C0-ATR-898

PLANNING FOR GROUNDWATER PROTECTION

Edited by

G. WILLIAM PAGE

Department of Urban Planning
and Center for Great Lakes Studies
University of Wisconsin–Milwaukee
Milwaukee, Wisconsin

1987

ACADEMIC PRESS, INC.

Harcourt Brace Jovanovich, Publishers

Orlando San Diego New York Austin
Boston London Sydney Tokyo Toronto

ACADEMIC PRESS, INC.
Orlando, Florida 32887

United Kingdom Edition published by
ACADEMIC PRESS INC. (LONDON) LTD.
24–28 Oval Road, London NW1 7DX

Library of Congress Cataloging in Publication Data

Planning for groundwater protection.

Includes index.
1. Water, Underground—Pollution—United States.
I. Page, George William, Date
TD426.P58 1987 363.7'394 86-22342
ISBN 0–12–543615–7 (alk. paper)

PRINTED IN THE UNITED STATES OF AMERICA

87 88 89 90 9 8 7 6 5 4 3 2 1

Contents

Preface ix

1 HYDROGEOLOGIC FRAMEWORK FOR GROUNDWATER PROTECTION

Mary P. Anderson

I. The Role of Groundwater in the Hydrologic Cycle 1
II. Transport of Chemicals through the Soil Zone to the Water Table 6
III. Physical, Chemical, and Biological Defense Mechanisms 9
IV. Groundwater Monitoring 13
V. Mathematical Simulation 19
VI. Areas of Uncertainty and Implications for Planning 21
References 24

2 THE INSTITUTIONAL FRAMEWORK FOR PROTECTING GROUNDWATER IN THE UNITED STATES

Timothy R. Henderson

I. Introduction 29
II. Overview 30
III. The State Role in Groundwater Protection 33
IV. Federal Laws and Institutions Protecting Groundwater 44
V. Concluding Remarks and Summary 64
References 66

3 DRINKING WATER AND HEALTH

G. William Page

I. What Microcontaminants Are Found in Water Supplies? 70
II. What Are the Health Implications of Toxic Contaminants in Drinking Water? 72
III. Cancer 73

IV. Cancer Processes 73
V. Dose–Response Functions 74
VI. Animal Studies 75
VII. Epidemiologic Studies 77
VIII. Cancer Risk in Perspective 79
IX. How Can We Protect the Health of the Public? 82
X. Conclusions 83
References 84

4 TECHNOLOGICAL APPROACHES TO REMOVING TOXIC CONTAMINANTS

Robert M. Clark

I. Introduction 89
II. Alternatives to Treatment 89
III. Treatment Options 91
IV. Conventional Treatment 93
V. Chemical Feed and Handling 99
VI. Ion Exchange 102
VII. Removal of Organics by Adsorption 105
VIII. Aeration 109
IX. Cost Comparisons 122
References 122

5 DATA AND ORGANIZATIONAL REQUIREMENTS FOR LOCAL PLANNING

Martin Jaffe

I. Introduction 125
II. Rediscovering the Past 126
III. Local Data Needs 130
IV. Identifying Priority Areas 147
V. Establishing a Local Protection Program 151
VI. Broadening the Local Protection Program 153
VII. Conclusion 154
References 155

6 LONG ISLAND CASE STUDY

Lee E. Koppelman

I. Physical Setting 157
II. The Groundwater System 160
III. Groundwater Problems 162
IV. Institutional Roles 175

V. Plan Recommendations 184
VI. Citizen Participation 188
VII. Implementation 195
References 201

7 DADE COUNTY, FLORIDA, CASE STUDY

Reginald R. Walters

I. Introduction 205
II. Wellfield Protection Study 216
III. Wellfield Protection Program 226
IV. Conclusion 238
References 239

8 WAUSAU, WISCONSIN, CASE STUDY

G. William Page

I. Background 241
II. Sequence of Events 243
III. Sources of Outside Assistance 248
IV. Protection Plan Elements 253
References 258

9 URBAN GROWTH MANAGEMENT AND GROUNDWATER PROTECTION: AUSTIN, TEXAS

Kent S. Butler

I. Introduction 261
II. The Edwards Aquifer and Associated Watersheds 262
III. Contributing Factors to Enactment of the Ordinances 267
IV. Watershed Development Standards 272
V. Current Events and Long-Range Issues 282
VI. Concluding Observations 284
References 286

10 PERTH AMBOY, NEW JERSEY, CASE STUDIES

G. William Page

I. Background 289
II. Nontoxics Problems at the Runyon Well Field 291

III. Toxics Contamination Problems 294
IV. Groundwater Protection Plan Components 296
References 298

11 SANTA CLARA VALLEY (SILICON VALLEY), CALIFORNIA, CASE STUDY

Thomas Lewcock

I. Overview, Past and Present, of the Santa Clara Valley 299
II. The Water System of the Santa Clara Valley 302
III. Contamination Alarm—A New Awareness 305
IV. Implementation of the Model Storage Ordinance 314
V. Who's in Charge—Unresolved Issues 318
VI. Conclusion 323
References 323

12 SOUTH BRUNSWICK, NEW JERSEY, CASE STUDY

G. William Page

I. The Municipality 325
II. Chronology of Toxics Pollution 327
III. Response to Well Closings 328
IV. Consulting Studies 329
V. Corporate Response to Toxics Contamination 331
VI. Government Roles in Developing Groundwater Protection Policies 333
VII. Protection Plan Elements 334
VIII. Effectiveness of Plan Components 339
References 340

13 BEDFORD, MASSACHUSETTS, CASE STUDY

Bonnie J. Ram and Harry E. Schwarz

I. Introduction 341
II. Background 342
III. Hydrogeology of the Tributary Watershed 342
IV. Institutional Structure 342
V. Groundwater Crisis 346
VI. State and Local Investigations 348
VII. Results of Bedford's Technical Investigation 351
VIII. Profile of Groundwater Supplies 352

IX. Significant Local Initiatives 354
X. Litigation Proceedings 356
XI. Bedford versus Advanced Metal Resource Corporation *et al.* 357
XII. Summary and Conclusion 358
XIII. Update to 1985 362
References 367

14 SUMMARY

G. William Page

I. Complexities of the Groundwater System 370
II. Fragmentation of Laws and Institutions 374
III. Political Support 377
IV. Significant Unknowns 379
References 381

Index *383*

Preface

Toxic substances contamination of groundwater is one of the most important environmental problems in the United States and other industrially developed nations. It is a problem about which we have only recently become aware as advances in a wide range of fields have alerted us to the presence of a great number of toxic substances in groundwater supplies and to the potential health risks caused by this contamination. The full dimensions of this problem are not known. Citizens and municipalities all over the country want to know if their drinking water contains toxic contaminants, yet are usually unable to have this legitimate concern answered because of the considerable expense of sampling and chemical analysis for toxic contaminants. In a nation where half of the population depends on groundwater for its source of drinking water, evidence from the limited sampling that has taken place indicates that many additional municipalities in all parts of the nation will soon discover that they have a groundwater contamination problem. Many of the approximately 16,000 known hazardous waste sites in the United States that are now closed will eventually begin to leak, as will the 2 million underground storage tanks for chemicals and petroleum products. These problems exist at a time when many surface water sources of drinking water are already polluted and few high-quality alternative sources of drinking water remain in proximity to human population centers.

This book describes selected groundwater contamination problems and the best available approaches for protecting groundwater. Our society has released large quantities of many pollutants including toxic contaminants into the environment. Many of these contaminants are harmful at even trace concentrations and have been found in our groundwater resources where they may be protected from the usual environmental degradation processes and where their movement is difficult to predict. Because of the slow movement of groundwater, its nonturbulent flow characteristics, its isolation from the atmosphere, and the very limited cost-effective removal or treatment options, contaminated groundwater is effectively lost as a source of drinking water for the foreseeable future. These factors make efforts for the protection of groundwater imperative.

On a national scale only a small portion of our total groundwater resources is contaminated with toxic substances, but often contamination occurs in areas where groundwater is the source of drinking water. Groundwater supplies about 50% of the U.S. population with drinking water and about 95% of the rural households. In light of the risks of contamination and the limited possibilities and high costs of reclamation, it is urgent that municipalities and water supply utilities take actions designed to protect their groundwater from contamination.

In the past 10 years, many instances of groundwater contamination with toxic substances have been reported. This proliferation of known problems is due in part to the fact that more samples are being collected and analyzed for toxic contaminants. In most areas of the country, local groundwater supplies have not been tested for contamination by any of a wide array of toxic chemicals. Industrial solvents are found in areas far from our urban industrial centers because of disposal practices and the decentralization of our industrial base. A variety of pesticides cause widespread groundwater problems in agricultural and suburban areas. Leaking storage tanks are problems throughout the nation. Groundwater protection will undoubtedly grow in importance as more groundwater contamination problems are identified in localities that have never suspected the existence of a groundwater contamination problem.

Policy to contend with groundwater contamination exists at all levels of government. In the United States, the Environmental Protection Agency released its National Groundwater Protection Strategy in August 1984. This national policy includes attempts to alleviate inconsistency and improve coordination among the 16 separate federal laws and other programs that have some impact on groundwater protection, and attempts to assist state and local government. The national strategy leaves the development, implementation, and management of specific policies to protect groundwater to local levels of government.

There is great variety in groundwater protection programs at the state and local levels of government. Some states have developed innovative and elaborate programs. Florida, Connecticut, Arizona, and Wisconsin are examples of states that have accomplished the most (see Chapter 2 for a more complete description of state approaches). Some states have not yet addressed the issue. In the United States, it is the local governments that must implement site-specific actions to protect groundwater from contamination. In most cases the municipality is the level of government that must take this action. In some cases the county or a regional government agency may take the lead role. For these reasons, this book focuses on municipal, county, and regional approaches to groundwater protection. Groundwater contamination affects both public and private wells. The focus of this book is on programs designed to protect public water supplies. Programs that protect groundwater will protect both public and private wells from contamination.

While there are some similarities among the contamination events, the variety and complexity of geological and hydrological conditions and of toxic contaminants produce extremely complex problems with great differences between groundwater contamination sites. These are new problems with which few municipalities have had experience and for which few municipal officials are prepared. This book attempts to provide the reader with an understanding of contamination processes and groundwater systems as well as options available in planning to protect groundwater and in planning to respond to groundwater contamination. Given past disposal practices, the prodigious quantities of toxic chemicals in use, and possibilities for accidental spills, all municipalities should take action now to prepare to minimize the potential for crisis if groundwater contamination occurs, and most importantly to plan a program tailored to local conditions that will help protect groundwater from toxic contaminants.

Every groundwater system is unique, as are the patterns of human activities that may cause contamination. Lists of approaches to protect groundwater are not too useful unless they are linked to the unique social, cultural, economic, and physical conditions in a community. This volume combines summary chapters that describe the background to groundwater contamination problems with detailed case studies. Some of the case studies document the discovery of serious groundwater contamination and the remedial and preventive actions taken in response to the resulting crisis. Other case studies describe preventive actions accomplished before serious groundwater contamination with toxic substances reached crisis proportions. All of the case studies describe how and why groundwater protection policies have been developed and tested in response to the local characteristics of the contamination threats, the physical environment, and the human environment.

Individual chapters provide an overview of the problem, describe the hydrogeological framework for groundwater protection, review the health risks associated with groundwater contamination, describe the instititutional framework for groundwater protection in the United States, propose the data and organizational requirements for developing a local groundwater protection program, discuss technological approaches to removing contaminants, and present detailed descriptions of approaches designed to protect groundwater. The series of chapters presenting detailed case studies from around the country comprise approximately half of the book. These case studies have been carefully selected to present the most effective and innovative approaches in the United States from areas experiencing a wide variety of hydrogeological conditions and toxic contaminant source problems. These case studies of groundwater protection policies used within diverse contexts will provide important information that will be very useful in the selection of management methods that are most appropriate to the unique circumstances of any community.

Protecting groundwater from toxic contaminants is not an easy task even when one is equipped with the best information and the political power to implement the most thorough of plans. Our knowledge of groundwater systems is incomplete; the chemistry of toxic chemicals in groundwater, sources of toxic chemicals, their health effects, and the efficacy of the different components of the groundwater protection plan are all incompletely understood. Because of the uncertainty in planning for groundwater protection, it is very important that groundwater protection plans be continuously reassessed. Each component should be evaluated to determine if it is having its intended effect. The process of protecting groundwater may have to be modified as new information is received. This book is designed to assist with the initial development of a plan and with the ongoing management of a program for groundwater protection.

G. William Page

1

Hydrogeologic Framework for Groundwater Protection

MARY P. ANDERSON
Department of Geology and Geophysics
University of Wisconsin–Madison
Madison, Wisconsin 53706

To understand the basic principles of groundwater occurrence and movement, it is necessary to appreciate the ways in which groundwater interacts with other kinds of water moving through the hydrologic cycle (Fig. 1). In particular, it is important to recognize that the division of water into different categories is artificial because water moving through the cycle readily crosses boundaries between zones. For example, a source of pollution at the land surface can cause contamination of soil water, which in turn may contaminate groundwater, which then may discharge to lakes, streams, and the ocean. For this reason, contamination of any part of the hydrologic cycle may lead to widespread problems. However, it is also important to recognize that there are natural mechanisms within the subsurface that can mitigate the effects of a pollutant source. In this chapter, we will explore the way in which pollutants enter the groundwater system as well as the natural defense mechanisms operative in the subsurface.

I. THE ROLE OF GROUNDWATER IN THE HYDROLOGIC CYCLE

The hydrologic cycle contains water in three states: liquid water, water vapor, and water temporarily locked up in glaciers as snow and ice. Our concern here is with liquid water which occurs in two forms: (1) subsurface water, which includes soil water and groundwater, and (2) surface water, which is found in streams, lakes, and the ocean.

Water moves continuously through the hydrologic cycle and for this reason it is difficult to isolate the soil zone, groundwater zone, or surface

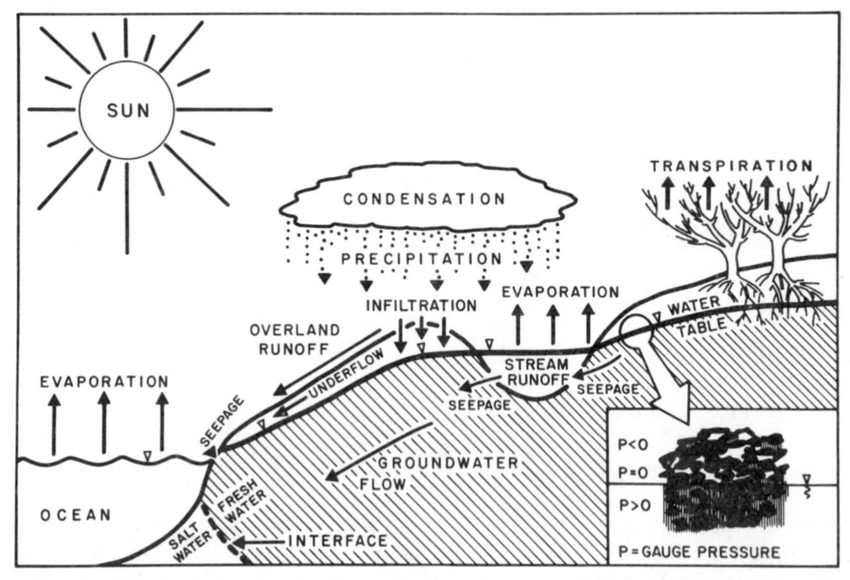

Fig. 1. The hydrologic cycle. Inset shows a close-up of the boundary between the unsaturated and saturated zones.

water zone for individual analysis; all three zones are connected. Therefore, the groundwater hydrologist must also be knowledgeable about processes in the soil zone as well as surface water processes.

A. Components of the Hydrologic Cycle

Processes operative in the hydrologic cycle are illustrated in Fig. 1 and are defined below.

Transpiration: transformation of liquid water to water vapor through the action of vegetation.
Evaporation: transformation of liquid water to water vapor.
Condensation: transformation of water vapor to liquid water.
Precipitation: the process of water falling to the land surface.
Infiltration: the entry of water into the soil zone.
Underflow: lateral movement of water through the soil zone.
Percolation: vertical movement of water through the soil zone.
Groundwater flow: movement of water through the groundwater zone.
Seepage: movement of water to or from the groundwater zone.
Overland runoff: precipitation that does not infiltrate but runs off along the land surface.

Stream runoff: that part of stream flow consisting of water derived from overland runoff or underflow. *Base flow* is that part of stream flow consisting of water derived from seepage.

It should be noted that the energy to drive the hydrologic cycle is derived from the sun.

B. Definition of Groundwater

The inset in Fig. 1 shows a portion of the subsurface in detail. The water table marks the top of the zone of saturation below which the porous material that makes up the subsurface is saturated with water. The water table can also be described as a surface along which pressure is equal to atmospheric pressure (zero gauge pressure). The water table is delineated in the field by measuring the elevation to which water will rise in a well that just penetrates the top of the saturated zone. Immediately above the water table is a layer of porous material that is saturated with water under tension. Here pressure is less than atmospheric but all the void spaces are filled with water. Water in this layer, which is known as the capillary fringe, is drawn up from the saturated zone by capillarity and by adsorption to soil particles. Above the capillary fringe is the unsaturated zone, in which water is under tension (pressure less than atmospheric) and where only some of the void space is filled with water. The other voids are filled with water vapor and with air. Water in the saturated zone is under pressure greater than atmospheric and will flow freely into wells, whereas water in the unsaturated zone will not since it is held within the voids by tension.

Therefore, we can define groundwater to be subsurface water that occurs below the water table in the saturated zone, whereas soil water is water that is found above the water table in the unsaturated zone.

C. Forces That Cause Groundwater Movement

Movement of groundwater through the saturated zone occurs in response to two forces: gravity and pressure. Gravity, of course, acts to pull water downward. However, pressure forces can cause water to move upward. Hence, it is not generally correct to say that "water moves downhill" when referring to subsurface water. Water has energy by virtue of its position in the earth's gravitational field and its position in a pressure field within the subsurface. This energy is known as potential energy because it originates by virtue of position, in contrast to kinetic energy, which arises by virtue of movement. The potential energy of groundwater is expressed in terms of head (also known as groundwater potential). Head is equal to the elevation of the water level in a well measured above an arbitrarily defined datum (Fig. 2). Total head (h) has two components: elevation head (z), which is an expression of the potential energy resulting from gravity, and pressure head (ψ), which is an expression of potential energy resulting from pressure. Water always flows from points of higher to lower total head. But water may flow from points of lower to higher pressure head or from points of lower to higher elevation head (Fig. 2).

The velocity of groundwater in any given direction, say, x, is expressed by a form of Darcy's Law:

$$v_x = \frac{-K_x}{n}(dh/dx) \tag{1}$$

where K_x is the permeability or hydraulic conductivity in the x direction, n is the effective porosity, and dh/dx refers to the hydraulic gradient in the x direction. Hydraulic gradient is the change in head (dh) over distance (dx). Velocity is a vector with direction as well as magnitude and v_x is one of its components. In a three-dimensional flow system it would be necessary to write analogous forms of Eq. (1) for the velocity components in the y and z directions.

Head is measured in the field by means of wells. Wells are open to a portion of the aquifer through a well screen or perforated casing or simply through an open hole below the casing. The water level in the well measures the average head in that portion of the aquifer open to the well. That is, a well with a short screen, on the order of one foot,

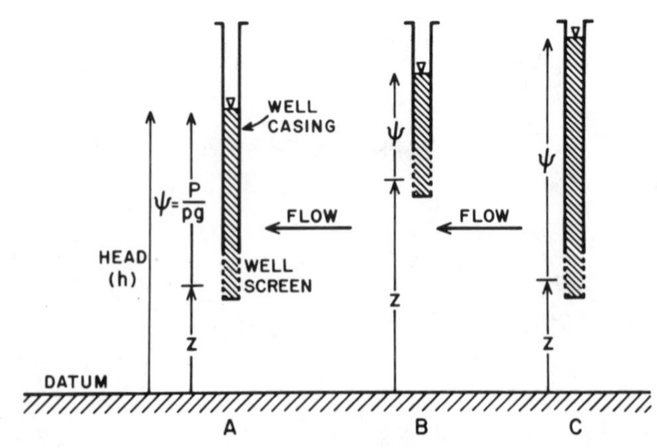

Fig. 2. Definition of head. Flow arrows are drawn assuming a horizontal flow system.

measures the head at essentially a point in the subsurface. This type of well is known as a piezometer. Wells used for water supply are open to larger sections of the aquifer and may have ten to thousands of feet in connection to the aquifer. Hence, information about well construction is critical in translating water level measurements to head values within groundwater flow systems such as the system shown in Fig. 3. Failure to appreciate the significance of well construction can give rise to erroneous conclusions regarding groundwater movement (e.g., see Saines, 1981).

Figure 3 shows how information about groundwater movement is obtained using nested wells—a set of two or more wells each open to a different portion of the aquifer. Water levels in nested wells are different when there is vertical movement of water in the aquifer. In flow system terminology, areas of vertically downward flow below the water table are known as recharge areas while areas of upward flow toward the water table are known as discharge areas. Water that enters the groundwater system in recharge areas may follow relatively long flow paths through the saturated zone. Hence, it is a common objective of groundwater management plans to identify and protect groundwater recharge areas.

In analyzing regional groundwater systems, it is usually assumed that the flow system is at steady state and heads are invariant with time. In general, fluctuations in head are small relative to the total head drop from recharge to discharge area. Moreover, seasonal fluctuations in

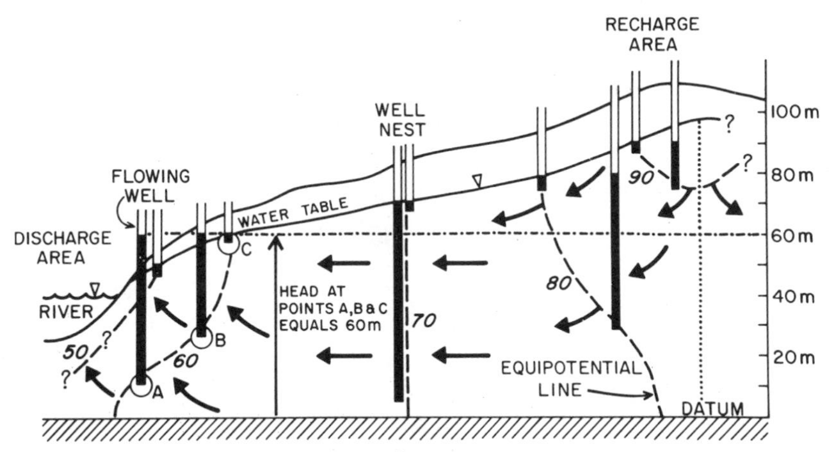

Fig. 3. Schematic representation of a regional groundwater flow system. Nested piezometers measure head at points. Equipotential lines connect points of equal head. Water flows from higher to lower head.

head may not affect the overall flow pattern. Hence, the steady-state assumption is generally acceptable. However, for certain regional problems and for some more localized analyses, it may be important to quantify groundwater fluctuations. Usually quarterly measurements of groundwater levels are sufficient to characterize seasonal fluctuations but more frequent measurements (monthly, for example) might also be needed depending on site-specific hydrologic conditions and the problem being addressed. The magnitude of seasonal fluctuations varies with geographic location as well as with hydrologic conditions.

II. TRANSPORT OF CHEMICALS THROUGH THE SOIL ZONE TO THE WATER TABLE

Chemicals are transported through the unsaturated zone dissolved in water that infiltrates at the land surface and percolates down to the water table. Hence, analysis of chemical movement requires an analysis of water movement in the unsaturated zone. The details of water movement in the unsaturated zone are complex and only a few highlights will be presented below.

A. Moisture Content

The first step in describing water movement in the unsaturated zone is to quantify the moisture content of the soil profile. Moisture content (θ) is defined to be volume of water divided by volume of porous material. Moisture content equals porosity when the volume of porous material in question is fully saturated. However, in general, moisture contents in the unsaturated zone are less than porosity because some of the void spaces are filled with air rather than water (see inset to Fig. 1). Moisture contents can be measured in the field. However, such measurements require sophisticated instrumentation (e.g., a neutron moisture logger) and must be made periodically to capture fluctuations in moisture contents that may occur daily or even hourly.

Pressure head (ψ) in the unsaturated zone is negative since water is held under tension. Soil moisture tension is expressed in units of pressure (e.g., mbar) while pressure head is expressed in units of length ($\psi = p/\rho g$, where p is soil moisture tension, ρ is density of water, and g is the constant of acceleration of gravity). Pressure head (or soil moisture tension) is related to moisture content by means of the soil moisture characteristic curve (Fig. 4a). Each soil can be expected to have a unique soil moisture characteristic curve. These curves are determined in the laboratory or field by experiments in which both ψ and θ are measured directly under controlled conditions.

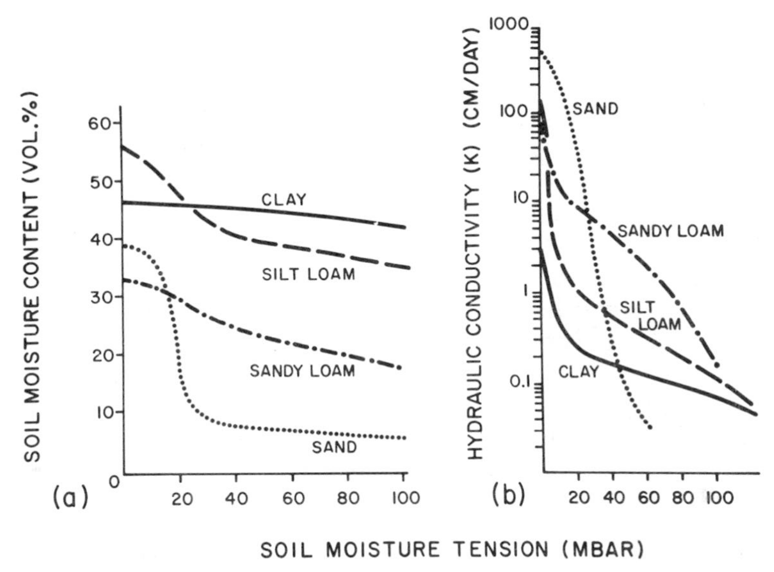

Fig. 4. Generic soil curves (adapted from Bouma *et al.*, 1974).

Flow of water in the unsaturated zone, as in the saturated zone, occurs from higher to lower total head. In general, water moves vertically downward through the unsaturated zone to the water table. Significant lateral movement or underflow (Fig. 1) would be expected only when the unsaturated zone is relatively wet. Hence, in predicting the movement of chemicals in the unsaturated zone, the velocity of a contaminant that does not undergo chemical reactions might be approximated as the velocity of water moving vertically through a column of unsaturated soil:

$$v_z = -K(\psi)/\theta \ (dh/dz) = -K(\psi)/\theta \ [(d\psi/dz)+1] \qquad (2)$$

where $h = \psi + z$ and $K(\psi)$ is the permeability of the soil. Permeability in the unsaturated zone is a function of pressure head (or moisture content). Permeability is higher under low tensions or high moisture contents. For example, permeability is high in the capillary fringe because this zone has high moisture content and pressure head close to zero. The presence of air at lower moisture contents acts to block water movement. The relationship between permeability and soil moisture tension can be measured in the laboratory or field for a given soil; Fig. 4b shows some typical curves.

It is evident that determination of the velocity distribution in a soil profile is not a simple process. It requires that soil curves such as those shown in Fig. 4 be generated for each soil in the profile. Furthermore, site- and time-specific measurements of moisture content (or pressure head) are needed to calculate the velocity distribution from Eq. (2). In practice, it is sometimes justified to assume steady-state conditions. Steady-state conditions exist when heads are constant in time. This implies that inflow to the profile (infiltration) equals outflow (groundwater recharge). Then a constant velocity can be approximated as the infiltration rate divided by an average moisture content known as "field capacity." Field capacity is defined to be the prevalent moisture content in the soil profile after drainage has occurred. However, the concept of field capacity has been criticized because conditions of complete drainage may never exist for very long under field conditions owing to the periodic occurrence of infiltration events. In any case, the assumption of a constant velocity is not appropriate when heterogeneities or layers are present in the soil profile.

B. Heterogeneities

Heterogeneities or layers in the soil profile imply the existence of two or more different soil types, each with different soil curves. Analysis of flow under these conditions is complex. For example, it is well known that under transient unsaturated conditions, zones of high permeability act as barriers to flow. Eventually, moisture contents may increase sufficiently at the top of such a high-permeability layer to force water to move from the smaller pore spaces in the low-permeability layer to the larger pore spaces below. When steady-state conditions are established in such a soil profile, the fluxes ($v_z\theta$) through each layer are equal but the velocities (v_z) are different. The presence of heterogeneities can profoundly affect the movement of water and chemicals in the subsurface. Some indication of the nature of layers present at a site can be inferred from soil cores. However, the areal distribution and configuration of heterogeneities are difficult to detect by analysis of a limited number of soil cores. Consequently, analysis of the movement of chemicals in the subsurface is fraught with uncertainty (Nielsen and Biggar, 1982). (Editor's note: see Chapter 11, Fig. 2, in the Santa Clara Valley case study, for an illustration of heterogeneities in the subsurface.)

C. Groundwater Recharge

The groundwater hydrologist is interested in the unsaturated zone because it yields the water that becomes groundwater recharge. If contamination is present at the land surface or in the unsaturated zone, groundwater recharge may also be contaminated. Groundwater recharge can be defined as water that flows downward across the water table.

The rate of groundwater recharge is difficult to estimate. Direct measurement of groundwater recharge is labor-intensive and few direct estimates are available (e.g., see Freeze and Banner, 1970; Steenhuis *et al*, 1985; Dreiss and Anderson, 1985; Stoertz, 1985). In the absence of direct measurements, it is standard practice to let recharge equal a fraction of the average annual precipitation. The ratio of recharge to precipitation can be expected to vary with the geographic location of the study area. For example, on Long Island, New York, groundwater recharge is commonly assumed to be 50% of annual precipitation (Steenhuis *et al.*, 1985). In Wisconsin, a ratio of one-third is commonly used. However, it is likely that the average annual groundwater recharge varies from close to zero in parts of eastern Wisconsin, where there are nearly impermeable soils, to perhaps as much as 50% of annual precipitation in the central and northern portions of the state, where sandy glacial deposits cover the surface. Moreover, it is likely that recharge varies spatially within a groundwater basin (e.g., Winter, 1983; Stoertz, 1985).

The groundwater recharge rate determines the amount of water reaching the water table. The amount of contaminant entering the groundwater system can be quantified as mass flux ($cv_z\theta$), where c is the concentration of the contaminant crossing the water table and $v_z\theta$ is the flux of water across the water table (i.e., the groundwater recharge rate). The concentration (c) of the contaminant crossing the water table will be different from the concentration in leachate emanating from the source because of dilution and degradation that occur in the unsaturated zone.

III. PHYSICAL, CHEMICAL, AND BIOLOGICAL DEFENSE MECHANISMS

Natural defense mechanisms in the subsurface mitigate the effects of a pollutant source by causing a reduction in the concentration of the contaminant in solution. These mechanisms include physical processes such as dilution of contaminated water with uncontaminated recharge water and mixing with uncontaminated water already present in the subsurface (dispersion). Chemical processes such as adsorption and precipitation can also remove chemicals from solution at least temporarily. Biological processes, as well as decay and hydrolysis, can permanently remove or alter chemical compounds.

A. Physical Processes

Contaminants entering the soil zone may be diluted during infiltration by mixing with precipitation. Some mixing also takes place by diffusion in the unsaturated zone, which consists of mixing on a molecular level as ions move from points of higher to lower concentrations. Within the saturated zone, contaminants will again be diluted by mixing with uncontaminated groundwater recharge that is added to the saturated zone downstream of the contaminant source, causing the contaminant plume to be depressed below the water table (Fig. 5).

Mixing and dilution with ambient groundwater occur by a process known as dispersion. Dispersion occurs on a microscopic scale as water moves around individual grains of the porous medium, causing flow lines to be contorted. This contortion of flow results in spreading and dilution of the plume of contaminated water. However, microscopic dispersion is relatively minor compared to macroscopic dispersion, which arises from the presence of heterogeneities in the subsurface. At a macroscopic scale, diversion of flow lines occurs as water moves with different velocities through areas of differing permeabilities causing the plume to spread (Fig. 6). It follows that to predict the movement of contaminants in the saturated zone, it is necessary to understand the nature of the heterogeneities present in the subsurface. Geologic well logs, analysis of soil cores, and geophysical measurements are helpful in detecting the presence of heterogeneities. However, there rarely is enough information at a site to allow heterogeneities to be characterized at the level of detail needed for a quantitative analysis of dispersion. This problem is at the root of one of the major obstacles in quantifying

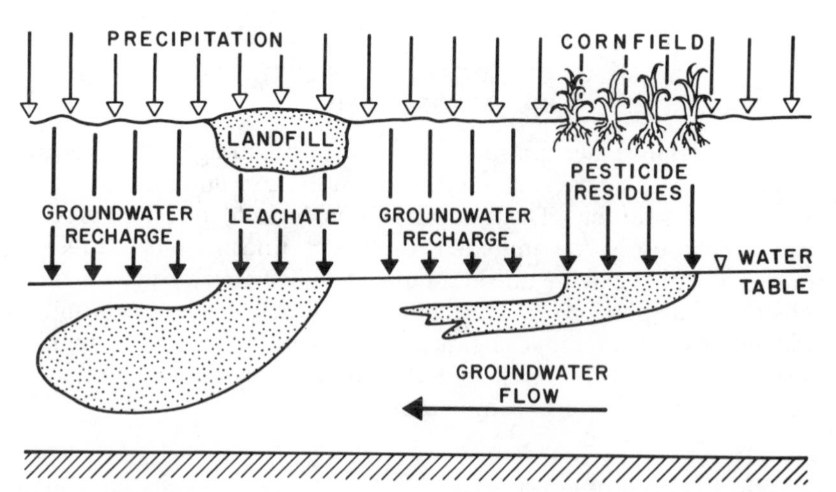

Fig. 5. Schematic representation of contaminant plumes in the subsurface showing the effects of dispersion as well as the depression of the plume below the water table downgradient of the source.

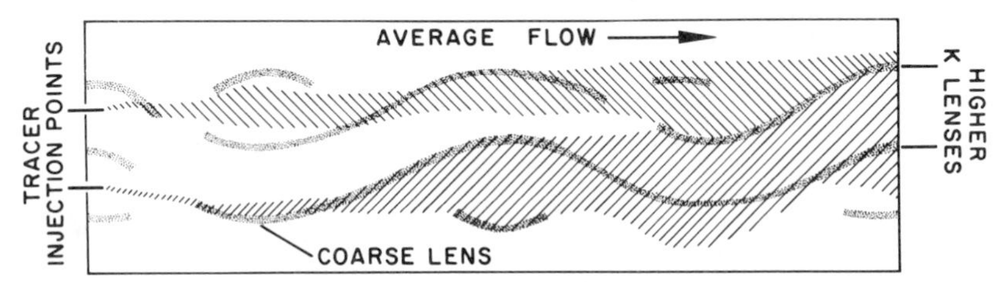

Fig. 6. Macroscopic dispersion as represented in a laboratory experiment (adapted from Skibitzkie and Robinson, 1963). "Higher K lenses" refers to lenses of high hydraulic conductivity.

dispersion for input to predictive models (Anderson, 1984; also see Section VI,A).

Diffusion also occurs in the saturated zone but its effects are generally small compared to those of macroscopic dispersion. However, dispersion is directly related to velocity and when velocities are small diffusion may become the dominant mixing process. Small velocities may occur because of the presence of material of low permeability, for example.

B. Chemical Processes

Chemical processes in the subsurface include precipitation, adsorption, and hydrolysis (sometimes represented as first-order decay). Precipitation results in the formation of a crystalline substance within the porous material, which may undergo dissolution if chemical conditions within the aquifer change. While precipitation reactions are important in describing the evolution of groundwater chemistry in regional flow systems, adsorption, hydrolysis, and decay are of greater interest in contaminant hydrology.

1. *Adsorption*

Adsorption refers to the removal of a chemical from solution by fixation onto the porous material itself. Adsorption is quantified by means of a distribution coefficient,

$$K_d = \bar{c}/c \tag{3}$$

where c is the concentration in solution and $\bar{c}$ is the mass adsorbed; c is expressed in units of gm/ml and $\bar{c}$ is expressed in units of gm of ion adsorbed per gm of porous material, so that the units of K_d are ml/gm.

A retardation factor (R) can also be defined such that

$$R = v/v_c \tag{4}$$

where v is the velocity of groundwater and v_c is the velocity of the contaminant. It can be shown that

$$R = 1 + K_d(\rho_b/n) \tag{5}$$

where ρ_b is the bulk density of the porous material and n is porosity.

Distribution coefficients are dependent on the ions involved in the reaction as well as on the porous material (Reardon, 1981) and can be measured in the laboratory by an experiment in which both c and $\bar{c}$ are measured directly. Distribution coefficients can also be measured in the field by means of a tracer test (e.g., Pickens *et al.*, 1981). For certain organic compounds in soils for which the organic content by weight is greater than 0.1%, it has been shown that K_d is directly proportional to the organic content of the porous medium and to the octanol water partition coefficient (e.g., see Cherry *et al.*, 1984).

The velocity of a contaminant can be calculated from Eqs. (4)–(5) provided the distribution coefficient and the velocity of groundwater are known and the bulk density and porosity can be estimated.

Use of the distribution coefficient and retardation factor implies that adsorption is reversible, indicating that the porous material eventually releases the contaminant and allows it to go back into solution. Hence, reversible sorption serves only to delay the arrival of the contaminant. However, the delay may be on the order of hundreds of years or even more, depending on the value of the retardation factor. If $R_d = 1$ there is no adsorption but if $R_d > 6$, generally the chemical is virtually immobilized.

2. *Hydrolysis/Decay*

Hydrolysis refers to the chemical reaction of a compound with water. For certain substances the hydrolysis reaction can be expressed as a first-order decay process involving a half-life, analogous to the half-life used in radioactive decay. However, in hydrolysis reactions the half-life is dependent on temperature and pH, whereas the half-life in a radioactive decay process is a constant.

For example, the pesticide aldicarb undergoes alteration and degradation by hydrolysis and oxidation. The parent product, aldicarb, is quickly oxidized to aldicarb sulfoxide. Sulfoxide is degraded by hydrolysis to sulfoxide oxime, which then may be oxidized to aldicarb sulfone. Sulfoxide and sulfone degrade to nontoxic oximes and nitriles (Hansen and Spiegel, 1983). The breakdown of aldicarb into nontoxic "daughter" products can be quantified by means of a half-life. The half-life of aldicarb in groundwater in Wisconsin is estimated to be 475–1800 days (Anderson, 1986).

C. Biological Processes

Bacteria have long been recognized to be important participants in chemical reactions in the soil zone and almost every organic compound can be metabolized by some organism. Bacteria also exist in relatively high abundance in the groundwater zone (Table I) and attention has recently focused on the role of bacteria in the groundwater system (Bitton and Gerba, 1984).

The degree of biodegradation of an organic compound is determined by the geochemical properties of the site-specific subsurface environment. Studies have shown that several halogenated aliphatics, alkyl benzenes such as toluene, chlorobenzenes and certain pesticides may undergo biodegradation (Newsom, 1985). Examples of the potential for biotransformation of selected organic compounds are given in Table II.

IV. GROUNDWATER MONITORING

Two types of groundwater monitoring activities are important for groundwater planning and protection: (1) to protect groundwater supplies from overdraft and (2) to protect groundwater from pollution.

Before discussing each of these objectives of groundwater planning,

TABLE I

Numbers of Organisms in the Subsurface Environment

Site	Depth to water table (m)	Subsoil[a]	Just above water table[a]	Just below water table[a]
Lula, Okla.[b]	3.6			
February 1981		6.8	3.4	6.8
June 1981		9.8	3.7	3.4
Fort Polk, La.[c]				
Borehole 6B	6.0	3.4	1.3	3.0
Borehole 7	5.0	7.0	1.3	9.8
Conroe, Texas[d]	6.0	0.5	0.3	0.6
Long Island, N.Y.[d]	6.0	—	—	36
	3.0	170	—	—
Pickett, Okla.[c]	5.0	—	—	5.2

Source: Wilson and McNabb (1983), published by the American Geophysical Union.

[a] In millions per gram dry material.

[b] Wilson *et al.* (1983).

[c] Ghiorse and Balkwill (1983).

[d] W. G. Ghiorse (personal communication, 1982).

TABLE II
Prospect of Biotransformation of Selected Organic Pollutants in Water Table Aquifers

Class of compounds	Aerobic water, concentration of pollutant (μg/liter) >100	Aerobic water, concentration of pollutant (μg/liter) <10	Anaerobic water
Halogenated aliphatic hydrocarbons			
Trichloroethylene	None	None	Possible[a]
Tetrachloroethylene	None	None	Possible[a]
1,1,1-Trichloroethane	None	None	Possible[a]
Carbon tetrachloride	None	None	Possible[a]
Chloroform	None	None	Possible[a]
Methylene chloride	Possible	Improbable	Possible
1,2-Dichloroethane	Possible	Improbable	Possible
Brominated methanes	Improbable	Improbable	Probable
Chlorobenzenes			
Chlorobenzene	Probable	Possible	None
1,2-Dichlorobenzene	Probable	Possible	None
1,4-Dichlorobenzene	Probable	Possible	None
1,3-Dichlorobenzene	Improbable	Improbable	None
Alkylbenzenes			
Benzene	Probable	Possible	None
Toluene	Probable	Possible	None
Dimethylbenzenes	Probable	Possible	None
Styrene	Probable	Possible	None
Phenol and alkyl phenols	Probable	Probable	Probable[b]
Chlorophenols	Probable	Possible	Possible
Aliphatic hydrocarbons	Probable	Possible	None
Polynuclear aromatic hydrocarbons			
Two and three rings	Possible	Possible	None
Four or more rings	Improbable	Improbable	None

Source: Wilson and McNabb (1983), published by the American Geophysical Union.

[a] Possible, probably incomplete.

[b] Probable, at high concentration.

we will first consider some general guidelines for groundwater monitoring.

A. General Guidelines

1. *Contamination or Pollution?*

In the discussion presented above, "contaminant" has been used in preference to "pollutant." A contaminant is here defined to be a solute that is present in the subsurface at a concentration above background levels. Pollution is defined to occur when contaminants are present to a degree that is judged to be harmful. Therefore, pollution carries a subjective connotation and implies the acceptance of standards against which to judge what is harmful. Often such standards are those set by the U.S. EPA or state enforcement agencies.

2. *Aquifers and Confining Beds*

When addressing the issue of groundwater contamination, most concern arises over the potential for pollution of an aquifer. An aquifer is defined to be a unit of porous material that yields economically significant quantities of water to wells. Like the definition of pollution, this definition is a subjective one. Wells in a low-permeability unit may yield quantities of water that are economically significant to homeowners, suggesting that the unit should be called an aquifer. However, the same low-permeability unit may not yield economically significant quantities of water for industrial purposes and under these circumstances the unit might not be considered an aquifer. Needless to say, such subjectivity may create problems when enacting enforcement guidelines.

Units of porous material that are not aquifers are called confining beds. Aquifers can be divided into confined aquifers (aquifers overlain by confining beds) and unconfined aquifers (aquifers not overlain by confining beds). The top of an unconfined aquifer is the water table, whereas the top of a confined aquifer is the base of the confining bed. The hydrologic surface of a confined aquifer that is analogous to the water table in an unconfined aquifer is the potentiometric surface. When describing an aquifer as confined or unconfined it is convenient to assume that there is no component of vertical flow in the aquifer, or, in other words, flow through the aquifer is horizontal. This approximation is here referred to as the "aquifer viewpoint." The aquifer viewpoint is useful mainly when dealing with problems involving the effects of localized pumping. In describing regional flow systems it is often more appropriate to view the system from the "flow system viewpoint" as shown in Fig. 3. Because all regional flow systems have recharge and discharge areas, there must be vertical flow occurring in the system and in the flow system viewpoint vertical gradients are not assumed to be negligible.

3. *Flow Direction and Geologic Characterization*

In all types of monitoring it is essential to determine the general groundwater flow direction within the aquifer of interest. The direction and magnitude of the horizontal head gradient in a given portion of the aquifer can be determined by measuring the heads in a minimum of three wells. However, it is desirable to use as many more wells as is economically practical given that deflections in the contour lines that describe the distribution of head in the aquifer are common owing to the presence of heterogeneities in the subsurface. Hence, linear interpolation of contours between well points may not yield an accurate representation of the head distribution. For this reason, it is desirable to obtain as much information as possible about the geology of the area under investigation so that geologic units can be defined and the presence of heterogeneities anticipated. Hydraulic conductivities of geologic units are routinely measured using the results of pumping tests or bailer tests (Freeze and Cherry, 1979, pp. 339–350). In addition, it is essential to determine the magnitude of vertical head gradients within the aquifer using nested wells as shown in Fig. 3.

B. Monitoring for Overdraft

Planning for adequate supplies of groundwater generally involves analysis of the effects of pumping one or more wells and delineating the cone of depression formed in the water table or potentiometric surface. The details of this type of analysis are complex and occupy a good deal of space in the groundwater literature. The interested reader is referred to Freeze and Cherry (1979, Chap. 8) for an overview.

The design of a network of monitoring wells is necessarily site and problem specific. However, a few comments on some of the issues related to planning concerns are offered below.

1. *Effects of Overpumping*

Effects of overpumping include the decline of groundwater levels and may also involve decline of surface water levels and draining of water out of the unsaturated zone. In coastal areas, overpumping may induce landward movement of the fresh water/salt water interface (Fig. 1), causing salt water intrusion of coastal water wells. For example, the history of groundwater development on Long Island, New York, and the resultant salt water intrusion forms a classic example in groundwater mismanagement (Heath *et al.*, 1983).

2. *Safe Yield*

A basinwide monitoring scheme may be implemented in an attempt to ensure that the "safe yield" of the aquifer is not exceeded. Safe yield is defined to be the amount of water that can be withdrawn from an aquifer annually without causing undesirable effects. Again we have a subjective definition. What is meant by "undesirable effects"? The concept of safe yield has been strongly criticized (e.g., Anderson and Berkebile, 1977; Gass, 1981; Bredehoeft *et al.*, 1982). However, the concept seems to be immortal despite repeated death blows. The term safe yield has even found its way into state regulations: "The amounts and timing of water appropriated shall be limited to the safe yield of the aquifer to the maximum extent feasible and practical" (Minnesota Dept. of Natural Resources, GMCAR 1.5052, Sect. 3a).

Other terms have been introduced in attempts to identify parameters that are more amenable to objective quantification. For example, Freeze and Cherry (1979, pp. 364–367) discuss the use of natural basin yield and maximum stable basin yield. Natural basin yield refers to the amount of water that discharges under steady-state conditions from a flow system, such as the one shown in Fig. 3, that has not been developed through the use of pumping wells. Pumping wells intercept groundwater flow and prevent it from reaching natural discharge areas. The resultant decline of the water table induces drainage from the unsaturated zone and temporarily may increase recharge to the aquifer. Pumping may also induce recharge to the aquifer from surface water bodies such as rivers. Hence, groundwater development through pumping wells may actually increase the amount of recharge to the aquifer. However, if pumping continues to increase after the maximum amount of recharge has been induced, there will come a time when water levels will begin to fall sharply as water is removed from storage in the aquifer. At this point the unsaturated zone can no longer deliver recharge to the aquifer and the maximum stable basin yield has been exceeded; the aquifer is now being mined. The implications of the problems involved in a safe yield determination are discussed in Section VI,B,2.

C. Monitoring for Pollution

Sources of pollution include point and nonpoint sources. Point sources include sanitary landfills, hazardous waste disposal facilities, high-level radioactive waste repositories, and other waste disposal operations. Monitoring of such localized sources requires a detailed network of nested piezometers so that the contaminant plume may be accurately delineated. Research studies in which numerous nested piezometers have been installed (e.g., Kimmel and Braids, 1980; MacFarlane *et al.*, 1983) make it clear that even with a large number of well points, there is still some measure of uncertainty regarding the configuration of the plume. Unfortunately, in most situations outside the research arena, it is rare to find sufficient well coverage to allow the contaminant plume to be characterized adequately.

Plumes from point sources sometimes develop as relatively thin zones of contaminated water occurring near the water table. The chromium plume described by Permutter and Lieber (1970) and the sewage plume described by LeBlanc (1984) are such plumes. Other plumes seem to sink a certain distance below the water table in response to either density contrasts between the leachate and the ambient groundwater or the existence of vertical hydraulic gradients in the subsurface. The landfill plumes described by Kimmel and Braids (1980) and by MacFarlane *et al.* (1983) are of this type (Fig. 7).

Nonpoint sources include urban runoff and infiltration contaminated by agricultural products. Monitoring for nonpoint source pollution is usually done at a regional scale, that is, a network of wells is established throughout one or more counties. In urban areas, contaminants of concern include chloride and sodium from road salt as well as trace metals

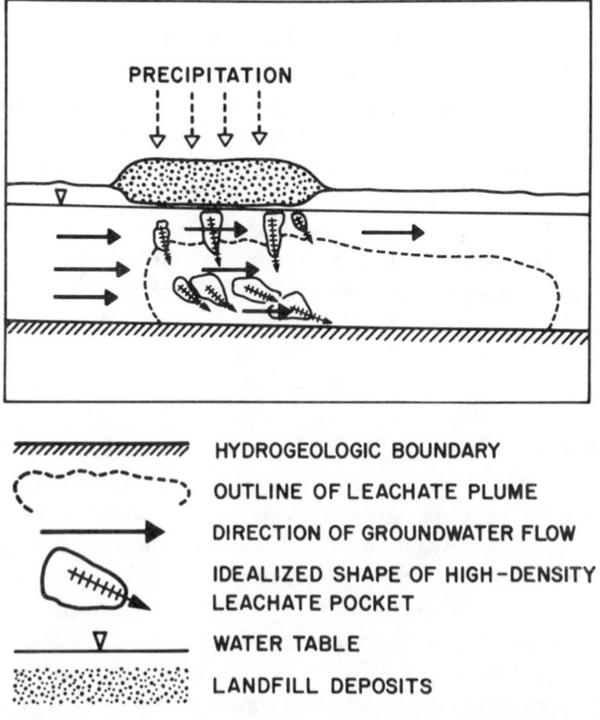

Fig. 7. Sinking leachate plume (adapted from Kimmel and Braids, 1980).

from road runoff (Eisen and Anderson, 1979; Kimmel, 1985). In agricultural regions, wells might be sampled periodically for selected pesticides as well as nitrate (NO_3) and ammonium (NH_4).

Sampling procedures are dependent on well construction and the chemical constituents of interest. Details on sampling protocol are given by Barcelona *et al.* (1986), Morrison (1983), and Scalf *et al.* (1981).

V. MATHEMATICAL SIMULATION

In Sections III and IV we discussed the parameters and variables used to characterize the hydrologic state of the soil and groundwater zones. Namely, head is the variable of interest and hydraulic conductivity is the key parameter. In addition, we need to know soil moisture distribution in the unsaturated zone and porosity in the saturated zone. Under transient conditions in the saturated zone we also need to quantify storativity, which describes the way in which water is released or taken up into storage in response to changes in head. Storativity is represented by a family of parameters that include storage coefficient, specific storage, and specific yield. The interested reader is referred to Freeze and Cherry (1979) for details.

While an inspection of system parameters together with information on the head distribution provide useful information, a more powerful analysis is possible when a mathematical model is used to quantify the movement of water and solutes in the subsurface.

A. Definition of a Mathematical Model

A mathematical model consists of a set of equations that incorporates the relevant physics that are thought to represent processes occurring in the subsurface. For example, water movement in the subsurface is described by a water balance equation for the system, which states that inflow to a representative element in the system minus outflow equals changes in storage with time. For the groundwater zone, Darcy's Law is used to substitute head for flow rate so that the resulting partial differential equation is written with head as the dependent variable. In the unsaturated zone a form of Darcy's Law known as Richard's Equation is used in the water balance equation.

A form of Darcy's Law written for flow in the x direction is:

$$q_x = v_x n \tag{6}$$

where q_x is the flow rate and v_x is given by Eq. (1). Richard's Equation written for flow in the z direction is:

$$q_z = v_z \theta \tag{7}$$

where v_z is given by Eq. (2).

The result of combining a water balance equation with Darcy's Law or Richard's Equation is a partial differential equation that describes water movement in the subsurface. Such an equation is known as the governing equation and it forms the heart of a mathematical model. Other equations that describe the hydrologic state of the boundaries of the system and the initial state of the system (e.g., the head distribution at the beginning of the simulation) complete the model. Input to the model includes parameters that characterize the system, e.g., hydraulic conductivity and storativity. The output of the model consists of the distribution of head in space and time. Velocities can be calculated from the head distribution using equations like Eqs. (1)–(2).

Velocities are input to another model that uses a different partial differential equation to describe the movement of contaminants. In Section III, we noted that the movement of contaminants is influenced by the physical processes of dilution and dispersion as well as by chemical and biological processes. The physics that are thought to describe these processes are used to derive an equation that attempts to quantify contaminant movement. This equation is generally known as the advection–dispersion equation or sometimes the convection–dispersion equation. The output of this model is the concentration of the contaminant in space and time.

B. Powers and Limitations

Scientists as well as planners and regulators are attracted to mathematical models because of the model's potential to predict the future. For example, different types of groundwater development schemes can be simulated using a model and probable results can be evaluated before an optimal scheme is selected. Similarly, different scenarios for waste disposal can be evaluated using a contaminant transport model.

However, the accuracy of such predictions depends on a number of factors including the accuracy and completeness of the field data used as input to the model and the skill of the modeler (Faust *et al.*, 1982). There is considerable potential for misuse of model results (e.g., see Anderson, 1983), and for this reason it is particularly important that the limitations of a model be appreciated and the assumptions used in a specific modeling exercise be clearly stated and understood: "applying a model is an exercise in thinking about the way a system works. Automating a modeling exercise to the extent that the model can be used by someone lacking the necessary background in groundwater hydrology destroys the essence of modeling. It is the thought process needed when applying a model that should lead to a decision, not necessarily and certainly not exclusively the answers generated by the model itself" (Anderson, 1983). In other words, planning decisions must be made using model results together with other information, fully recognizing the limitations and uncertainties inherent in every modeling exercise.

C. Models for Planning

The literature is packed with examples of models that potentially are useful in making planning decisions. However, very few models are sufficiently well documented to allow even the specialist to use them with ease. There are a number of computer codes that solve relatively sophisticated models that are well documented and used frequently by specialists. These include the groundwater flow models developed by Prickett and Lonnquist (1971) and McDonald and Harbaugh (1984) and the contaminant transport models by Prickett *et al.* (1981) and Voss (1984). Useful listings of available models can be found in Thomas *et al.* (1982) and Van der Heijde *et al.* (1985).

There has been some impetus within the U.S. EPA to select relatively simple models and document them in such a way that they can be used by persons having no background in modeling. An attempt to document a contaminant transport model in this way was presented by Donigian *et al.* (1983), who reduce the model to a work sheet and a nomograph.

Models that attempt to track contaminants from the land surface through the unsaturated zone to the water table include EPA's Pesticide Root Zone Model (PRZM) (Carsel *et al.*, 1984) and Cornell University's MOUSE program (Method of Underground Solute Evaluation), which has been developed in a user-friendly form for the IBM-PC (Steenhuis *et al.*, 1985).

VI. AREAS OF UNCERTAINTY AND IMPLICATIONS FOR PLANNING

A. Uncertainties

Major areas of uncertainty include quantifying the nature of heterogeneities present in aquifers (Section III,A) and in soil profiles (Section II,B), quantifying groundwater recharge rates (Section II,C), identifying the kinds of chemical and biological reactions that occur in the subsurface (Sections III,B and C), and quantifying these reactions for input to mathematical models (Section V,A). In addition it should be recognized that application of the principles in Sections I–V to fractured porous media is a new science. For example, it is still unclear whether mathematical models (Section V) derived for continuous porous media can be adapted to fractured rock systems.

B. Implications for Planning

1. *Identification of Recharge Areas*

Recharge areas as defined in Section I,C can be identified from a basinwide analysis of vertical gradients at the water table. However, it is rare to have sufficient information from nested wells (Fig. 3) to allow an analysis of basinwide vertical gradients. Attempts to locate recharge areas have been made using water table or potentiometric surface maps and the aquifer viewpoint (Section IV,A,2). These maps may help identify the origin of water flowing in the aquifer. However, analysis of such maps for the purpose of delineating recharge areas can be misleading because vertical gradients that cause recharge to enter an aquifer are ignored in the aquifer viewpoint.

Recent attention to the recharge process focuses on detailed analyses of flow in the unsaturated zone and vertical gradients at the water table. These studies may eventually lead to a practical approach to basinwide analysis that will allow recharge areas to be delineated more accurately (Stoertz *et al.*, 1985).

2. *The Safe Yield Myth*

It is clear from the discussion in Section IV,B,2 that determination of the true aquifer safe yield (the maximum stable basin yield) involves an analysis of the interaction between the saturated and unsaturated zones and between groundwater and surface water zones. Hence, determination of maximum stable basin yield is not a simple calculation. Traditionally, safe yield has been approximated to be equal to some fraction of the net annual precipitation using the flawed rationale that safe yield is equal to groundwater recharge. While such an approach may be justified on a regional scale to get a rough estimate of this type of parameter for comparison purposes (e.g., Bloyd, 1975; Fetter, 1981), the weaknesses inherent in the approach should be fully recognized. Specifically, groundwater recharge is very difficult to estimate accurately (Section II,C); equating recharge to some fraction of precipitation is at best only a rough approximation of the actual groundwater recharge. Furthermore, it is physically impossible to capture 100% of the natural groundwater recharge by pumping. Finally, the true basin yield depends on interaction of the unsaturated and surface water zones with the groundwater zone.

It may be more appropriate to plan groundwater development using the concept of optimal yield (Freeze and Cherry, 1979, pp. 364–365). Optimal yield is determined after consideration of the effects of a number of different groundwater management schemes on groundwater flow conditions within a specific aquifer. These effects might be predicted using a mathematical modeling approach (Section V). The optimum scheme best meets a set of economic and social objectives. Under certain circumstances the optimum scheme may call for groundwater development in excess of the groundwater recharge rate. For example, in arid regions there may be virtually no groundwater recharge under predevelopment conditions. However, it may be appropriate to develop the aquifer if recharge can be induced from surface water bodies or it may be appropriate to mine the aquifer. On the other hand, in coastal areas it may not be advisable to capture any of the groundwater recharge because of concern over salt water encroachment.

3. *Controlled Degradation versus No Degradation*

The aim of contaminant transport analysis in a planning context is to determine the concentration and travel times of potential pollutants at critical discharge points, such as springs, wells, or surface water bodies. One may plan to ensure that the concentration arriving at a discharge point will be below some predetermined value such as a standard set by a regulatory agency. Or one may plan to ensure that contaminant flow paths do not intersect critical discharge points. When the discharge point of interest is a well, this approach may utilize the concept of well protection zones which involves delineating the cone of depression surrounding the well (Section IV,B) and the recharge area (Section I,C) for the well. The area so defined is then zoned to prevent or restrict waste disposal (e.g., Hennings *et al.*, 1985; Zaporozec, 1985). Both of these planning strategies invoke a philosophy of controlled degradation. Alternatively, one may choose to plan that no contaminants at all enter the groundwater system. This planning strategy follows a no degradation philosophy.

While it may be desirable to apply the no degradation philosophy to certain critical and sensitive aquifers, it is unrealistic to apply it universally. Given that the subsurface is the preferred disposal medium for certain municipal wastes (e.g., in sanitary landfills), as well as hazardous and high-level radioactive wastes, and given that all engineered barriers are subject to failure, a certain amount of groundwater degradation is inevitable. It has been suggested that for certain types of municipal wastes it may be appropriate to plan for controlled degradation and therefore to spend a minimum on engineered barriers and controls. However, for hazardous and radioactive wastes, protection by natural site geology as well as engineered structures is mandatory in order to safeguard society against toxic chemicals.

Engineered structures include canisters, liners, covers, drains, grouting curtains, removal wells, and hydraulic barriers created through the use of injection wells. Natural site geology is also important in planning disposal sites. For example, porous materials with low hydraulic conductivity, such as unfractured shale and clay, are favored disposal media owing to the likelihood of slower groundwater movement and greater time for dilution (Section III,A) and chemical and biological degradation (Sections III,B and C). Similarly, high-permeability material such as sandstone and sand, as well as rocks characterized by networks of well-connected fractures, should be avoided. In particular, it should be recognized that the analysis of flow through fractured porous media is hindered by uncertainties (Section VI,A).

As a conclusion, I feel compelled to issue a warning. Recent attention to groundwater protection by government agencies, the media, and concerned citizens has drawn groundwater hydrologists into closer communication with planning and regulatory personnel. Groundwater hydrology is a relatively young science and prior to 1970 few students selected this field as a profession. Consequently, there are relatively few experienced groundwater hydrologists available to meet society's recent concerns. As a result, some persons in related areas in the physical sciences have begun to practice groundwater hydrology. Some of these professionals have managed to acquire the skills necessary to address groundwater problems effectively while others have not. Planners involved in planning for groundwater protection should be cautious when seeking advice.

REFERENCES

Anderson, M. P. (1983). Groundwater modeling—The Emperor has no clothes. *Ground Water* **21**(6), 666–669.

Anderson, M. P. (1984). Movement of contaminants in groundwater—Groundwater transport—Advection and dispersion. *In* "Groundwater Contamination," Chapter 2, pp. 37–45. National Academy Press, Washington, D.C.

Anderson, M. P. (1986). Field validation of groundwater models. *ACS Symp. Ser.* **315,** 396–412.

Anderson, M. P., and Berkebile, C. A. (1977). Discussion of Hydrogeology of the South Fork of Long Island, New York, by C. W. Fetter. *Geol. Soc. Am. Bull.* **88,** 895–896.

Barcelona, M. J., Gibb, J. P., Helfrich, J. A., and Garske, E. E. (1986). "Practical Guide for Ground-Water Sampling." Ill. State Water Surv., Champaign.

Bitton, G., and Gerba, C. P., (eds). (1984). "Groundwater Pollution Microbiology." Wiley (Interscience), New York.

Bloyd, R. M., Jr. (1975). Summary appraisals of the nation's ground-water resources—upper Mississippi region. *Geol. Surv. Prof. Pap. (U.S.)* **813-B,** 1–22.

Bouma, J., Baker, F. G., and Veseman, P. L. M. (1974). Measurements of water movement in soil pedons above the water table. *Inf. Circ.—Wis. Geol. Nat. Hist. Surv.* **27,** 1–114.

Bredehoeft, J. D., Papadopulos, S. S., and Cooper, H. H., Jr. (1982). Groundwater: The water-budget myth. *In* "Scientific Basis of Water-Resources Management," pp. 51–57. National Academy Press, Washington, D.C.

Carsel, R. F., Smith, C. N., Mulkey, L. A., Dean, J. D., and Jowise, P. (1984). "Users' Manual for the Pesticide Root Zone Model (PRZM) Release 1," EPA-60013-84-109. U.S. Environmental Protection Agency, Washington, D.C.

Cherry, J. A., Gillham, R. W., and Barker, J. F. (1984). Contaminants in groundwater: Chemical processes. *In* "Groundwater Contamination," Chapter 3, pp. 46–64. National Academy Press, Washington, D.C.

Donigian, A. S., Jr., Yo, T. Y. R., and Shanahan, E. W. (1983). "Rapid Assessment of Potential Ground-water Contamination under Emergency Response Conditions," EPA-600/8-83-030. U.S. Environmental Protection Agency, Washington, D.C.

Dreiss, S. J., and Anderson, L. D. (1985). Estimating vertical soil moisture flux at a land treatment site. *Ground Water* **23**(4), 503–511.

Eisen, C. E., and Anderson, M. P. (1979). Effects of urbanization on ground-water quality—A case study. *Ground Water* **17**(5), 456–462.

Faust, C. R., Silka, L. R., and Mercer, J. W. (1981). Computer modeling and ground-water protection. *Ground Water* **19**(4), 362–365.

Fetter, C. W. (1981). Interstate conflict over ground water. Wisconsin-Illinois. *Ground Water* **19**(2), 201–213.

Freeze, R. A., and Banner, J. (1970). The mechanism of natural ground-water recharge and discharge. 2. Laboratory column experiments and field measurements. *Water Resour. Res.* **6,** 138–155.

Freeze, R. A., and Cherry, J. A. (1979). "Groundwater." Prentice-Hall, Englewood Cliffs, New Jersey.

Gass, T. (1981). The safe-yield dilemma. *Water Well J.* p. 37.

Ghiorse, W. C., and Balkwill, D. L. (1983). Enumeration and morphological characterization of bacteria indigenous to subsurface environments. *Dev. Ind. Microbiol.* **24.**

Hansen, J. L., and Spiegel, M. H. (1983). Hydrolysis studies of aldicarb, aldicarb sulfoxide and aldicarb sulfone. *Environ. Toxicol. Chem.* **2,** 147–153.

Heath, R. C., Foxworthy, B. L., and Cohen, P. (1983). The changing pattern of ground-water development on Long Island, New York. *In* "Environmental Geology" (R. N. Tank, ed.), pp. 443–454. Oxford Univ. Press, London and New York.

Hennings, R., Thompson, M., and Wilson, T. (1985). "Don't Pollute our New Wells." Wis. Geol. and Nat. Hist. Surv., Madison.

Kimmel, G. E. (1985). Nonpoint contamination of groundwater on Long Island, New York. *In* "Groundwater Contamination," Chapter 9, pp. 120–126. National Academy Press, Washington, D.C.

Kimmel, G. E., and Braids, O. C. (1980). Leachate plumes in ground water from Babylon and Islip landfills, Long Island, New York. *Geol. Surv. Prof. Pap. (U.S.)* **1085,** 1–38.

LeBlanc, D. R. (1984). Sewage plume in a sand and gravel aquifer, Cape Cod, Massachusetts. *Geol. Surv. Water-Supply Pap. (U.S.)* **2218,** 1–28.

McDonald, M. G., and Harbaugh, A. W. (1984). "A Modular Three-dimensional Finite-difference Ground-water Flow Model." U.S. Geol. Surv., Reston, Virginia.

MacFarlane, D. S., Cherry, J. A., Gillham, R. W., and Sudicky, E. A. (1983). Migration of contaminants in groundwater at a landfill: A case study. *J. Hydrol.* **63,** 1–29.

Morrison, R. M. (1983). "Ground Water Monitoring Technology." Timco Mfg., Inc., Prairie du Sac, Wisconsin.

Newsom, J. M. (1985). Transport of organic compounds dissolved in ground water. *Ground Water Monit. Rev.* **5**(2), 28–36.

Nielsen, D. R., and Biggar, J. W. (1982). Implications of vadose zone to water resources management. *In* "Scientific Basis of Water-Resources Management," pp. 41–50. National Academy Press, Washington, D.C.

Perlmutter, N. M., and Lieber, M. (1970). Dispersal of plating wastes and sewage contaminants in ground-water and surface water, South Farmingdale-Massapequa Area, Nassau County, New York. *Geol. Surv. Prof. Pap. (U.S.)* **1879-G,** *1–67.*

Pickens, J. F., Jackson, R. E., Inch, K. I., and Merritt, W. F. (1981). Measurement of distribution coefficients using a radial injection dual-tracer test. *Water Resour. Res.* **17**(3), 529–544.

Prickett, T. A., and Lonnquist, C. G. (1971). Selected digital computer techniques for groundwater resource evaluation. *Bull—Ill. State Water Surv.* **55,** 1–66.

Prickett, T. A., Naymik, T. G., and Lonnquist, C. G. (1981). A random-walk solute transport model for selected groundwater quality evaluations. *Bull—Ill. State Water Surv.* **65,** 1–103.

Reardon, E. J. (1981). Kd's—Can they be used to describe reversible ion sorption reactions in contaminant migration? *Ground Water* **19**(3), 279–286.

Saines, M. (1981). Errors in interpretation of ground-water level data. *Ground Water Monit. Rev.* **1**(1), 56–61.

Scalf, M., McNabb, J., Dunlay, W., Cosby, R., and Fryberger, J. (1981). "Manual of Ground Water Sampling Procedures." National Water Well Assoc., Worthington, Ohio.

Skibitzkie, H. E., and Robinson, G. M. (1963). Dispersion in ground-water flowing through heterogeneous materials. *Geol. Surv. Prof. Pap. (U.S.)* **386-B,** 1–5.

Steenhuis, T. S., Van der Marel, M., and Pacenka, S. (1985). A pragmatic model for diagnosing and forecasting ground water contamination. *In* "Conference on Practical Applications of Ground Water Models," National Water Well Assoc., Worthington, Ohio.

Steenhuis, T. S., Jackson, C. D., Jung, S. K. J., and Brusaert, W. (1985). Measurement of groundwater recharge on Eastern Long Island, New York, U.S.A. *J. Hydrol.* **79,** 145–169.

Stoertz, M. W. (1985). Evaluation of groundwater recharge in the central sand plain of Wisconsin, M.S. Thesis, University of Wisconsin, Madison.

Stoertz, M. W., Bradbury, K. R., and Faustini, J. M. (1985). Delineating groundwater recharge areas using water table maps and mass balance calculations. *Eos* **66,** 883.

Thomas, S. D., Mercer, J. W., and Ross, B. (1982). "NUREG/CR-2781, "Model summary report for repository siting. Nuclear Regulatory Commission.

Van der Heijde, P., Bachmat, Y., Bredehoeft, J., Andrews, Holtz, D., and Sebastian, S. (1985). Groundwater Management: The use of numerical models. *Am. Geophys. Union Monogr.* **5,** 1–127.

Voss, C. I. (1984). "SUTRA: Saturated-Unsaturated Transport." U.S. Geol. Surv, Reston, Virginia.

Wilson, J. T., and McNabb, J. F. (1983). Biological transformation of organic pollutants in groundwater. *Eos* **64,** 505.

Wilson, J. T., McNabb, J. F., Balkwill, D. L., and Ghiorse, W. C. (1983). Enumeration and characterization of bacteria indigenous to a shallow water table aquifer. *Ground Water* **21**(2), 134–142.

Winter, T. C. (1983). The interaction of lakes with variably saturated porous media. *Water Resour. Res.* **19**(5), 1203–1218.

Zaporozec, A., ed. (1985). "Groundwater Protection Principles and Alternatives for Rock County, Wisconsin," Spec. Rep. No. 8. Wis. Geol. Nat. Hist. Surv., Madison.

2

The Institutional Framework for Protecting Groundwater in the United States

TIMOTHY R. HENDERSON
Office of Regional Counsel, Region II
Waste and Toxic Substances Branch
*U.S. Environmental Protection Agency**
New York, New York 10278

I. INTRODUCTION

A complex set of laws and institutions attempt to protect groundwater from contamination by toxic chemicals. In contrast to the programs established to respond to air and surface water pollution, no one federal law sets the standard for preventing and responding to groundwater contaminant discharges. The Safe Drinking Water Act, the Resource Conservation and Recovery Act (RCRA), and the Comprehensive Environmental Response, Compensation, and Liability Act (CERCLA or Superfund) laws implemented by the Environmental Protection Agency (EPA), each have major groundwater pollution control components; none alone is comprehensive, nor is the combination of all three. State efforts to protect groundwater are spread out among an even larger number of laws and agencies. Typically, pollution control agencies regulate industrial discharges and waste disposal (often implementing the federal programs through delegation), natural resource management agencies regulate contamination caused by withdrawal and use, departments of agriculture regulate contamination from farm practices, departments of health regulate drinking water supplies, and departments of public works regulate the construction, maintenance, and closure of

* The ideas and opinions expressed are the author's; they do not represent the views and positions of the agency.

wells. This patchwork of federal, state, and local laws fragments implementation responsibility among many federal, state, and local agencies.

In spite of the fragmentation, certain common characteristics run through these laws, implemented by different agencies, at different levels of government. Groundwater contamination regulations fall into two basic categories: (1) cleanup and (2) prevention. While the amount of groundwater already contaminated is small relative to the volume of clean water, the actual volume of groundwater contaminated or threatened with contamination is large and growing. Thousands of surface impoundments holding industrial waste, thousands of operating and abandoned industrial and municipal landfills, and millions of septic systems leak tremendous quantities of leachate to groundwater each year. A mammoth effort by federal, state, and local governments will be required to remedy these existing problems. However, groundwater once contaminated can rarely be cleaned up completely and only at great expense. Therefore prevention ultimately will be the most effective solution to groundwater contamination.

This chapter describes the existing patchwork of federal and state groundwater quality protection laws and institutions and provides a brief overview of efforts to improve their protection of this important resource.

II. OVERVIEW

A number of specific approaches or techniques for preventing groundwater contamination have been developed by states and by the federal government. EPA and many states have set drinking water standards that if exceeded should lead to the shutting down of wells. Some states have set discharge standards for problem contaminants. Many states have devised groundwater classification systems that provide greater protection for regions with high-quality aquifers, such as recharge areas or important sources of water, than for regions where the groundwater is of poor quality, inaccessible, or under land relatively impervious to leaching. EPA has established a loose classification system to guide the implementation of the agency's groundwater responsibilities (U.S. Environmental Protection Agency, 1984). Most states have well construction laws that require well drillers to use materials and covers designed to prevent the seepage of contaminants to aquifers through the well and that prescribe well closure requirements to prevent old wells from becoming conduits for contaminants. Many states, and now EPA with the recent amendment of the Resource Conservation and Recovery Act, prescribe construction and monitoring standards for underground tanks storing hazardous materials or waste. Hazardous waste disposal is regulated by EPA and the states under RCRA through the imposition of construction standards for and monitoring of waste disposal facilities, and through land disposal bans for contaminants shown to pose serious threats to human health and safety. Most states regulate the siting and use of private septic systems.

A smaller range of programs emphasizes cleanup. Superfund, administered by EPA, is the premier groundwater cleanup program. Groundwater contamination is the major environmental problem addressed by most Superfund cleanup efforts. In addition, any significant spill of a hazardous material must be reported and a cleanup effort initiated. Many states have developed state cleanup programs to supplement Superfund. RCRA also requires permitted disposal sites to clean up groundwater contaminated by leaking storage sites when leaks are uncovered through monitoring. Unresolved questions underly any cleanup effort. When is the job complete? What standard governs? Should the groundwater be cleaned up to background purity or some lesser standard? Finding the answers to these questions presently occupies the energies of many federal and state policymakers.

In spite of all these laws, implemented by all these agencies, some significant forms of groundwater contamination go unregulated, e.g., brine pits from oil and gas operations, abandoned hard rock mines, surface impoundments used to store industrial wastes considered nonhazardous under RCRA but too polluted to discharge to surface water, and tailing ponds from mines. Many lawmakers, experts, and members of the public see a need for additional, more complete laws and regulations to protect the valuable groundwater resource.

Why is the regulation of groundwater so complex and why is it less complete, or at least less mature, than protections for air and surface water? A major reason is the extent, or limited extent, of our understanding of the resource. The science of groundwater hydrogeology is in its infancy. Only recently have scientists developed an understanding about how groundwater moves through subsurface materials. Scientists only recently have learned how contaminants move and disperse with this flow. Predictions about the speed and pattern of dispersion in different geologic formations remain inexact. Consequently, policymakers have either ignored the groundwater implications of pollution control policies or devised systems that reflect the limits of scientific knowledge. The diversity of laws and institutions governing groundwater contamination arises in large part because of the diversity of sources and the multiple pathways for contaminants to reach groundwater. Virtually every form of waste disposal, from landfilling to surface water discharges, from septic systems to incineration, generates contaminants that may find their way to groundwater. Leaks and spills of chemicals and oil, as well as the planned application of road salts, fertilizers, and pesticides, also contribute contaminants. Any disturbance of land increases the opportunities for contaminant seepage, from mining to construction to the sinking of wells. And in recharge areas, any placement of contaminants creates a high probability of seepage. Perhaps most perplexing for law and policymakers, however, is the fact that contaminants flow in plumes, and slowly, in very hard to reach places.

Policymakers and legislators have turned for guidance to the well-entrenched air and surface water pollution control programs that depend on the natural processes of dispersal, dilution, and absorption to mitigate the impacts of pollutant discharges. The goal has been to find the carrying capacity of the environment beyond which unacceptable negative impacts on human health and the environment occur. With groundwater contamination, however, scientists tell us that contaminant plumes disperse very little; seldom attach to the underground materials (with the exception of the by-products of biological wastes, i.e., sewage); and, worse, tend to flow toward points of withdrawal (i.e., wells for drinking and other purposes).

Developing groundwater pollution discharge standards from the mind-set of air and surface water pollution leads to the inexorable conclusion of zero discharge standards for dangerous materials. Zero discharge standards, however, can be very expensive, and people raise the legitimate objection that not all groundwater is equally important, or equally at risk. They also beg the question of what to do about existing plumes of contaminants created by past discharges. The obvious solution is to cleanse the contaminated water, but this is an expensive proposition and it is unclear who should pay. Complicating the picture is the difficulty of identifying the sources of the contaminants, the logical group from which to exact money for cleanup. Members of Congress, of state legislatures, and of federal and state agencies are grappling with these issues in attempting to craft new laws and regulatory programs that work in a coordinated fashion to protect groundwater from contamination.

Between 1980 and 1986, federal and state legislatures and administrative agencies have worked hard to develop new and comprehensive groundwater protection programs. EPA published its Groundwater Strategy in 1984 after two false starts (U.S. Environmental Protection Agency, 1984). Congress amended RCRA to include a set of strict groundwater protections, established the National Groundwater Commission, and amended the Safe Drinking Water Act with deadlines for

EPA to set drinking water standards for many of the unregulated contaminants typically found in groundwater. Senator David Durenburger of Minnesota is working on national groundwater legislation. A few states such as Florida, Arizona, and Wisconsin have enacted groundwater quality protection legislation.

In the meantime, the nation's groundwater relies on the protections afforded by the federal and state laws and programs, described in the following sections of the chapter.

III. THE STATE ROLE IN GROUNDWATER PROTECTION

States traditionally have exercised use, management, and protection authority over groundwater. Until relatively recently, groundwater, unlike air and surface water, which flow freely and visibly across state lines, has been viewed as a resource contained within the borders of a state. While in many cases this is true, some aquifers extend under many states (e.g., the Ogallala Aquifer) and many activities that threaten groundwater are conducted by companies and individuals operating on a multistate basis. Consequently federal sensitivity to groundwater concerns has grown during the last decade, as has the reach of federal jurisdiction. Congress adopted pollution controls aimed primarily at addressing groundwater and soil contamination in the form of RCRA and Superfund. The Supreme Court ruled that groundwater is a commodity in interstate commerce subject to the restrictions and protections of the Commerce Clause of the Constitution and therefore subject to the legislative authority of Congress (*Sporhase* v. *Nebraska,* 1982, 455 U.S. 935, 12 ELR 20749). Nevertheless, states still bear major responsibility for controlling groundwater pollution.

A. The Common Law

Until recently, the focus of state groundwater management efforts has been on the access and use of the resource. Common law rules governing groundwater use evolved from English law to fit the characteristics of different areas of the United States. Not surprisingly the allocation rules that evolved for the water-plentiful eastern states differed from those of the dry western states. The English rule of "absolute ownership" gave landowners the right to withdraw as much groundwater as they wished without regard to the impact on their neighbors. This rule operated many years with little modification in the states east of the Mississippi. Today, however, the courts in most of these states have modified the English rule to place reasonable limits on wasteful and harmful withdrawals and uses. In dry western states, however, the rule of absolute ownership had shortcomings that immediately caused problems. A set of "prior appropriation" rules evolved in the western territories that gave persons with the first or earliest claims to water priority during times of shortage. These rules limited the claims to the amount of water needed for the use intended. Most western states have since incorporated these common law rules into laws that establish strict rules for the buying and selling of water rights. Groundwater appropriation rules did not always parallel the surface water rules in the western states (or the eastern states). In many states with limited rainfall and limited surface water supplies the underground water supplies seemed unlimited and therefore rules to allocate and conserve use seemed to make little sense. During the last few decades, however, people have learned that these underground sources are being depleted in much the same way that oil fields are depleted. The laws on groundwater extraction have been changed accordingly.

Four groundwater allocation doctrines govern access and use to groundwater in the states today: (1) the "English rule"; (2) the "American rule" of reasonable use; (3) the "correlative rights rule" of shared access; and (4) the "prior appropriation rule" establishing defined property rights.[1] The English rule was devised when little was known about the flow and movement of groundwater and in application provides little recognition of the interconnections of ground and surface waters. As discussed above, most states that followed the English rule eventually re-evaluated the strict application of the rule leading to the creation of the "reasonable use rule" (only Texas and some of the New England states still follow the English rule). Though the doctrine takes different form in different states, basically it places limits on withdrawals that harm the wells and springs of others through wasteful withdrawals. The correlative rights rule or the "California rule" builds on the "reasonable use rule" by providing equal access to those owning land overlying aquifers during periods of drought. The "prior appropriation systems" typically restrict groundwater withdrawals to beneficial uses and establish use priority for times of water scarcity. Beneficial uses are defined broadly but for no more water than is needed for the use. Priorities are set according to the overall benefit to the state of an activity and the date when persons first submitted their applications for a particular use. Prior appropriation systems often are administered through permit systems.

[1] For an in-depth discussion of these rules see *Restatement (2d) of Torts* (1979), Chapter 41, Note 4, pp. 254–258; and *State* v. *Michaels Pipeline, Inc.*, 63 Wis.2d 278, 217 N.W.2d 339 (1974).

These detailed state systems governing the access and use of groundwater provide little in the way of protection from or remedies for contamination. None of the four allocation doctrines includes the right to pollute as one of the prerogatives of use; on the other hand, these doctrines do not provide remedies to redress the harm caused by contamination that may occur as a result of use. If a use of groundwater contaminates the source aquifer, you can make a good argument that the use is an "unreasonable use" in states following the reasonable use doctrine, or in states with appropriation rights systems, that the use interferes with the appropriative rights of a senior appropriator. Judges, however, are reluctant to accept such claims and prefer claims under the traditional tort remedies of nuisance, negligence, and strict liability (Davis, 1984).[2]

Nuisance is one of the most widely applicable bases of liability for groundwater pollution and other types of pollution. A plaintiff must establish that the defendant's action caused a "substantial and unreasonable interference" with his land (private nuisance) or a property right held in common by the public (public nuisance). In reaching a decision on nuisance claims, judges balance the interests of the plaintiff, the defendant, and society at large. A shortcoming of nuisance actions to redress groundwater pollution is that defendants may avoid liability if they can prove that the contaminating activity makes their land productive (Grad, 1983; Rodgers, 1977).[3]

Negligence actions must establish the defendant's creation of an "unreasonable risk of harm" to others, and that the defendant failed to exercise "due care" to avoid the foreseeable harm. The key is whether the defendant knew or should have known that his actions could pollute groundwater. For example, courts are likely to find a defendant negligent where he discharges waste oil or solvents down abandoned wells and contaminates the underlying aquifer. Courts look to administrative, legislative, or other policy-making bodies for guidance in determining whether a defendant's conduct falls short of the requisite standard of care (Grad, 1983; Prosser, 1971; Rodgers, 1977).[4]

Strict liability is imposed for activities that are considered "abnormally dangerous." Courts hold defendants liable regardless of whether they act reasonably or with due care if the harm claimed by the plaintiff occurs as a result of the abnormally dangerous activity. Conducting the activity establishes the grounds for liability (Davis, 1984; Grad, 1983).[5]

[2] See *Restatement (2d) of Torts* (1979), §849, pp. 205–206.

[3] *Restatement (2d) of Torts* (1979), §821.

[4] *Restatement (2d) of Torts* (1979), §282.

[5] *Restatement (2d) of Torts* (1979), §§519 and 520.

Plaintiffs seeking relief under common law for damages to groundwater have a difficult task. Winning a suit in negligence, nuisance, or strict liability requires the plaintiff, at a minimum, to establish a causal connection between the contamination and an action of the defendant (Henderson *et al.*, 1984; Grad, 1983). Until recently our understanding of hydrology made it very difficult to determine whether a release of contaminants on the surface was the source of the same contaminants found in water drawn from wells in the nearby areas. Though knowledge about the nature of contaminant flow has increased, it remains difficult to determine, with the precision demanded by many courts, whether the contaminants released by a defendant are the source of those fouling the plaintiff's water supply. Typically, numerous potential sources of the contaminant are located in the same area. Recent trends in the evolution of common law indicate that future cases will expand liability for groundwater contamination.

Increasingly, courts have found companies using, producing as by-products, or manufacturing toxic chemicals liable for adverse health consequences attributed to exposure. Workers, their families, and those exposed by virtue of being neighbors have successfully sued for damages caused by exposure to dangerous chemicals such as asbestos, of which the health effects are well known.[6] More and more frequently courts are finding companies liable for exposure of plaintiffs to toxic chemicals where the link between exposure and health problems is somewhat murky and where the procedures followed by the defendant were the established or accepted method of the industry as a whole.[7] Two cases illustrate how these expanded theories of liability have been applied by courts for plaintiff claims of groundwater contamination. *Wood* v. *Picillo*, 443 A.2d 1244, 12 ELR 21000 (RI 1982) and *Branch* v. *Western Petroleum, Inc.* 657 P.2d 267, 13 ELR 20362 (Utah 1982).[8]

In *Picillo* the Rhode Island Supreme Court found that the defendant's hazardous waste disposal site constituted a public and private nuisance because it contaminated and threatened the continued contamination of groundwater (and surface waters through groundwater discharge). Substantial evidence presented to the trial court established that the plain-

[6] See *Borel* v. *Fibreboard Paper Products*, 493 F.2d 1076 (5th Cir. 1973) cert. denied, 419 U.S. 869 (1974) and its progeny.

[7] See *Ashland Oil* v. *Miller Oil Purchasing Co.*, 678 F.2d 1293, 12 ELR 20845 (5th Cir. 1982); *Branch* v. *Western Petroleum*, 657 P.2d 267, 13 ELR 20362 (Utah 1982); and *State DEP* v. *Ventron Corp.*, 94 N.J. 254, 483 A.2d 893, 13 ELR 20837 (1983).

[8] See also *Miller* v. *Cudahy Co*, 592 F. Supp. 976, 15 ELR 20051 (D. Kan. 1984), where the court held the defendants liable under the theory of public and private nuisance for punitive damages for polluting the aquifer that farmers depended on for irrigation through years of salt mining.

tiffs and the public were harmed by the seepage of toxic contaminants through the groundwater. The site was on a hill 800 feet above wetland headwaters for important state waters and above the plaintiff's land. Experts testified that contaminants had percolated from the defendant's landfill to the groundwater that supplied both the plaintiff's water supply and the public waters. In reaching its decision, the court emphatically rejected the prevailing nuisance rule in Rhode Island. That rule, established by the Rhode Island Supreme Court in the 1934 case of *Rose* v. *Socony-Vacuum,* 54 R.I. 411, ruled that nuisance actions claiming contamination of public or private waters by pollutants percolating through the soil and traveling underground must show not only actual harm but also that the defendant's actions were negligent. The facts of the 1934 case were similar to those in *Picillo,* but the court dismissed the case because the plaintiff failed to allege that the defendant oil company's actions were negligent. The court found the defendant's disposal practices to be standard business practice, and that the policy of the state was to encourage the survival of industry. The 1982 Rhode Island Court emphasized that the world has changed since 1934. Not only has the knowledge of groundwater improved, but "decades of unrestricted emptying of industrial effluent into the earth's atmosphere and waterways has rendered oceans, lakes, and rivers unfit for swimming and fishing, rain acidic, and air unhealthy. Concern for the preservation of an often precarious ecological balance, impelled by the spectre of "a silent spring," has today reached a zenith of intense significance." The court, in the forefront of state courts, has paved the way for the expansion of common law liability for pollution of "subterranean waters."

In *Branch,* an appellate court in Utah, citing the *Picillo* court as precedent, expanded its application of the strict liability doctrine to groundwater contamination cases. According to the record established by the trial court, the defendant's use of an evaporation pit for dumping formation water—a wastewater produced by oil wells containing oil, gas, and high concentrations of salt and chemicals—caused 66% of the contamination of the plaintiff's well water; the rest was caused by other unspecified parties. Two aquifers, one shallow and one deep, were involved. The plaintiff's well to the deep aquifer had provided clean water for drinking and culinary purposes. Both experts agreed that the water was fouled and that a new well produced bad water. They disagreed about the defendant's role in contaminating the deep aquifer. The jury believed the more experienced expert hired by the plaintiff. The dispute on appeal was whether strict liability was the proper cause of action. The defendant argued that such cases normally base "liability for pollution of subterranean waters on either negligence, nuisance, or trespass." The court disagreed, citing a long list of cases finding defendants strictly liable for harm caused by such abnormally dangerous activities as oil mining. In fact, the precedents cited by the court had never explicitly deemed that mining activities were abnormally dangerous because of the threat they pose to groundwater supplies. Going even further with its groundbreaking decision, the court broke its own procedural rule of refusing to consider causes of action not raised by the trial court and offered a suggested cause of action for future plaintiffs, that of negligence per se. It explained that negligence per se "is in reality just another term for strict liability."

B. State Groundwater Protection Programs

States have differing objectives and priorities for protecting their groundwater from contamination. Some rely totally on groundwater to supply their freshwater needs, others have only a moderate dependence. The groundwater resource varies considerably from state to state in terms of underlying geology, depth of aquifers, and abundance and quality of the water. Contamination threats also vary from state to state. Heavily industrialized states in the Midwest and East worry more about contamination from the release of the hazardous chemicals produced, handled, and disposed by industry. Arid states worry more about the chemical by-products of mining or the salts leached from irrigated farmland reaching their groundwater drinking water supplies. States in the northern snowbelt worry about the leaching of road salts used to melt ice and snow. In response to this diversity, state groundwater protection laws, policies, strategies, and institutions span a range of differences. However, they also have many features in common.

Concern about hazardous chemicals has generated a heightened interest in groundwater quality protection among states. Many have adopted management programs to complement their common law protections and federal groundwater management requirements (discussed in Section IV). These programs include contaminant standards for selected chemicals, aquifer classification, monitoring, containment and cleanup, well construction restrictions, storage tank standards, and septic systems controls. States in the forefront of program development have adopted such features as discharge permit systems (New Jersey and Wisconsin), compensation systems to replace contaminated supplies, and high-quality aquifer protection systems. Though they differ widely in design, state groundwater management programs consist of three basic components: (1) an overall protection policy; (2) a management strategy; and (3) protection techniques (adapted from Henderson *et al.*, 1984).

1. *Protection Policies*

A state's groundwater protection policy defines the level of protection its groundwater program seeks to provide. Existing and proposed state groundwater protection policies fall along a continuum from maintaining groundwater quality at its natural background levels to setting levels only as stringent as needed to protect the current and anticipated uses of specific aquifers or groundwater regions. A study by the Environmental Law Institute provides a good construct for describing this continuum by identifying three types of policies: nondegradation, limited degradation, and differential protection (Henderson *et al.*, 1984).

Nondegradation, the most protective, seeks to protect all groundwater at natural levels. Few states, if any, follow an absolute nondegradation policy. Florida, Minnesota, and Wisconsin come close but their programs acknowledge the inevitability of some degradation.

Limited degradation policies aim to maintain groundwater at as high a quality as possible, but allow contamination up to protection levels or standards. New Mexico's groundwater program operates under such a policy.

Differential protection policies protect groundwater only as needed to satisfy present and future uses. Such policies distinguish between aquifers on the basis of geologic characteristics, population dependency, depth, and vulnerability to contamination. The most popular groundwater protection policy among the states, differential protection, allows states to focus pollution prevention resources on high-quality drinking water or high-yield aquifers and to apply less stringent controls in other areas to accommodate industrial activities. For example, Wyoming, Connecticut, New York, Oregon, and Arizona all follow a form of differential protection.

The strongest argument for nondegradation policies is that they protect against our lack of knowledge about future drinking water needs and about the health effects of many toxic contaminants. The strongest argument for differential protection policies is their realistic evaluation of groundwater resources, i.e., that not all groundwater resources of usable quality are equally valuable or equally vulnerable to contamination.

2. *Management Strategies*

State groundwater management strategies dictate how much of a state's groundwater will receive the level of protection called for in the protection policy. They guide the formulation of specific protection techniques and serve to coordinate all state agencies and programs with groundwater protection responsibilities toward a common objective. Existing and proposed state programs follow one of three types of management strategy: uniform protection, contaminant or source classification, or aquifer classification (Henderson *et al.*, 1984).

Aquifer classification is the most widely followed strategy. Aquifers are classified according to the threat of contamination, importance of use, type of use, or water quality. With five classes distinguished on the basis of use, Connecticut's system is one of the most comprehensive (Connecticut Department of Environmental Protection, 1981). More restrictive protection techniques are applied to areas with higher classifications. For example, a total ban on hazardous waste disposal applies in special aquifer zones.

Contaminant and source classification systems single out sources and chemicals that cause serious pollution problems for priority attention. They mimic the discharge limitation by source approach of the Clean Air and Clean Water Acts. California's program emphasizes a contaminant management strategy by identifying common groundwater contaminants and setting safe drinking water levels for each (Henderson *et al.*, 1984). New Jersey follows a source control strategy by including groundwater contamination discharge limits in the surface water pollution discharge permits it issues to point sources (New Jersey Department of Environmental Protection, 1982).

Uniform protection gives all groundwaters an equal level of protection. It is the most practical strategy for states with uniformly high quality or highly important groundwater. Wisconsin's groundwater law, for example, emphasizes uniform protection (Wisconsin Department of Natural Resources, 1984).

3. *Protection Techniques*

States use a host of tools, protection techniques, to bring their groundwater policy and management strategies to life. Comprehensive state programs utilize as many techniques as needed to address their contamination problems. The suitability of the different techniques depends on the nature of a state's contamination problems and its hydrogeology. How the techniques are implemented depends on a state's legal/institutional establishment. In general, however, states use four basic techniques: standards implemented through permits; facility design requirements; well construction and use restrictions; and land use controls (Henderson *et al.*, 1984).

Groundwater quality standards specify maximum concentrations of contaminants allowed in ambient groundwater or describe minimum levels of groundwater quality. They are applied through permit programs; exceedances trigger enforcement actions. EPA estimates that 32 states use groundwater quality standards (EPA, 1985). New Mexico's program (New Mexico Water Quality Control Commission, 1982), for example, covers 35 chemical groups by modifying and extending federal Safe Drinking Water Act maximum contaminant limits (discussed in Section IV). Standards add certainty for groundwater users, purveyors, dischargers, and enforcement agencies. Once adopted they are relatively quick and easy to apply. Because they require extensive scientific analysis and debate, however, standards are expensive to set. In addition, standards are reactive. Once exceedances are detected the groundwater is already polluted. Given these conflicting costs and benefits, the adoption of standards by states and the federal government has been very controversial. (Editor's note: see Chapter 3 for a more thorough discussion of setting standards.)

Permit programs set discharge zone boundaries, or effluent discharge boundaries, for regulated facilities past which contaminant levels are not allowed to exceed the standards. The New Mexico program, for example, requires virtually any entity likely to discharge listed contaminants to groundwater to develop a discharge plan (New Mexico Water Quality Control Commission, 1982). The plan must show that the discharge will not exceed the applicable standard. New Jersey regulates discharges to groundwater as part of its New Jersey Pollution Discharge Elimination System program for surface water (New Jersey Department of Environmental Protection, 1982). Applicants must demonstrate that facility discharges will not contaminate groundwater, and they must comply with monitoring and recordkeeping procedures. Permits and standards fit in well with the federal pollution control programs that many states already oversee under the Safe Drinking Water Act, the Resource Conservation and Recovery Act, and the Clean Water Act (see Section III,C).

Facility design requirements specify materials, dimensions, and technologies for problem facilities to prevent the release of groundwater contaminants. The two types of facilities most commonly subject to design requirements are land disposal sites for wastes and underground chemical storage tanks. State hazardous waste programs (modeled after the federal Resource Conservation and Recovery Act requirements) typically require new landfills and surface impoundments to use plastic liners and to install leachate collection wells. States such as Maryland and South Carolina and counties on Long Island, New York, have long recognized the problem of leaking underground storage tanks and have required tank owners to comply with coating and wrapping, tank material, and double-walled storage design restrictions (Henderson *et al.*, 1984). Some programs such as the new California program require tank replacement (Editor's note: see Chapter 13.) As discussed in Section IV on federal programs, Congress has set the stage for all states to regulate underground storage tanks in this fashion through the 1984 amendments to the Resource Conservation and Recovery Act (the Hazardous and Solid Waste Amendments of 1984 (HSWA), P.L. 98-616, Title VI).

Well construction and use restrictions address the contaminant problems caused by improperly designed, built, or closed wells and those associated with overdrafting. Most states license well drillers; constructing wells without a license is illegal. To obtain a license, individuals must learn the installation, siting, and closure requirements. Well construction regulations prescribe the material, weight, and dimension of well casings and closure procedures. State well regulations also govern the withdrawal of water from the wells. These regulations allow states to keep track of volume reductions and to monitor areas suffering from drought or overdrafting. A few state laws, such as those of Florida, North Carolina, and Texas, explicitly prohibit the construction and use of wells that lead to the pollution of one aquifer by another (Henderson *et al.*, 1984).

User-oriented controls are a quick, inexpensive, and effective way for states to reduce a major source of groundwater contamination. However, the techniques are most suited to new well construction. Also, they tend not to apply to the rural users of small volumes of groundwater that dig their own wells.

Land use controls attempt to limit contamination of groundwater by restricting uses of the overlying land. They have the greatest utility for restricting activities on identifiable, sensitive areas, such as groundwater recharge zones. Land use controls for groundwater protection include all the traditional controls: critical watershed planning, zoning, siting, and building regulations and public acquisition programs. They typically rely on local government implementation through the application of the "best management practices" (DiNovo and Jaffe, 1984).

Few states rely on land use controls for groundwater protection. Vermont's groundwater program is an exception in relying solely on the development and implementation of land use controls by municipalities (Vermont Agency of Environmental Conservation, 1982). The state provides technical assistance and maps aquifer recharge zones. As discussed in later chapters of this book, many localities in many states have adopted groundwater protection features into their land use control programs.

Land use controls are the best technique for addressing pervasive problems caused by nonpoint sources, particularly in those states where groundwater contamination is caused by land use practices over large areas, e.g., the application of pesticides and fertilizers on agricultural lands. However, land use controls generally only offer protection against future uses of land, and by virtue of the fact that they are implemented mainly by local governments, they run into problems in providing consistent protection to aquifers that underlie many local jurisdictions.

C. State Primacy under the Federal Regulatory Programs

Most of the federal pollution control statutes discussed in Section IV follow the same basic pattern in shaping federal–state relationships: Congress gives a federal agency the lead role for developing the necessary scientific data, technical standards, and regulatory framework and gives states the opportunity to develop their own programs, modeled after the federal framework, in lieu of the federal program. Typically, the laws provide states with financial assistance to set up the programs. States choose to assume primacy voluntarily; those making the choice must devise programs that meet federally mandated minimum requirements. It is the job of the designated federal agency to make sure that the proposed state program adequately reflects the federal requirements before approving the program and to track the state's efforts to ensure the programs are implemented as promised. Technically, though this very rarely happens, the federal agency may take back implementation responsibilities from a state that does a poor job. This pattern of state assumption of major implementation responsibilities is followed by most of the federal environmental protection statutes with groundwater protection features, including the Resource Conservation and Recovery Act (RCRA), the Safe Drinking Water Act (SDWA), the Comprehensive Environmental Response, Compensation, and Liability Act (CERCLA or Superfund), the Surface Mining Control and Reclamation Act, and the Clean Water Act (CWA).

Many states have assumed primacy over pollution control programs authorized by these statutes. For example, the majority of states have received interim authorization to administer hazardous waste programs under RCRA, and many of these have received final authorization. The 1984 RCRA Amendments have complicated the state authorization process. States were given an additional year (until January 1986) to obtain final authorization, but they also were given new responsibilities to incorporate into their programs. Under CERCLA many states have entered into cooperative agreements with EPA to implement cleanup activities at hazardous waste sites. Almost all states and territories have assumed implementation responsibilities for the drinking water protection programs of the SDWA. For example, all the states have applied for primacy over the underground injection control (UIC) program: thirty-three have received full program implementation authority and five have received partial implementation authority (EPA Office of Drinking Water, State Programs Division, personal communication, February, 1986). Planning funds granted by EPA to states under Sections 208 and 303 of the Clean Water Act have supported the development of groundwater protection programs in a number of states, including Arizona, New Jersey, and Idaho. (Editor's note: see Chapter 6 for a case study of the use of these funds.)

IV. FEDERAL LAWS AND INSTITUTIONS PROTECTING GROUNDWATER

Many federal laws and agencies manage groundwater through regulation, technical assistance, research, funding, and land management. In 1983 the Department of Interior published a Directory of Groundwater Programs listing forty-four federal programs administered by such diverse agencies as the Environmental Protection Agency, the U.S. Geological Survey, the Office of Surface Mining, the National Science Foundation, and the Nuclear Regulatory Commission (NRC) (U.S. Department of Interior, 1983). Despite their numbers and variety, these federal programs have provided an incomplete and poorly coordinated system for protecting groundwater from contamination. Federal groundwater pollution control responsibilities are spread among at least eight federal laws: the Safe Drinking Water Act, 42 U.S.C. §§300f *et seq.*; the Resource Conservation and Recovery Act, 42 U.S.C. §§6901 *et seq.*; the Comprehensive Environmental Response, Compensation, and Liability Act, 42 U.S.C. §§9601 *et seq.*; the Clean Water Act, 33 U.S.C. §§1251 *et seq.*; the Federal Insecticide, Fungicide, and Rodenticide Act, 7 U.S.C. §§136 *et seq.*; the Toxic Substances Control Act, 15 U.S.C. §§2601 *et seq.*; the Atomic Energy Act, 42 U.S.C. §§2011 *et seq.*; and the Surface Mining Control and Reclamation Act, 30 U.S.C. §§1201 *et seq.*

Since EPA administers most of the federal regulatory programs protecting groundwater (the NRC and Office of Surface Mining administer the others), it has assumed the responsibility of piecing together a more coherent, complete, and coordinated system of federal groundwater protection. In 1984 the agency published the EPA Ground-Water Protection Strategy and set up an Office of Groundwater to oversee the implementation of the Strategy (U.S. Environmental Protection Agency, 1984). Progress, though slow, is being made by the office in convincing the rest of the agency to bring the strategy to life.

This section lays out the basic groundwater protection features of the eight laws and then describes the EPA Ground-Water Protection Strategy.

A. The Safe Drinking Water Act

Enacted in 1974, the Safe Drinking Water Act (42 U.S.C. §300f *et seq.*) is designed to ensure that public water supplies are safe to drink. Its jurisdiction covers groundwater aquifers when they are the source of drinking water supplies, which is very common since groundwater serves as the primary source of drinking water for over 50% of the population. The law establishes three programs: a system of national standards and treatment technologies that provide end-of-the pipe protection; a system that regulates the disposal of wastes through underground injection, the underground injection control (UIC) program; and a program to protect aquifers that are the primary source of drinking water for a community, the sole source aquifer program. EPA was given the responsibility to set up the programs, but states implement them.

Congress amended the act in June of 1986 (Safe Drinking Water Amendments of 1986, P.L. 99-339), adding significant new requirements to each of the three programs, and a new program designed specifically for groundwater supply protection, the "well-head protection area" program. States are directed to develop programs to protect the surface and subsurface areas surrounding wells that supply the public water systems from contamination. As originally enacted and as amended, states are given the primary implementation responsibility, while EPA establishes the standards and guidelines. As it did with the RCRA amendments of 1984, Congress sets a tight schedule within which EPA and the states must implement the new SDWA responsibilities.

1. *Drinking Water Standards*

The SDWA gives EPA broad authority to adopt both primary (health-based) and secondary (welfare-based) standards (42 U.S.C. §300g-1) for public water systems. The law calls for phased adoption by EPA of interim and revised primary standards for public water systems, defined as one with a minimum of 15 service connections or regular service to at least 25 people at least 60 days a year. A primary standard for a substance assigns a maximum contaminant level (MCL) that if exceeded establishes grounds for enforcement actions. The law directs EPA to set an MCL for a substance as close as possible, after considering costs of treatment technologies, to the concentration level which would have "no known or anticipated adverse effects on health" and which builds in an adequate margin of safety.

EPA efforts to fulfill this mission have proceeded slowly. Interim standards were proposed for basic drinking water pollutants such as coliform bacteria, metals, and salts in 1975 and radionuclides in 1976. While this basic list has been added to lately (for vinyl chloride, trichloroethylene, and pesticides), it retains its interim status. Less formal standards called Health Advisories have been developed for many other compounds such as trichloroethylene, PCBs, benzene and EDB because their potential health effects are not known. Many of these compounds are the ones that frequently contaminate groundwater (Henderson *et al.*, 1984). (Editor's note: see Chapter 3 for more discussion of the health effects and relationship to MCLs, and Chapter 14 for proposals for establishing additional MCLs.)

The basic regulatory purpose served by the interim MCLs is to set the minimum standard for public water supplies. When a designated state agency (or EPA in those states without any enforcement capability) discovers a compound in the water supply in excess of the MCL they either shut the water supply off or require treatment back to the MCL. EPA recommends the same response when health advisories are exceeded, but the law does not compel that result.

Standards set under the SDWA, both the interim MCLs and the Health Advisories, have taken on a larger role in groundwater protection than that assigned by the Act. Many of the states and localities that have recently adopted groundwater protection programs either reference the standards or actually adopt them. The level of cleanup required under the Superfund is often calculated in reference to the SDWA standards when background levels are difficult to determine. Many factors explain this phenomenon. Setting standards requires a high level of technical expertise and money, commodities state legislatures and agencies typically do not have. "How clean is clean" is the great unanswered question among groundwater policymakers grappling with hazardous waste cleanups. Reliance on one of the few standards around is logical, especially since the standards are set for protecting public health. Critics, however, point out that the drinking water standards are minimum standards, set assuming that prior to consumption the water will be treated using available treatment technologies. They argue that relying on such standards for cleanup or for protection may allow unnecessary contamination and generate unwarranted belief among users that the water meeting the standard is safe to drink without treatment. Such reliance is a special concern in rural areas where wells provide drinking water. After a Superfund cleanup well water may meet the applicable MCL yet, because it receives no treatment on withdrawal, may be unfit to drink.

The 1986 amendments greatly expanded the scope of the drinking water standards program. SDWA, §1412. Congress requires EPA to establish final national primary drinking water regulations for all the chemicals that presently have interim standards or health advisories (all 83 chemicals EPA has identified in the Advanced Notice of Proposed Rulemakings in 47 Fed. Reg. 9352 and 48 Fed. Reg. 45502). Within the next 3 years (by June 19, 1989), EPA must develop final standards for all 83 chemicals according to a phased schedule (9 the first year, 40 the next and 34 the final year). EPA is given the discretion to substitute 7 contaminants for those on the list if it finds that regulating the substitutes will provide better protection. In addition, EPA must set up an advisory working group including members of the National Toxicology Program, EPA's Offices of Drinking Water, Pesticides, Groundwater, and any others the agency deems appropriate, to identify additional chemicals that may contaminate drinking water. The group must produce a list "not later than January 1, 1988" and every three years thereafter. In turn, EPA is directed to develop standards for the chemicals identified and to ensure that standard development for these newly identified chemical contaminants does not go as slowly as it did under the 1974 requirements, Congress has required EPA to set regulatory standards for 25 chemicals from the list published in 1988 by 1991 and 25 more every three years thereafter.

Standards set under the new amendments will consist of a maximum contaminant level goal and a primary drinking water regulation standard designed to come as close as possible to achieving the goal. The amendments direct EPA to set the maximum contaminant level goals "at the level at which no known or anticipated adverse effects on the health of persons occur and which allows an adequate margin of safety." §1412(b)(4). The regulatory standard must be set as close to the goal as is feasible, which the law defines as the level that can be attained by "the best technology, treatment techniques [available] . . . after examination for efficacy under field conditions and not solely under laboratory conditions," and after taking cost into consideration. §1412(b)(5). To illustrate what it means by feasible technology, Congress points out that granular activated carbon is feasible for the control of synthetic organic chemicals. The maximum contaminant level goal therefore is the ideal, and the regulatory standard is the level end-of-the-pipe treatment technologies can achieve.

Potentially, the new standard-setting requirements offer expanded protection for groundwater. First, maximum contaminant level goals and protection standards will be set for many of the chemical constituents that frequently contaminate underground drinking water supplies.

Secondly, as said before, many other regulatory programs, state and federal, use the drinking water standards as guides for cleanup and protection. By ensuring that EPA develops many more standards according to a set schedule, especially the maximum contaminant level goals, the amended public drinking water supply section of the SDWA has the potential for achieving far-reaching improvement in groundwater protection. Given EPA's past history of missing statutory deadlines, however, especially in this era of limited federal funding for EPA program development, the safe prediction is that the agency will miss the new SDWA deadlines. (*See* generally, EESI & ELI, 1985). The more important question is whether the new requirements will stimulate the SDWA program to develop a substantial amount of new drinking water standards in relatively few years in contrast to its history of developing very few standards over a 12-year period. Hopefully, by the beginning of the next decade, the answer will be yes.

2. *The Underground Injection Control Program*

The underground injection control program (UIC) regulates by permit the deep-well injection of waste that might endanger underground sources of drinking water [42 U.S.C. §300h(d)(2)]. Injection wells "endanger" drinking water supplies under the SDWA if they contaminate the water to a level that requires the water to be treated to meet the MCL for the pollutant involved. EPA regulations, adopted in 1980 and amended in 1981, 1982, 1983, and 1984, set engineering, performance, monitoring, and information submission guidelines for different classes of disposal wells to prevent endangerment. Injection of most types of wastes requires a permit, but certain types of existing injection wells were automatically granted permit status if the owners or operators submitted specified information (including injections of brines for oil and natural gas production). States may assume authority to implement the UIC program if they adopt programs that follow the guidelines set by EPA.

The 1986 Amendments of the SDWA require EPA to monitor more closely the potential for contamination of underground drinking water supplies by underground injection of hazardous and nonhazardous waste. By December 19, 1987, EPA must amend its Class I injection well-monitoring requirements. Class I wells inject hazardous waste deep below the surface, often under high pressure, but no closer than 1/4 mile of a drinking water well. Recent studies raise concerns about the threat of pollution associated with deep well injection. (Brown, 1986; Gordon and Bloom, 1985). The new monitoring regulations must identify monitoring techniques and methods that "provide the earliest possible detection of fluid migration into, or in the direction of, underground sources of drinking water from such wells," §1426(a), and require Class I well permit holders and applicants to adopt such monitoring techniques. By implication, the regulations that EPA develops probably should direct that when the monitoring systems indicate that use of an injection well poses the "potential for fluid migration from the injection zone that may be harmful to human health or the environment," §1426(a), injection of the wastes should cease. In addition, no later than September 1987 EPA must send a report to Congress that recommends "design, construction, installation, and siting requirements" for injection wells that discharge nonhazardous waste (Class V wells) into or above underground sources of drinking water, §1426(b). The report must examine the number and extent of class V wells that discharge near underground drinking water sources, identify the primary contamination problems posed by the wells, and conclude with recommended solutions.

The 1984 amendments to RCRA [P.L.98-616, §405(a)] further limit the underground injection of waste by adding 42 U.S.C. §6979 (a). The provision prohibits the disposal of hazardous waste through underground injection (1) into a formation which contains an underground source of drinking water within one-quarter mile of the well used for such underground injection or (2) above such a formation. The prohibition took effect in May 1985.

3. *Sole Source Aquifers*

The sole source aquifer program authorizes EPA to designate and protect aquifers, on petition from states or at its own discretion, that serve as the principal drinking water supply for an area [42 U.S.C. §300h - 3(e)]. Designated aquifers are protected through restrictions on the federal spending for projects in the aquifer recharge zone. EPA reviews such projects to determine whether they may contaminate the aquifer and create a significant hazard to public health. Negative determinations result in the withholding of federal funds. According to agency officials, while EPA has made very few negative determinations, many projects have been redesigned so that the aquifers are better protected (personal communication, November 1985).

The 1986 amendments to the SDWA enhance the protections afforded to groundwater by the sole source aquifer provisions, §1427. Within one year of enactment (by June 19, 1987), EPA must develop regulations that set criteria for identifying critical aquifer protection areas. The regulatory criteria must, at a minimum, consider the vulnerability of the aquifer to contamination, the number of persons that rely on the underground source for drinking water, and the costs and benefits associated with protection versus degradation, §1427(d).

States, municipal or local governments or regional planning entities with jurisdiction over an area would then apply to EPA to have selected areas within or encompassing sole source aquifers designated as aquifer protection demonstration areas. Applications from local and regional government must be accompanied by a joint application filed by the state. Presumably, Congress wanted to give the states an opportunity to prioritize the designations sought within their boundaries. To aid EPA in its decision, the application needs to define the boundary of the area, designate the lead planning agency, establish procedures for public participation in the development and implementation of the project, assess the surface and groundwater resources in the area, and contain a "comprehensive plan" of protection including schedules for implementation. After receipt of the application, EPA has 120 days to grant its approval or disapproval. Approval is conditional on whether the application satisfies the regulatory criteria. If approved, EPA and the applicant enter into a cooperative agreement to set up the demonstration project, and EPA may grant up to 50% of the project cost up to a limit of $4,000,000 per aquifer.

The amendments explain in some detail what a complete comprehensive plan must include and may include. For example, it shall identify "existing and potential point and nonpoint sources of ground water degradation," §1427(f)(1)(B); "[a]n assessment of the relationship between activities on the land surface and ground water quality," §1427(f)(1)(C); "[s]pecific actions and management practices to be implemented in the critical protection area to prevent adverse impacts on ground water quality," §1427(f)(1)(D); and the legal authority and funding to be relied on to implement the plan, §1427(f)(1)(E). Examples of what a comprehensive plan may comprise include: "[a]ctions in the special protection area which would avoid adverse impacts on water quality, recharge capabilities, or both," §1427(f)(2)(E); "[c]onsideration of specific techniques, which may include clustering, transfer of development rights, and other innovative measures sufficient to achieve the objectives of this section, "§1427(f)(2)(F); and "[p]ollution abatement measures, if appropriate," §1427(f)(2)(I).

The driving force for the program is the federal funding. Congress authorized yearly program funding amounts ranging from $10,000,000 in 1987 to $17,500,000 in 1991 (the last year of the program). Except for the inducement to obtain the federal matching grants, the amendments include nothing to encourage or require states to participate in the program. Given the limited success of the sole source aquifer program set up in 1974, also a volunteer program, serious questions remain about the potential for success of the sole source aquifer demonstration program. As of September 1986, only 21 sole source aquifers had been designated nationwide, and approximately 18 are pending (Personal communication, EPA Office of Drinking Water, September 1986), twelve years after the program began. These 39 areas, consisting of both those designated and those pending, represent the locations from which local governments will select sole source aquifer demonstration program areas. Perhaps with the heightened awareness of the health threats associated with contaminated groundwater among states and EPA, the new sole source aquifer protection program will have greater success. Only time will tell.

4. *Wellhead Protection Areas*

In concept, wellhead protection concentrates regulatory effort on the surface and subsurface area subject to the influence of a well. With the 1986 SDWA Amendments, Congress has devised a system to protect such areas from contamination. Wellhead protection areas are defined by the amendments to generally include "the surface and subsurface area surrounding a well or wellfield, supplying a public water system, through which contaminants are likely to move toward and reach such well or wellfield," §1428(e). States are required to develop programs designed to identify "wellhead areas" §1428(a), and to protect the areas "from contaminants which may have an adverse effect on the health of persons" §1428(e), within 3 years of enactment or by June 19, 1989. EPA must issue technical guidance by June 19, 1987 to help states identify the extent of wellhead protection areas. EPA is also given the role of reviewing and approving state programs, §1428(c). States must submit their programs to EPA for approval. If EPA approves a program the state may receive between 50% and 90% federal funding support. Congress authorized the appropriation of $20,000,000 for the purpose in 1987 and 1988, and $35,000,000 for the years 1989–1991.

To ensure consistent protection of wellhead areas nationwide, Congress directs state programs to at a minimum:

- specify the duties of the agencies and local governments assigned to develop and implement the programs;
- define the wellhead protection area for each wellhead (in accordance with the guidelines EPA develops) using available groundwater flow, recharge and discharge information;

- define all "potential anthropogenic [derived from human activities] sources of contaminants" that may threaten human health;
- set up a program that includes all the tools needed to protect the water supply within the well head protection area (e.g. financial assistance, "control measures," education);
- establish contingency plans for providing alternative water supplies; and
- require that decisions to locate new wells evaluate the effect of potential contaminant sources in the expected wellhead area.

A key element of each state program will be how they identify and define their groundwater protection areas. Recognizing this, Congress gives EPA some guidance on what factors the agency's wellhead protection area identification guidelines for states should contain. Section 1428(e) says that the guidance "may reflect" a host of factors that relate to the hydrogeologic conditions of the well location, such as the extent of the well's influence on the groundwater flow patterns in an area, the geology of the formation in which the well operates, and the rate of travel of contaminants in such formations given the hydrologic conditions (reflecting the amount and rate of water flowing in the ground and at the surface). The overriding consideration should be "the likelihood of contaminants reaching the well or wellfield."

How the specific features of the wellhead protection area program will evolve provides fruitful ground for speculation. Arguably a state must define the wellhead protection area for each drinking water well that serves a public water supply. For states that rely on underground supplies this represents a major undertaking. Complicating the task is the requirement that all potential "anthropogenic" sources of health-threatening contaminants be identified. Defining the universe of sources covered will be a controversial effort. One message that comes across loud and clear for anyone who reads the amendments is that Congress contemplated the development by states of comprehensive programs designed to protect groundwater water wells serving public drinking water supplies from contamination by controlling the use of the land around the wells.

B. The Resource Conservation and Recovery Act

The Resource Conservation and Recovery Act regulates the "life cycle" of hazardous and other solid wastes from cradle to grave (42 U.S.C. §6901 *et seq.*). The 1984 Amendments (HSWA, P.L. 98-616) added significant new requirements to govern the storage of hazardous substances in underground tanks and to minimize the land disposal of hazardous wastes. Theoretically, RCRA is designed to prevent our waste disposal practices from threatening the environment through unregulated discharges, releases, or seepage to the environment (including the air, surface water, soil, and groundwater) from landfills, dumps, pits, ponds, lagoons, and new storage tanks. In practice, RCRA has protected groundwater insofar as the law eliminates the widespread, uncontrolled disposal of wastes. Hazardous waste treatment, storage, and disposal facilities must be permitted and the permit requirements include groundwater monitoring. Once implemented, the underground storage provisions should stanch much of the leaks of chemicals from storage tanks and pipes. Spills, leaks, and seepage of solid or hazardous wastes to groundwater that threaten public health or the environment are regulated by the imminent hazard and corrective action provisions [§§3004(u), 3008(h), and 7003] that authorize EPA, or states with approved RCRA programs, to compel containment or cleanup.

Enacted in 1976 (amending the Solid Waste Disposal Act), RCRA distinguished between "hazardous waste" and "other solid wastes." Most of the public focus and interest has been on the "hazardous waste" control provisions found in Subtitle C, because hazardous wastes pose a greater threat to public health and the environment. "Other solid wastes" are covered by Subtitle D.

1. *Hazardous Waste (RCRA Subtitle C)*

The law governs hazardous waste from production to disposal (from cradle to grave). Key protections for groundwater include: site-specific performance standards for treatment, storage, and disposal (TSD) facilities, including use of liners to prevent leaks and leachate collection and removal systems; monitoring systems for both new and existing TSD facilities; concentration limits or protection standards built into TSD permits based on MCLs developed under the SDWA; and corrective action procedures for all spills, releases, or other accidents by generators, transporters, and TSDs that call for containment and cleanup. RCRA Subtitle C has spawned a confusing array of regulations.

A brief historical survey of the law's implementation will aid the understanding of the law's key groundwater protection provisions, those applicable to TSD facilities using surface impoundments, waste piles, or landfills. EPA published the first set of RCRA regulations in May 1980. These set forth the hazardous waste listing criteria and a list of many hazardous constituents (40 CFR Part 261 adopted pursuant to §3001), the generator and transporter regulations (40 CFR Parts 262 and 263 adopted pursuant to §§3002 and 3003), the state delegation guidelines explaining how states could obtain both interim and final program authorization (40 CFR Part 266 adopted pursuant to §3006), and those explaining to TSDs how to obtain interim status and final permits (40 CFR Parts 264 and 265 adopted pursuant to §§3004 and 3005). The specific storage, disposal, and treatment regulations for different types of facilities have come out in several packages, the last and most difficult to produce being the land disposal regulations. All generators, transporters, and TSDs were required to notify EPA that they were handling hazardous waste within 90 days of the May 1980 issuance of the regulations for listing hazardous wastes. At that point all TSDs were required to apply for and obtain a Part A permit. Since 1983, EPA has exercised its authority to call in Part B permit applications from all TSD facilities for final permit processing. The idea was to bring all existing TSDs under the more stringent permit requirements, including strict groundwater monitoring and facility design standards to protect groundwater.

Congress further complicated the RCRA regulatory scheme with the enactment of the HSWA Amendments in November of 1984. HSWA Section 213(a), which amended RCRA §3005(e), required all existing TSD facilities to submit their Part B applications for final permits by November 8, 1985, and required those facilities with land disposal units (e.g., surface impoundments, landfills, and waste piles) to certify that they were in compliance with the groundwater monitoring requirements by that date. Those facilities choosing to close, instead of submitting their Part B permit application and compliance certification, were required to initiate closure procedures that ensure no contamination of the environment including groundwater and soils.

RCRA's groundwater monitoring requirements for TSDs with land disposal units, coupled with the removal and treatment requirements for contamination detected through monitoring, are the principle groundwater protection features of the law. EPA regulations for implementing groundwater monitoring set out a phased sequence of increasingly stringent monitoring requirements that lead to cleanup and correction.

Interim status regulations required TSDs to set up a groundwater monitoring system to detect any releases from their facilities to groundwater and, if so, to characterize the contaminants. The regulations direct TSD facilities operating under interim status to sink three monitoring wells, one up-gradient of the groundwater flow and three down-gradient. Samples are analyzed for changes against background of four indicator parameters (pH, specific conductance, total organic carbon, and total organic halogen). If changes are detected which indicate problems, assessment monitoring commences. The first step in assessment monitoring is to determine whether hazardous waste constituents have migrated from the facility to groundwater. If so, the facility must sink additional wells to define the vertical and horizontal extent of the plume. If the interim status groundwater assessment monitoring establishes that hazardous waste constituents have migrated from the facility, then either cleanup and containment strategies are incorporated as conditions of the facility Part B permit (interim status groundwater monitoring results are to be submitted as part of the TSD's Part B permit application) or, where no Part B application for a final permit is submitted, EPA brings an enforcement action under RCRA §§3008(h) and 7003, or CERCLA §106, to remedy the problem.

The TSD permit regulations set up a three-stage program to detect, evaluate, and correct groundwater contamination. The first stage, detection monitoring, requires permittees to monitor for indicator parameters specified in the permit. The indicator parameters are tailored to the wastes managed at the permitted facility. Similar to the detection monitoring under interim status, the system is intended to serve as a trigger for more comprehensive monitoring if problems are detected. Monitoring results showing changed indicator levels trigger the more rigorous compliance monitoring. Those interim status TSD facilities that have already initiated assessment monitoring launch right into compliance monitoring. Compliance monitoring requires the TSD to set up a monitoring system for a broad array of hazardous constituents. The compliance monitoring system must contain clusters of wells similar to those required for interim status assessment monitoring. Where monitoring results from these wells show that the concentration limits in the permit are exceeded, the permittees are supposed to set up a corrective action program. Corrective actions must reduce contamination below the concentration limits listed in the permit.

Another important groundwater protection feature of RCRA is the design and operating standards for land disposal facilities. As a result of HSWA, all new and existing land disposal facilities (surface impoundments, waste piles, and landfills) eventually will be held to comply with double liner and leachate collection and removal systems (42 U.S.C. §6925). The 1984 Amendments changed the requirements that only new facilities install liners by directing existing facilities to retrofit, though the exact features of the liner/leachate collection system vary according to the type of facility and whether it is new or existing. The requirements for existing facility retrofits will be phased in over the next few years.

An overall thrust of the 1984 RCRA Amendments is to minimize the amount of hazardous waste generated by industry and reduce its disposal on land to a minimum by 1990. Congress built a graduated series of constraints and deadlines into the law to achieve these goals. HSWA §224 amended RCRA §3002(b) to require hazardous waste generators, as of September 1985, to certify that they have a program in place to reduce the volume and toxicity of their wastes. HSWA §201 amended RCRA §3004(c) to set deadlines banning the placement of liquids in landfills (by February 1986) and in any TSD facility (i.e., storage containers, etc.) that presents a risk of contaminating underground drinking water supplies (by November 1985). HSWA §201 also amended RCRA §§3004(d) and (g) to set up a system that could lead to the extinction of land disposal as a viable method of hazardous waste disposal by May 1990. Essentially the law prohibits the land disposal of any hazardous waste that *may* present a threat to human health or the environment *for as long as the waste remains hazardous.* EPA has the burden of justifying exceptions, taking into account the persistence, toxicity, mobility, and propensity to bioaccumulate for the life of the waste in the landfill. In combination, these new provisions bring RCRA closer to being the federal law that is designed to *prevent* the contamination of groundwater by hazardous waste disposal practices.

2. *Other Solid Wastes (Subtitle D)*

EPA governs the disposal of wastes other than those fitting the definition of hazardous waste, both liquid and solid, by providing states with technical and financial assistance to regulate nonhazardous solid waste facilities, such as municipal landfills and open dumps. The Subtitle D regulations encourage the development by states of environmentally sound methods of solid waste disposal. Each state is called on to develop a plan to implement the guidelines and to assume enforcement responsibility. However, EPA has no legal authority to require states to follow the guidelines.

The Subtitle D guidelines contain a few specific groundwater protection measures. They prohibit solid waste facilities from contaminating current or potential underground drinking water sources beyond the solid waste disposal site boundary, or an alternative boundary selected by a court.

3. *Leaking Underground Storage Tank Controls*

The provisions included in HSWA for leaking underground storage tanks (LUST) add a potent new weapon to the federal regulatory arsenal for fighting groundwater contamination (HSWA §601 which added Subtitle I, §§9001-9010 to RCRA). They require owners and operators of the storage tanks within the jurisdiction of the law to notify state agencies and to comply with release detection, prevention, and correction regulations. Congress assigned states the major implementation responsibilities and EPA the responsibility for drafting regulations and guidance. Jurisdiction of the LUST provisions extends to any tanks, or combinations of tanks, including pipes storing "hazardous materials" that have 10% or more of their bulk underground. "Hazardous materials" includes all substances covered by CERCLA in addition to oil and oil byproducts. Substances covered by RCRA Subtitle C (hazardous wastes) are exempt because they are already addressed by the TSD regulations. The number of tanks that will be brought under the jurisdiction of the law is staggering. EPA estimates that two million tanks are subject to the law, that 100,000 leak, and that 350,000 more will spring leaks in the future (U.S. Environmental Protection Agency, 1984).

By May of 1986 owners of such underground storage tanks must notify the appropriate state or local agency that they own the tank and give its age, size, type, location, and uses (§9002). The notification requirement extends to all tanks in use since January 1, 1974, even if they have been taken out of operation. All owners of tanks built after May 1986 have 30 days to notify the designated agency about the tank's existence, size, and intended use.

The release detection and prevention features of the law require EPA to develop regulations that require tank owners and operators to set up leak detection and correction systems; to set out tank design, construction, installation, and compatibility standards; and to establish financial capabilities to clean up spills and properly close tanks (§9003). The law gives EPA the discretion to make distinctions in the regulations reflecting both the nature of the tank (e.g., tank type and size, kinds of materials stored, and ages) and the nature of the land where the tanks are located (including climate, soil type, water table, and hydrogeology).

C. Comprehensive Environmental Response, Compensation, and Liability Act of 1980 (CERCLA)

CERCLA, or Superfund, authorizes EPA to initiate removal and/or cleanup of hazardous substance disposal sites, to seek compensation for the cost of cleanup or other corrective actions from responsible parties, and to initiate cleanup, abatement, and enforcement actions to minimize the threat to health and the environment from spills of hazardous substances (42 U.S.C. §§9601 *et seq.*). Groundwater is included in the statutory definition of the "environment" that Superfund is intended to protect.

CERCLA establishes three basic programs. First, the Hazardous Substances Response Fund, or the "Superfund," finances government containment or cleanup responses to actual or threatened releases of substances that may harm human health or the environment, including groundwater. It is sustained mainly by taxes on petroleum products and chemical feedstocks. Second, the liability provisions authorize EPA to hold polluters liable for the expenses of removal, cleanup, and containment, as well as to force the responsible parties to undertake such actions at their own expense. Third, the response provisions obligate private and government entities to report spills of hazardous substances in excess of specified quantities to EPA or the Coast Guard.

Groundwater cleanup and contaminant containment has been a major EPA concern in implementing Superfund. Of the 541 sites listed, as of January 1986, for priority attention, over 400 were selected, at least in part, because of existing or threatened groundwater contamination. Many contaminant soil removal actions have worked to prevent leaching of the material to groundwater. Also, the National Contingency Plan, which describes the way the federal government will respond to chemical pollution incidents, contains extensive provisions for minimizing the health and environmental hazards posed by releases to soil and groundwater. (Editor's note: see Chapter 8, the Wausau, Wisconsin case study, for discussion of CERCLA involvement.)

However, though the law protects groundwater from some hazardous waste sites and waste spills, it is designed primarily to remedy existing contamination rather than to prevent such problems. CERCLA provides relatively limited funds to throw at a rather large universe of abandoned hazardous substance disposal sites and EPA resources are stretched thin implementing CERCLA and five other major environmental laws. While the law's liability provisions do set up a system for making those responsible for a site pay for cleanup, the responsible parties for many sites cannot be found. Even where responsible parties have been identified, cost recovery has proven difficult in terms of both legal maneuvering and calculating the proper amount to seek to recover.

D. The Clean Water Act

While the Clean Water Act (CWA) focuses mainly on protecting surface water quality, it addresses groundwater quality protection in limited ways. Section 303 gives EPA the authority to require states to promulgate groundwater quality standards where a clear "hydrologic nexus" has been established between ground and surface waters. Even though some experts argue that the language supports a more expansive reading, EPA interprets the authority to apply in very limited circumstances. As a result, the agency has not used this section to encourage states to adopt groundwater standard programs. Some states, such as Pennsylvania, however have chosen to include groundwater quality provisions in their water quality standard regulations (33 U.S.C. §1313).

The planning provisions of the law require EPA to oversee the development of comprehensive management plans by states for "nonpoint sources" of pollution, such as erosion from agricultural lands and construction sites and storm water discharges to both ground and surface water. Section 208, for example, requires designated state and local agencies to plan for "disposal of pollutants on land or in subsurface excavations . . . to protect ground and surface water quality" (33 U.S.C. §1288). Many states have used §§208 and 303 planning funds to develop groundwater management area protection plans and a few have used the money to develop regulatory programs. New Jersey, for example, used CWA planning funds to help offset some of the costs of developing its comprehensive groundwater permit program. The EPA Groundwater Protection Strategy (U. S. Environmental Protection Agency, 1984) identifies planning funds from §205(j), or 33 U.S.C. §1285(j), as one of the principal sources of support for state program development.

Section 201, which governs federal grants for the construction of sewage treatment plants, directs EPA to establish grant conditions to protect groundwater (33 U.S.C. §1281). Projects employing land application techniques to reuse and recycle nutrients must protect groundwater for present and projected future uses, based on present quality.

E. The Federal Insecticide, Fungicide, and Rodenticide Act and the Toxic Substances Control Act

Under the Federal Insecticide, Fungicide, and Rodenticide Act (FIFRA, 7 U.S.C. §§135-135k and 136-136y) and the Toxic Substances Control Act (TSCA, 42 U.S.C. §2601 *et seq.*), EPA regulates a variety of chemicals at the production, distribution, and application end of commerce. Many of these may contaminate groundwater through spills or even through normal use (e.g., pesticide applications).

FIFRA authorizes EPA to control pesticide use through registration. The agency has developed registration and testing guidelines for determining the potential for pesticides to leach to groundwater through normal patterns of use. Pesticide manufacturers may be asked to monitor the experimental or actual use of pesticides suspected of contaminating groundwater. For example, EPA has the authority to limit potential damage by banning the use of a pesticide in recharge areas where monitoring shows that groundwater contamination has occurred.

Under TSCA, EPA regulates the manufacturing, processing, use, and disposal of toxic chemicals through notification. Theoretically, where a chemical has the potential to contaminate groundwater and create an "unreasonable risk" to health or the environment, EPA may: (1) place restrictions on the use of the chemical; (2) require warning labels; (3) mandate that users adopt application procedures to control pollution; or (4) require the development of special disposal plans. For example, the agency's PCB regulations contain restrictions designed to prevent and clean up spills that could enter groundwater. Landfills permitted by EPA to receive PCB wastes are required to conduct groundwater monitoring pursuant 40 CFR §761.75. In practice, however, EPA rarely exercises its TSCA authority for the purpose of groundwater protection.

F. Other Federal Laws

Other federal agencies administer regulatory programs that offer some protection of groundwater quality. The Nuclear Regulatory Commission (NRC) regulates the leaching of contaminants from uranium mill tailings under the Atomic Energy Act (42 U.S.C. §2022). The Department of Interior's Office of Surface Mining oversees the Surface Mining Control and Reclamation Act of 1977 (SMCRA, 30 U.S.C. §1265). SMCRA contains strict provisions to prevent groundwater contamination and disruption of groundwater flow patterns from coal mining. The law, however, exempts hardrock mining for minerals and metals.

G. The EPA Groundwater Strategy

After many false starts, EPA issued a Ground-Water Protection Strategy in August of 1984. The first draft policy had been prepared during the Carter Administration. An additional two drafts were prepared during the Reagan Administration's first term, the latter becoming the version that won approval.

Why the need for a policy? EPA's authority to regulate activities with a potential for contaminating groundwater is fragmented among six statutes. Groundwater protection is not the main mission of any one statute. A number of different program offices have implementation responsibilities. As a result, implementation of groundwater protection at EPA has suffered from an overall lack of coordination. In addition, many activities with a significant potential for contaminating groundwater are not addressed by these statutes. EPA's Ground-Water Protection Strategy seeks to remedy these problems.

EPA's stated goal in crafting its policy was to provide officials at all levels of government with a "common reference" as their responsible institutions "work toward the shared goal of preserving, for current and future generations, clean water for drinking and other uses, while protecting the public health of citizens who may be exposed to the effects of past contamination" (U.S. Environmental Protection Agency, 1984). In line with the agency's mission to preserve and enhance environmental quality, the strategy focuses on managing groundwater quality as opposed to water quantity. It includes four major components: (1) State Program Development Assistance; (2) Studies of Unregulated Threats; (3) Guidelines for Consistent EPA Management of Groundwater under the Different Laws; and (4) Programs to Improve Intra- and Interagency Institutional Organization. An Office of Ground-Water Protection was established to work with the agency program offices and the ten EPA regions to further refine and implement the strategy.

Helping states set up groundwater protection programs. State assistance, the agency believes, is the key to ensuring nationwide protection of groundwater. "EPA believes that the most effective and broadly acceptable way to increase national institutional capability to protect ground water is to strengthen State programs" (U.S. Environmental Protection Agency, 1984). The strategy commits the agency to help states develop groundwater protection goals and to coordinate efforts of the various institutions given implementation responsibilities; to identify legal and institutional barriers to comprehensive programs; to design source- or contamination-specific groundwater protection programs; to create data management systems on quantity and quality; and to provide technical assistance to help solve specific groundwater problems. Funding for state assistance efforts is to come from the grant programs of the existing laws [e.g., §§106, 205(g), 205(j) of the Clean Water Act, §1443(b) of the Safe Drinking Water Act, and §3011 of the Resource Conservation and Recovery Act]. Federal assistance will be given solely for the development of new programs, however, not for implementation.

States have criticized the strategy on two counts. First, states with programs in place criticize the agency's decision to focus its state assistance resources on program development to the exclusion of supporting routine operations or implementation. They argue that the strategy offers to support states that have lagged behind in setting up groundwater protection programs. EPA takes the position, however, that the limited funds available will do the most good in helping set up programs in states where there are none. Second, states have criticized the choice to rely on funds from existing grant authorities. They point out that using the money for groundwater protection leaves less for uses more directly in line with the statutory purpose of the authorizing statutes.

Assessing inadequately addressed groundwater problems. As discussed above, many sources of groundwater contamination go unregulated. Some, such as surface impoundments holding legally nonhazardous wastes but still potentially dangerous substances, appear to pose substantial threats of contamination. But a lack of information makes it difficult to know the extent and seriousness of the problems associated with such sources. EPA, recognizing the problem of limited information and the preliminary evidence that sources such as existing suface impoundments and underground storage tanks holding toxic substances pose serious threats of groundwater contamination, made the evaluation of the threats from these sources and the development of controls one of the priorities of the strategy. Upon publication of the strategy, EPA initiated studies of underground storage tanks, surface impoundments, and contamination by pesticides and fertilizers from agricultural uses. (Congress, by enacting HSWA, closed a few of these loopholes with the provisions governing the use of underground storage tanks and existing surface impoundments.) The strategy also committed the agency to set up a groundwater monitoring system to help define the nature, extent, and severity of contamination from sources such as septic tanks, mining, unregulated drilling, natural gas pipelines, and sinkholes.

Creating a policy framework for guiding EPA programs. The strategy sets out a three-tiered classification system to guide agency implementation of those laws with groundwater protection provisions. The goal is to ensure consistency in EPA groundwater protection programs. As roughed out in the strategy, the classification guide attempts only to provide a broad definition of each class and the basic criteria to be used in classifying specific groundwater resources. Class I areas are "Special Ground Waters." They encompass those aquifers that are highly vulnerable to contamination because of their hydrogeological characteristics (e.g., a high rate of exchange or high volume flow of water) coupled with their importance for drinking water and/or for sustaining an ecologically important area. Steps to be taken to protect Class I areas include discouraging and eventually banning the siting of new hazardous waste land disposal facilities and emphasizing the immediacy of threats to Class I areas as an important factor to consider in selecting Superfund sites.

Class II is the classification that the "majority of usable ground water in the United States" should fall into. Relative to Class I areas, Class II areas encompass "[a]ll other ground water currently used or potentially available for drinking water and other beneficial use . . ." They will receive levels of protection consistent with those now provided groundwater under the agency's existing programs. "This means that prevention of contamination will generally be provided through application of design and operating requirements based on technology, rather than through restrictions on siting" (U.S. Environmental Protection Agency, 1984).

Class III areas are those that are not a potential source of drinking water and of limited beneficial use. To be so classified, the groundwater not only must be saline, or otherwise contaminated beyond usable levels, but also must not be connected to Class I or Class II aquifers or to surface waters. While a Class III designation will provide the area the same level of protection from further contamination as presently provided by law, the agency would be willing to consider less stringent RCRA or CERCLA cleanups or remedial actions in such areas. According to the agency, the three-tiered classification system is the "basic framework" for the strategy. It is also the most controversial.

The controversy stems from several unrelated factors. Potentially, the framework requires major changes in the way existing EPA programs are implemented, including the solid and hazardous waste, drinking water, and surface water programs. This appears to program managers as an arrogation of their authority; it threatens a shift in priorities and possibly dollars from what they see as their main mission to groundwater protection. The fact that the strategy omits the details of defining and implementing the classification guidelines compounds the problem. Both inside and outside the agency, people argue whether the classification proposed by the strategy is an organizing principle that will generate only minor adjustments in existing programs or is a call for a major restructuring of program implementation that will lead ultimately to the designation of hazardous waste zones and hazardous waste-free zones.

Strengthen internal groundwater organization. The agency established the Office of Ground-Water in 1984 to oversee implementation of the strategy. The office works closely with the other federal agencies, EPA program offices, EPA regional offices, and the states to help coordinate their many and diverse efforts to protect groundwater under different statutory authorities.

During its first year of existence the Ground Water Office initiated a number of studies and had some success in further defining the classification framework. In March 1985 the office published a two-volume Overview of State Ground-Water Program Summaries (U.S. Environmental Protection Agency, 1985) that gives a good overview of the groundwater programs of the fifty states as they appear on paper. A groundwater monitoring strategy was developed in 1985 and draft guidelines for the groundwater classification system should be issued for public comment in the Federal Register sometime in 1986.

V. CONCLUDING REMARKS AND SUMMARY

The institutional framework for protecting groundwater from contamination by toxic chemicals is labyrinthine. So many agencies and branches of agencies at every level of government run so many programs under the authority of so many laws that experts and lay persons alike become confused. As a result, you find some experts proclaiming that no underlying framework exists and others that claim the framework has far too many supporting pillars. Clearly, the nation's groundwater does not go unprotected from pollution by toxics, but just as clearly the system could benefit from an overhaul that emphasizes simplicity and efficiency.

During the last few years, EPA, Congress, and states have taken positive steps to improve the situation. EPA's Ground-Water Protection Strategy recommends important steps for streamlining and coordinating state and intra-EPA groundwater management efforts. States continue to develop innovative groundwater protection programs—e.g., Wisconsin, Florida, Arizona, and Connecticut—that plug gaps in the federal system and that serve as examples for other states of how programs can be put into practice to yield positive results. Congress has actively sought to strengthen groundwater protections. The 1984 Amendments to the Resource Conservation and Recovery Act provide powerful new tools to EPA and the states to use in preventing leaks, seepage, and spills of hazardous wastes to groundwater; the Safe Drinking Water Amendments of 1986 strengthen the protection of underground drinking water supplies. In November of 1985, members of both the House (HR. 3808) and Senate (S. 1836) introduced comprehensive groundwater protection bills. Despite these positive efforts many problems and questions remain unresolved.

Overall the system still cries for coordination among different levels of government, different agencies, and different divisions within agencies. At the state and federal level, many agencies have implementation responsibilities, and in agencies such as EPA, many program offices claim jurisdiction over groundwater protection efforts. With so many responsible, overlap and inefficiency are inevitable.

Some contamination threats remain ungoverned. For example, thousands of surface impoundments used to store dangerous but nonhazardous materials go unregulated in most states, as do pesticides and fertilizers that leach from farmland. Many chemicals that analysts have found in drinking water lack standards. New programs seem justified, but proposals to develop new programs to regulate these uncontrolled threats have generated raging controversy.

A number of difficult questions face policymakers attempting to improve national groundwater protection efforts. One fundamental choice is whether programs designed to protect public health alone are enough. Will natural systems depending on groundwater supplies be destroyed if our regulations extend solely to groundwater supplies that serve human needs? Will EPA's groundwater classification system set up a scenario where only groundwater resources used by humans are protected?

The 1984 RCRA Amendments highlight another difficult dilemma. The Amendments, by calling for the phaseout of land disposal of hazardous waste by the mid-1990s, rely on treatment technologies. According to some critics, these technologies remove the pollutant threat from groundwater only to send it to the air, or surface water—e.g., through incineration or wastewater discharge. How can new groundwater protections be devised so that such multimedia implications are taken care of?

The present federal strategy relies heavily on state implementation and leadership. Is this reliance based on sound judgment? Do states have the resources and the resolve? Will states use money from other pollution control programs to set up and run groundwater protection programs? EPA's Ground-Water Strategy suggests as much in directing states to seek §106 Clean Water Act funds to set up groundwater programs.

Cleanup efforts under Superfund and corrective actions at RCRA-regulated facilities place an emphasis on technological solutions such as double plastic liners, leachate control systems, and aeration. RCRA, as amended, also sets store in liners and leachate collection for existing land disposal units. Yet experts all seem to agree that all storage systems are temporary; even the most impermeable liner eventually leaks. Will so-called secure landfills, storage sites, and "cleaned-up" Superfund sites remain clean and remain secure, or will they become the Superfund sites of the future?

Determining when a cleanup or corrective action is complete is one of the most difficult decisions to be made on a site-specific basis. To date, no clear set of standards or guides has been developed. EPA policy recommends following, where available, the standards for hazardous constitutents developed under such statutes as the Safe Drinking Water Act, Comprehensive Environmental Response and Recovery Act, the Clean Water Act, or any other federal statute with standards. Unfortunately, many common groundwater contaminants have no standard or protection level under any law. Also, in most cases the standards developed under these laws were not developed with groundwater exposure in mind. Until a set of uniform standards or parameters is developed, decision-makers will decide "how clean is clean" on a case-by-case basis, an expensive proposition and one that leads to different outcomes in similar cases.

Decision-makers at all levels of government that are attempting to improve the institutional framework for protecting groundwater grapple with these and other difficult questions on a daily basis. Hopefully, they will be successful; clearly, success will not be achieved overnight.

REFERENCES

Brown, M. (1986). Lower depths: Underground injection of hazardous wastes. *Amicus J.* **7**(3), 14.

Connecticut Department of Environmental Protection (1981). "A Handbook for Connecticut's Water Quality Standards and Criteria." Conn. DEP.

Davis, P. N. (1984). Groundwater pollution: Case law theories for relief. *Mo. Law Rev.* **39**, 117.

DiNovo, F., and Jaffe, M. (1984). *"Local Groundwater Protection: Midwest Region."* American Planning Association, Washington, D.C.

Environmental and Energy Study Institute and the Environmental Law Institute (1985). "Statutory Deadlines in Environmental Legislation: Necessary but Need Improvement." EESI, Washington, D.C.

Gordon, W. (1984). "Citizen's Handbook on Groundwater Protection." Natural Resources Defense Council, New York.

Gordon, W., and Bloom, J. (1985). "Deeper Problems: Limits to Underground Injection as a Hazardous Waste Disposal Method." Natural Resources Defense Council, New York.

Grad, F. (1983). "Treatise on Environmental Law," Vol. 1, Sect. 3.05, pp. 3-432 to 3-442. Matthew Bender and Co., New York.

Henderson, T., Trauberman, J., and Gallagher, T. (1984). "Groundwater: Strategies for State Action." Environmental Law Institute, Washington, D.C.

New Jersey Department of Environmental Protection (1982). "Ground Water Discharge Permits—Program Status and Issues." N.J. DEP, Bureau of Ground Water Discharge Permits.

New Mexico Water Quality Control Commission (1982). "New Mexico Administrative Regulations. Part 3. Water Quality Control." N.M. WQCC.

Prosser, W. (1971). "Law of Torts," pp. 190–204. West Publishing Co., Minneapolis, Minnesota.

Rodgers, W. (1977). "Environmental Law," pp. 107–133. West Publishing Co., Minneapolis, Minnesota.

Restatement (2d) of Torts (1979). Sections 282, 519, 520, 821, and 849, and Chapter 41, Notes 3 and 4, pp. 214–258. West Publishing Co., Minneapolis, Minnesota.

U.S. Congress, Office of Technology Assessment (1984). "Protecting the Nation's Groundwater from Contamination," 2 vols. U.S. Government Printing Office, Washington, D.C.

U.S. Department of Interior (1983). "Directory of Groundwater Programs." USDOI, Office of Water Policy, Washington, D.C.

U.S. Environmental Protection Agency (1984). "A Groundwater Protection Strategy for the

Environmental Protection Agency," pp. 1–54. USEPA, Office of Ground-Water Protection, Washington, D.C.

U.S. Environmental Protection Agency (1985). "Overview of State Ground-Water Program Summaries," Vols. 1 and 2. USEPA, Office of Ground-Water Protection, Washington, D.C.

Vermont Agency of Environmental Conservation. (1982). "Vermont Ground Water Protection Strategy," Vermont AEC.

Wisconsin Department of Natural Resources. (1984). "Groundwater Report." Wis. DNR.

3

Drinking Water and Health

G. WILLIAM PAGE
Department of Urban Planning
and Center for Great Lakes Studies
University of Wisconsin–Milwaukee
Milwaukee, Wisconsin 53201

The human health effects of consuming drinking water containing toxic organic and inorganic substances are poorly understood and the subject of considerable controversy. The general public and public officials want clear standards, which define for each toxic chemical the concentration that will not cause adverse health effects, for each chemical potentially present in our drinking water supplies. The scientific basis for establishing definitive human health-based standards does not exist. This chapter describes the issues and controversies resulting from the presence of toxic contaminants in drinking water supplies. This chapter does not address the many human health-related issues caused by nontoxic contaminants in drinking water. Most nations face problems much more immediate and life threatening than toxic contaminants in providing a source of drinking water to their populations. Most nations must concentrate their efforts on providing water and on controlling bacteria, human parasites, and viruses in water supplies (Page, 1987). Toxic contamination of drinking water supplies troubles not only the economically developed nations, for Third World countries as well as the more economically developed countries presently have microcontaminants in their water supplies and should start planning to protect their groundwater and surface water from continued and more serious contamination.

Improvements in chemical instrumentation permit us to detect a wide variety of organic and inorganic chemicals in water supplies. Often these substances are found at concentrations in the low parts per billion (ppb) and, for some chemicals, the parts per trillion (ppt) range. These concentrations are near the lower limit of analytic chemists' ability to detect chemical compounds in water. Because of these extremely low concentrations, these substances are often referred to as microcontaminants. The health issues concerning microcontaminants addressed in this chapter are: (1) What microcontaminants are found in water supplies? (2) What are the health implications of ingesting microcontaminants in drinking water? and (3) How can we protect the health of the public from the risks of exposure to toxic chemicals in drinking water?

I. WHAT MICROCONTAMINANTS ARE FOUND IN WATER SUPPLIES?

There has been a chemical revolution in the past four decades. Insights to the structure of organic molecules led to the synthesis of vast numbers of not naturally occurring organic chemicals. United States industries registered more than two million new compounds in the period 1965–1972. The United States produced less than one billion pounds of synthetic organic chemicals in 1941, but production increased to 172 billion pounds by 1978 for just 50 of the most common synthetic organic chemicals (Council on Environmental Quality, 1980). The most common contaminants of drinking water supplies are volatile organic compounds (VOCs) that are halogenated hydrocarbon solvents, aerosol propellants, and refrigerants. They are frequent contaminants in water supplies because of the large quantities produced, decentralized use patterns, chemical and biological stability, and negligible adsorption to soils and sediments.

In a survey of the literature, Shackelford and Keith (1977) reported 1259 different organic compounds identified in water samples. Chemists are able to identify only about 10% of the organic compounds in water, with little success in identifying nonvolatile compounds (Rohlich, 1978; Kool *et al.*, 1982). As an example of the dimensions of the identification problem, the 89 toxic organic compounds identified in New Orleans drinking water represent about 2% by weight of the total organic chemicals in that water (Guinan *et al.*, 1978). Advances in analytic methods of chemical identification are stimulating research on the public health effects of microcontaminants in drinking water. Public health concerns became a national issue in the 1970s because of the identification of volatile organic chemicals and other toxic substances made possible by the development of the combined gas chromatograph/mass spectrometer. In the 1980s, the development of solid-state Carbon-13 nuclear magnetic resonance spectroscopy, extended mass range spectrometry, and new ionization techniques, including fast atom bombardment and combined liquid chromatography/mass spectrometry, make possible the detection of an extended number of microcontaminants in drinking water.

These analytic advances extend research capabilities in three primary areas: (1) the mechanism of the aquatic humic acid/aqueous chlorine reaction, (2) the relationship of chlorination by-products to mutagenic activity, and (3) the distribution and concentration of high-molecular-weight polar substances and N-chloroorganic compounds (Christman *et al.*, 1985).

Water supplies also contain inorganic contaminants that are toxic. Heavy metals contaminate many drinking water supplies and increasingly acidic precipitation causes higher concentrations of many inorganic contaminants in water. For example, lead levels exceeded the Maximum Contaminant Level (MCL) of 50 micrograms per liter (μg/liter) in 16.1% of the 2654 U.S. households sampled (Francis, 1984). Asbestos is another inorganic contaminant released in greater concentrations to water supplies from natural sources and man-made products such as concrete–asbestos water pipes as a result of acid precipitation (McDonald, 1985).

There are a variety of processes by which our drinking water may become contaminated with toxic chemicals. Microcontaminants can pollute our water supply sources, they can be created in the water treatment process, and they can originate in the drinking water distribution system (National Academy of Sciences, 1982). The most serious problems are with our sources of drinking water. Some microcontaminants such as pesticides and herbicides are intentionally released in the environment. Others result from transportation accidents and accidental chemical spills. Even small spills of these toxic chemicals can cause large groundwater contamination problems. Microcontaminants escape landfills, effluent lagoons, and petroleum storage and distribution facilities. Many household products contain toxic chemicals that escape to the environment.

The process of treating and distributing potable water can introduce organic microcontaminants into drinking water. Chlorination, used to disinfect water, combines with naturally occurring humic and fluvic acids to create a class of organic microcontaminants known as trihalomethanes. They are the most widely identified toxic organic chemicals in U.S. water supplies. The U.S. Environmental Protection Agency water supply survey of finished drinking water from both ground and surface sources showed that 95 to 100% of the systems surveyed contained chloroform, one of the trihalomethanes (U.S. Environmental Protection Agency, Office of Toxic Substances, 1975). Chloroform and other trihalomethanes are suspected human carcinogens (National Academy of Sciences, 1977). Other research has identified organic microcontaminants that originate in the water pipes of the water supply distribution system possibly from microbial action on the walls of the pipes or from the coal tar used to line some pipes (Schwartz *et al.*, 1979).

Contamination of groundwater with toxic chemicals is an especially serious problem. Traditionally, we assumed groundwater was pure. Even in developed countries, groundwater often is not treated. Despite this assumption, a study found groundwater at least as contaminated with organic and inorganic microcontaminants as surface water, and where contamination exists the study found higher concentrations of toxic contaminants in groundwater than in surface water (Page, 1981). Contaminants move through an aquifer as a discrete plume that is not subject to rapid dilution because of its slow movement and laminar flow. Once below the water table, microcontaminants are not in contact with the atmosphere. Even volatile organic chemicals cannot escape solution, and aerobic degradation processes are highly restricted. A plume of contaminants in groundwater may move slowly or quickly in unpredictable directions for years before being pumped from the ground into a private or public water supply.

II. WHAT ARE THE HEALTH IMPLICATIONS OF TOXIC CONTAMINANTS IN DRINKING WATER?

The health implications of exposure to organic microcontaminants in drinking water are the subject of considerable controversy. There is no doubt that exposure to large doses of many of these chemicals causes acute illness. However, concentrations large enough to cause these health effects are rarely found in water supplies. The concern for drinking water supplies is to know what are the health effects of consuming water contaminated with low concentrations of microcontaminants over the course of a lifetime. In earlier centuries, human populations often suffered serious epidemics caused by bacteria, viruses, or parasites that originated in distant lands and to which they had not developed biological resistance nor cultural adaptation. Most of the organic microcontaminants of concern in water supplies are man-made chemicals. Humans have evolved without prior exposure to these substances and lack defense mechanisms, acclimation capabilities, or excretion pathways to contend with these chemicals once they have entered the body. When consumed, many microcontaminants contained in water and food concentrate in fat cells faster than they are degraded or eliminated. The long-term health implications are unknown. There is widespread agreement that they are adding to the disease burden in a significant, although as yet not precisely defined way (Department of Health and Human Services, 1980).

III. CANCER

The greatest concern from the long-term exposure to organic microcontaminants in drinking water is cancer. As early as the 1960s, some scientists warned that microcontaminants in our drinking water supplies were animal carcinogens and that the threat of cancer from these pollutants was a rapidly increasing danger to humans (Hueper, 1960; Hueper and Payne, 1963). Epidemiologic studies implicate microcontaminants in water supplies with elevated cancer mortality rates. Many of the organic microcontaminants found in drinking water are animal carcinogens and on that basis are considered "suspected human carcinogens." Most of the organic microcontaminants identified in samples of drinking water have not yet been tested for carcinogenicity. We have virtually no information on potential synergisms among the large numbers of microcontaminants that often are present in water supplies.

IV. CANCER PROCESSES

We have learned much of the biological mechanism of cancers, but important parts of the process remain unknown. Three steps in the process are initiation, promotion, and progression (Farber, 1981). The initiation step starts with an irreversible lesion in the DNA. A promoter agent can attack an initiated cell and can accelerate that cell's progression to cancer. If the same promoter attacks an uninitiated (normal) cell, the resulting damage is thought to be reversible. Each of these three steps involves some unknown components.

The initiation step may result from the "hit" of a toxic chemical on a DNA molecule in the cell of an organ or body site. Toxic chemicals that attack the DNA are genetic carcinogens. Epigenic agents such as asbestos do not attack the DNA, but can initiate cancer through other cellular controls. The "hit" of a genetic carcinogen on the DNA of a cell results in a damaged portion of the DNA known as an "adduct." The human body has an excision–repair process that can often remove the adduct and replace it with a new segment of DNA and thus reverse the cancer-forming process (Ricci and Molton, 1985). When the body's natural defenses are not able to remove the adduct and repair the DNA, then the cancer process is likely to move to the progression stage, leading to the uncontrolled growth of a somatic cell and eventually to cancer. Research has found that individuals with defective DNA repair mechanisms have a sharply higher incidence of cancer (Cairns, 1981).

Is there a threshold exposure to a carcinogen, below which cancer does not result? This is one of the most important and contentious issues in the debate over toxic contaminants in drinking water. Many diseases other than cancer have proven thresholds, for instance gastrointestinal-radiation syndrome (National Academy of Sciences, 1977). Exposure below a determined dose has a zero probability of producing the effect. Scientists favoring a threshold theory for carcinogens in drinking water argue that the human body's immune defenses and gene repair mechanisms are able to protect us from exposure to low doses of carcinogens. If scientific evidence can establish a threshold for a specific carcinogen, then a government agency can establish that threshold exposure as an environmental standard to protect human health.

The argument against a threshold contends that a single hit to the DNA can produce cancer. Scientists in favor of this position recognize that human gene repair mechanisms exist, but believe that different humans have different capacities to repair damaged DNA. Different people, therefore, may have different thresholds, and some individuals may have no threshold. If these assumptions are correct, then there can be no single threshold exposure to a carcinogen that can be used as a standard to protect humans from cancer. These scientists also argue that even if thresholds do exist, they occur at sufficiently low doses that it would require massive, expensive, and impracticable experiments to establish verifiable thresholds (National Academy of Sciences, 1977).

V. DOSE–RESPONSE FUNCTIONS

The issue of a threshold is part of the issue of determining the relationship between the level of exposure to a carcinogen and the resulting likelihood of getting cancer. Scientists attempt to define this relationship mathematically as a dose–response function. A graph of the dose–response function for an animal carcinogen plots the percentage of animals with tumors or the average number of tumors per animal against the dose of the carcinogen (see Fig. 1). These functions are gross simplifications of the complex processes by which exposure is modified by absorption, distribution, membrane permeability and enzyme binding, metabolism, and excretion of the carcinogen. Ricci and Molton (1985) briefly describe the four most commonly used dose–response functions for cancer. The one-hit model assumes that one hit leads to cancer. The multistage model assumes several stages that are dependent on the dose that results in exposure. The multihit model assumes that some number of hits to the sensitive tissue are required to initiate a cancer. The Weibull model is a variation of the multihit model. It assumes that the hits occur in a single cell line and that different cell lines compete independently in producing a tumor.

All these dose–response models assume a linear relationship between

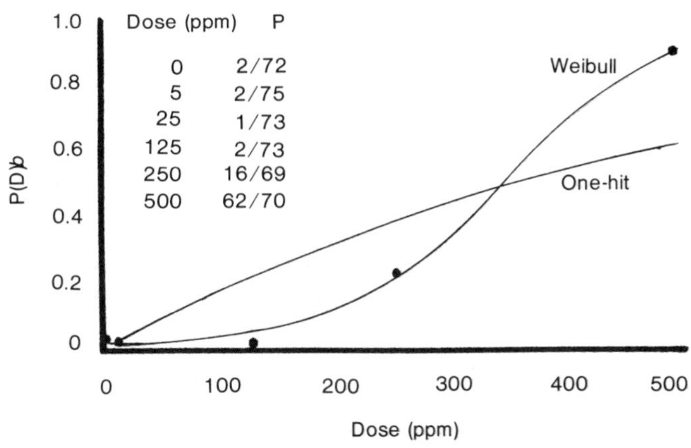

Fig. 1. Dose–response functions relating exposure to probability of adverse health effect. [Reprinted with permission from Ricci and Molton (1985). Copyright 1985 American Chemical Society.]

dose and hits at concentrations typically encountered in drinking water. The one-hit model predicts the greatest risk of cancer for a given exposure of these four dose–response models. The multistage, Weibull, and multihit models are much closer in their predictions, with the Weibull and multihit models flatter at low doses (see Fig. 1) and therefore predictors of lower risk than the multistage model (Brown and Koziol, 1983). The National Academy of Sciences (1980a,b) originally recommended the use of the multistage model to extrapolate risks from waterborne chemicals, but has since decided that it cannot recommend a single approach. The form of the dose–response function at low doses is the critical issue in estimating risk and developing policy to protect humans from potential damage resulting from microcontaminants in drinking water. Microcontaminants are common in drinking water supplies at low concentrations. The form of the dose–response function at these very low concentrations (doses) determines our best estimates of the risk of cancer associated with different concentrations of microcontaminants in water. The dose–response function selected produces more uncertainty in estimating the health risk of exposure to microcontaminants in drinking water than any other component in the risk analysis process (Schneiderman, 1980; Ricci and Cirillo, 1985).

VI. ANIMAL STUDIES

Experiments testing the effects of potential carcinogens on animals are used to help estimate the human health effects of exposure to these substances. Animals are used in these experiments because of the universal opinion that using humans for experimentation is unethical. Animal experiments conducted with exposure to microcontaminants at concentrations comparable to those in drinking water are prohibitively expensive because of the thousands and even millions of animals needed for such experiments. As a result, animal experiments are conducted with higher dosage levels than humans would be exposed to in drinking water. Experiments typically involve two species with several groups of about 50 males and 50 females. Different groups are exposed to one of three dosages, plus the no-exposure control groups. The dosage levels include the maximum tolerated dose and two fractions of it. At the end of the experiment, the animals are sacrificed and examined for evidence of cancer.

Scientists must extrapolate the cancer risks determined from animal studies at high-dose levels to cancer risks in humans at the low-dose levels generally encountered in drinking water. This procedure requires two extrapolations: (1) from high-dose levels in the animal experiments to low-dose levels and (2) from the animals studied to humans. The extrapolations to cancer risks associated with exposures to chemicals outside the range of exposures used in the experiments are made based on assumptions of the shape of the dose–response function at low doses. Scientists consider the four dose–response functions described and the many variations of these basic models "linear at low dose" and assume that there is no threshold (National Academy of Sciences, 1977). They use the models to help estimate the risk of cancer associated with exposure to the low concentrations of carcinogens that may exist in drinking water supplies.

The second extrapolation from animal data to humans is supported by large bodies of experimental data showing that exposure of substances that are carcinogenic to animals are carcinogenic to humans. There are, however, differences in the susceptibility to cancer between different animal species, between different strains of the same species, and between individuals of the same strain. Scientists use several approximations of metabolic pathways to account for metabolic differences between species. Approaches include milligrams per square meter of body surface per day and milligrams per kilogram of body weight per lifetime (Du Muchel and Harris, 1983). With the possible exception of arsenic and benzene, all known carcinogens in man are also carcinogens in some animal species, although not in all species (National Academy of Sciences, 1977).

Besides cancer, microcontaminants in drinking water also represent other threats to human health. Many of the organic microcontaminants found in water supplies are also suspected human teratogens and mutagens. The organic microcontaminants that are teratogens can produce birth defects or congenital abnormalities (Berg, 1979; Bloom, 1981). Mutagens increase the genetic load of recessive mutant alleles, which can lead to an increasing incidence of genetic diseases in future generations (Hollaender, 1973). The risk of genetic damage may manifest itself in the grandchildren of those exposed today to mutagens present in drinking water.

VII. EPIDEMIOLOGIC STUDIES

Considerable epidemiologic research investigating a link between water supplies containing microcontaminants and increased cancer mortality exists. This epidemiologic research has included both ecological studies and case control studies. These epidemiologic studies are among the most useful and common approaches for determining within and between which population groups a disease occurs (see Table I). The ecological studies compare cancer rates to aggregate measures of exposure, usually with the county as the unit of analysis. There have been many ecological studies since Page *et al.* (1976) first reported elevated cancer mortality in New Orleans, presumably caused by microcontaminants in water supplies.

Many of the epidemiologic studies assessing the relationship between cancer and drinking water quality tested for the effects of chlorination by-products in water. The trihalomethanes (THMs), chloroform, bromodichloromethane, dibromochloromethane, and bromoform are the most common chlorination by-products in water supplies. Studies have used historical information on water sources and chlorination practices as surrogate measures of exposure to trihalomethanes. The National Academy of Sciences (1980b) reviewed the descriptive studies conducted through 1978. Of the thirteen studies reviewed, three found associations between cancer mortality and current levels of THMs, and nine of ten studies found associations between historical patterns of chlorination and either cancer incidence or cancer mortality. These epidemiologic studies all suffer from the inability to adequately control for all potentially confounding variables, such as diet, smoking, drug use, migration, and lifetime occupational histories (Greenberg and Page, 1981).

Analytic epidemiologic studies, including case control studies and cohort studies, include information on exposure, disease, and potential confounding variables for each study participant. Analytic epidemiologic studies are better able to assess and control confounding bias than the more descriptive ecological epidemiology studies. The case control studies relate exposure levels experienced by people with the disease

TABLE I
Epidemiologic Methods in Risk Assessment

Study type	Unit of analysis	Disease measure	Remarks
Ecological correlations (spatial or temporal)	Groups of individuals	Prevalence, incidence, or mortality rates	Helpful for developing hypotheses between exposure and effect
Cross-sectional	Single individuals	Prevalence rates	Useful in the study of diseases with latency or that vary with time
Case control	Single individuals	Proportion of cases to controls	Useful in forming and establishing etiological hypotheses; exposure is difficult to establish; controls can be doubtful; cases may be unrepresentative
Cohort (retrospective or prospective)	Groups with and without exposure are followed over time	Incidence rate; frequency of disease in exposed to frequency in the unexposed[a]	Appropriate for testing causal hypotheses; diseases are determined from initial development; confounding can be avoided and rare diseases accounted for
Disease measures Rate[b]			

Prevalence = (Cases in a specific population)
– (Population); measured at time.
Incidence = (New cases in a specific population)
– (Population); measured over a period of time.
Mortality = (Total deaths in a specific population)
– (Unit of population); measured over a period of time.

[a] Commonly used measure of excess risk is the "relative risk," which is defined as the ratio of observed cancers to expected cancers.

[b] For a specific disease, alternative specifications of the denominator are feasible, as are adjustments for age (specificity and standardization), sex, race, and socioeconomic characteristics.

(cases) to exposure levels experienced by people without the disease (controls). Cohort studies follow each individual selected because of exposure status to determine morbidity or mortality. Analytic studies calculate relative risk or an exposure-odds ratio to measure the association of water quality to cancer. Several reviews of the analytic studies of the relationship between drinking water and cancer exist (Shy and Struba, 1980; Hoel and Crump, 1981; Crump and Guess, 1982; Cantor, 1983; Craun, 1985).

Almost all the epidemiologic studies have found significant statistical associations between use of contaminated drinking water and several body site cancers. The strongest and most consistent associations have been with rectal, bladder, and colon cancer. The most convincing of these studies involve colon and bladder cancer. Researchers found a weak to moderate association between water chlorination and colon cancer in an elderly population (Cragle *et al.*, 1985). The association was strongest in individuals exposed to chlorination for more than 15 years. Cantor *et al.* (1985) found a moderately strong relationship between chlorinated surface water and bladder cancer in nonsmokers who had used chlorinated water for 60 or more years.

Both the ecological and the analytic epidemiologic studies have consistently found statistically significant associations, however, this evidence suggests that microcontaminants in drinking water are not responsible for a large proportion of cancers. Rectal, bladder, and colon cancer are the cancers most strongly associated with toxic chemicals in water and these body site cancers account for a small proportion of total cancer mortalities. Despite shortcomings in controlling potentially confounding variables, the consistency of findings across many studies supports the hypothesis that microcontaminants in drinking water cause an increase in cancer.

The carcinogenic, teratogenic, and mutagenic effects from exposure to contaminated drinking water may be caused by microcontaminants other than those now under suspicion. Some researchers suggest that nonvolatile organic compounds, which have not yet been identified in water and often coexist with trihalomethanes and the common volatile organic chemicals, may cause these effects (Hileman, 1984). Evidence exists that many mutagenic compounds, isolated from water but not identified, may be polar and nonvolatile (Zoeteman *et al.*, 1982; Kool *et al.*, 1982).

VIII. CANCER RISK IN PERSPECTIVE

The cancer risk associated with drinking water containing microcontaminants is probably one of the smaller cancer risks commonly experienced in modern society. The potential cancer risk from exposure to carcinogens in tobacco and alcohol, in the workplace, in the air we breathe, and in the food we eat may be substantially greater for most people than the risk associated with exposure to carcinogens in drinking water.

Efforts to understand the relative magnitude of cancer risks associated with the different sources of human exposure are increasing. Estimating these risks requires measurement of pollutant concentrations in the air, water, food, and in contact with the skin. This approach is known as "total human exposure" (Ott, 1985). Researchers collect a detailed record of a person's exposure to contaminants as a function of time thoughout the day. This record is called an exposure profile. Individual exposure profiles are aggregated into a frequency distribution of exposures to characterize the total exposure of the population of a city or other area. Researchers can develop exposure profiles for individual contaminants or a group of contaminants.

Researchers using the total human exposure approach have studied volatile organic chemicals, which are of great concern in drinking water supplies. They measured concentrations of 20 volatile organic chemicals in the air, drinking water, food, and in the breath of a representative random sample of the population of eight U.S. cities for 12-hour periods (Wallace *et al.*, 1982, 1984). Each person in the study noted their activities in a detailed diary and carried a specially designed, miniaturized pump connected to a 6-inch Tenax cartridge. The researchers analyzed the Tenax cartridge for the concentrations of volatile organic chemicals each person encountered during the 12-hour exposure period. This study found that levels of 11 important organic compounds were significantly higher in the indoor atmosphere than outdoors (Ott, 1985). Sources inside homes include paint, solvents, furniture, drapes, carpets, construction materials, clothing, and spray cans. Previous research approaches focused on contaminant releases in the environment and as a result overemphasized outdoor sources of exposure. The total human exposure approach may assist policymakers in allocating scarce resources to more effectively protect the public's health from environmental contaminants. (Editor's note: see Chapter 11, the Santa Clara Valley case study, for a description of a cross-media study of risks due to air, water, and other resources.)

Microcontaminants in drinking water may produce health risks through a variety of exposures in addition to ingestion of the contaminated water. Adsorption of chemicals present in water through the skin while bathing can present a risk to health. Some research suggests that this means of exposure represents a risk to human health comparable in magnitude to the risk associated with direct ingestion of the contaminated water (Brown *et al.*, 1984). Volatilization of contaminants in water brought into a home can produce air pollutants that also represent a risk to health. Researchers estimate that air exposures to volatile organic chemicals and radon in a home using contaminated drinking water are six times greater than the exposure resulting from the ingestion of an assumed two liters of contaminated water per day (Andelman, 1985). Showers allow large quantities of volatile organic chemicals or radon present in water to escape to the atmosphere. Showering in water for a total of one hour a week using water contaminated with trichloroethylene produces an exposure that by itself is comparable to ingesting the contaminated water (Andelman, 1985). Water temperature and drop path of the shower water produced variation in exposure. These exposures are additive to the other sources of exposures inside the home.

The cancer risk associated with consuming carcinogens in the diet may exceed the risk associated with consuming carcinogens in drinking water. Epidemiologic studies have identified dietary practices as one of the most important causes of cancer risk (National Research Council, 1982). Many plants common in human diets synthesize large quantities of toxic chemicals as a defense against bacterial, fungal, insect, and other animal predators. many of these substances are mutagens, teratogens, and carcinogens (Ames, 1983). In addition to naturally synthesized toxic substances in plant material, alcohol, rancid fat, molds, burnt and browned material from heating protein, nitrosamines formed from nitrate and nitrite, and other carcinogens are added to our diets or formed by the processes of food storage and preparation. Fortunately, human diets contain a variety of anticarcinogens, such as vitamin E, beta-carotene, selenium, ascorbic acid, and others, which help protect humans against cancer (Ames, 1983).

Strongly divergent conclusions exist concerning the relative importance of water as a cancer risk. Humans are exposed to such a large quantity of naturally occurring carcinogens that man-made chemicals may represent an insignificant additional cancer risk. Every common beverage and every meal contains such a large quantity of naturally occurring carcinogens that the microcontaminants in drinking water may represent a trivial risk. Some of the most contaminated groundwater, containing 2800 ppb of trichloroethylene, "is at least 1,000 times less hazardous than an equal volume of cola, beer, or wine" according to some highly respected scientists (Ames, 1986). Other respected scientists performing risk assessments using conservative assumptions conclude that most U.S. municipal water supplies should be banned as "unfit for human consumption" because of toxic contaminants (Crouch *et al.*, 1983).

IX. HOW CAN WE PROTECT THE HEALTH OF THE PUBLIC?

There are several distinct approaches to protect the health of the public from the risks posed by microcontaminants in drinking water. Intervention to keep these pollutants out of the sources of water supply is by far the best approach (Page, 1984). Unfortunately so many toxic chemicals have been released to the environment from so many diverse sources that there are precious few water supply sources that do not already have microcontaminants present. Most of the microcontaminants are present in extremely low concentrations, and the most effective intervention is to keep additional toxic chemicals out of water supply sources, especially groundwater sources. There are so many water supply sources and so many sources of toxic chemicals that this approach is unlikely to be completely successful.

Another intervention approach is to establish standards for the maximum allowable concentration of each microcontaminant in drinking water. In the United States, the Environmental Protection Agency has established Maximum Contaminant Levels (MCLs) for a small number of organic and inorganic microcontaminants. The National Interim Primary Drinking Water Regulations contain Maximum Contaminant Levels for six organic microcontaminants plus total trihalomethanes (40 CFR parts 141 and 142). EPA has initiated the formal process for establishing Maximum Contaminant Levels for an additional eight volatile organic chemicals (50 FR 46902; Nov 13, 1985).

The process of establishing Maximum Contaminant Levels is extremely contentious and time-consuming because the long-term human health effects are unknown (Page and Greenberg, 1982). Maximum Contaminant Levels are enforceable standards. The Safe Drinking Water Act (42 U.S.C. 300f, *et seq.*) prescribes the process. The greatest period of time is spent developing and assembling the scientific evidence to support a standard. The EPA Office of Drinking Water provides Health Advisories for each toxic substance as the scientific evidence becomes available. The Health Advisories, which are updated as new information becomes available, are not enforceable standards but provide guidance to the states and interested parties. During 1985, EPA wrote 52 Health Advisories.

Establishing Recommended Maximum Contaminant Levels (RMCLs) is the first step of the official standard setting procedure. Recommended Maximum Contaminant Levels are nonenforceable health goals set at a level that, with an adequate margin of safety, should produce no known or anticipated adverse health effects. At the same time Recommended

Maximum Contaminant Levels are issued, EPA announces Proposed Maximum Contaminant Levels. After a period for public comment and public hearings, Maximum Contaminant Levels are established. Maximum Contaminant Levels are enforceable standards. The Environmental Protection Agency must set Maximum Contaminant Levels as close as feasible to Recommended Maximum Contaminant Levels. Feasible means that the Maximum Contaminant Level can be achieved using available and affordable technology. (Editor's note: see Chapter 14 for discussion of proposals to quickly establish many additional MCLs.)

The last intervention approach to be discussed is technological in nature. Alternative water treatment processes to disinfect water are being tested in an attempt to avoid the creation of trihalomethanes (National Academy of Sciences, 1980). While there are alternatives to the use of chlorine as a disinfectant, none are widely considered to be better than chlorine because each alternative has known or potential problems (National Academy of Sciences, 1980). Researchers are evaluating other technological approaches as a means of removing organic microcontaminants from water. The present water treatment techniques such as filtration, coagulation and settling, and disinfection are not effective in removing organic microcontaminants. Air stripping towers are effective at removing volatile organic compounds. Activated carbon has been extensively studied and does remove many, but not all, microcontaminants from water (National Academy of Sciences, 1980a). This technology is not widely used because of the high operating expenses required to regenerate the activated carbon at frequent intervals. (Editor's note: for a more thorough discussion of technological approaches, see Chapter 4.)

X. CONCLUSIONS

The consequences of failing to keep our drinking water supplies free of microcontaminants are unknown. Because there is usually an interval of 20–30 years between exposure to a carcinogen and the incidence of the disease in humans, it is possible that the widespread exposure to microcontaminants in water supplies now taking place may in the future produce substantial increases in cancer incidence and mortality. The best information available suggests that this will not happen. Risk assessments of consuming water with low concentrations of carcinogens and other toxicants estimate that this exposure will produce some cancers that would not otherwise have occurred, but this number of additional cancers may not be great compared to other causes of mortality (National Academy of Sciences, 1980b; Greenberg and Page, 1981; Crouch *et al.*, 1983). Present risk assessment methods are designed to be conservative; however, they are not able to include the unknown risks caused by synergistic effects of the mixtures of many different microcontaminants that may simultaneously be present in drinking water. The risks associated with consuming water containing microcontaminants are different in some respects from other health risks. The risks from smoking are much greater than the risks from drinking water containing microcontaminants. Tobacco use causes 30% of the cancer deaths in the United States and in Great Britain (National Research Council, 1982). While the health risk from tobacco use is much greater than that from contaminated drinking water, it may cause much less concern in large segments of the population because it is a voluntary risk (Slovic *et al.*, 1979). Drinking water is usually a public service provided by or licensed by local government. It is government responsibility to provide safe drinking water. People often are upset when exposed to risks over which they have no control and when that risk is caused by a government agency charged with the task of protecting them from such a risk.

There are ways to protect the public from the risks caused by microcontaminants in drinking water supplies. This book describes methods used to plan for the protection of water supplies from toxic contamination. This book also describes technologies that can remove microcontaminants. These technologies can be expensive, especially for small water utilities (Page, 1985). However, researchers found the use of drinking water treatment to remove microcontaminants inexpensive using both the net benefit and the cost per life saved approaches in a comparison with other public health measures (Clark *et al.*, 1984). Drinking water is not a unique source, nor is it a major source of exposure to carcinogens, but it is one of the most easily and cost-effectively controlled. With so many uncertainties about microcontaminants in water and about cancer, conclusions are likely to be modified or change, but at this point it seems prudent to take prompt action to protect human health.

REFERENCES

Ames, B. (1983). Dietary carcinogens and anticarcinogens. *Science* **221,** 1256–1264.

Ames, B. (1986). Interview with Ames test inventor. *Environ. Rep.* **16,** 1813.

Andelman, J. B. (1985). Inhalation exposure in the home to volatile organic contaminants of drinking water. *Sci Total Environ.* **47,** 443–460.

Berg, K., ed. (1979). "Genetic Damage in Man Caused by Environmental Agents." Academic Press, New York.

Bloom, A. D., ed. (1981). "Guidelines for Studies of Human Populations Exposed to Mutagenic and Reproductive Hazards." March of Dimes Birth Defects Found., White Plains, New York.

Brown, C. C., and Koziol, J. A. (1983). Statistical aspects of the estimation of human risk from suspected environmental carcinogens. *SIAM Rev.* **25,** 151–181.

Brown, H. S., Bishop, D. R. and Rowan, C. A. (1984). The role of skin absorption as a route of exposure for volatile organic compounds in drinking water. *Am. J. Public Health* **74,** 479–484.

Cairns, J. (1981). The origin of human cancers. *Nature (London)* **289,** 353.

Cantor, K. P. (1983). Epidemiological studies of chlorination by-products in drinking water: An overview. *In* "Water Chlorination: Environmental Impacts and Health Effects" (R. L. Jolley, W. A. Brungs, J. A. Cotruvo, R. B. Cumming, J. S. Mattice, and V. A. Jacobs, eds.), Vol. 4, pp. 1381–1397. Ann Arbor Sci. Publ., Ann Arbor, Michigan.

Cantor, K. P., Hoover, R., Hartge, P., Mason, T. J., Silverman, D. T., and Levin, L. I. (1985). Drinking water source and risk of bladder cancer: A case–control study. *In* "Water Chlorination: Chemistry, Environmental Impact, and Health Effects" (R. L. Jolley, R. J. Bull, W. P. Davis, S. Kate, M. H. Roberts, Jr., and V. A. Jacobs, eds.), Vol. 5, pp. 143–49. Lewis Publishers, Chelsea, Michigan.

Christman, R. F., Norwood, D. L., and Johnson, J. D. (1985). New directions in oxidant by-product research: Identification and significance. *Sci. Total Environm.* **47,** 195–210.

Clark, R. M., Goodrich, J. A., and Ireland, J. C. (1984). Cost and benefits of drinking water treatment. *J. Environ. Syst.* **14,** 1–30.

Council on Environmental Quality (1980). "Drinking Water and Cancer; Review of Recent Findings and Assessment of Risks." U.S. Govt. Printing Office, Washington, D.C.

Cragle, D. L., Shy, C. M., Shuba, R. J., and Siff, E. J. (1985). A Case–control study of colon cancer and water chlorination in North Carolina. *In* "Water Chlorination: Chemistry, Environmental Impact, and Health Effects" (R. L. Jolley, R. J. Bull, W. P. Davis, S. Kate, M. H. Roberts, Jr., and V. A. Jacobs, eds.), Vol. 5, pp. 151–157. Lewis Publishers, Chelsea, Michigan.

Craun, G. F. (1985). Epidemiologic considerations for evaluating associations between the disinfection of drinking water and cancer in humans. *In* "Water Chlorination: Chemistry, Environmental Impact, and Health Effects" (R. L. Jolley, R. J. Bull, W. P. Davis, S. Kate, M. H. Roberts, Jr., and V. A. Jacobs, eds.), Vol. 5, pp. 131–141. Lewis Publishers, Chelsea, Michigan.

Crouch, E., Wilson, R., and Zeise, L. (1983). The risks of drinking water. *Water Resour. Res.* **19,** 1359–1375.

Crump, K. S., and Guess, H. A. (1982). Drinking water and cancer: Review of recent epidemiologic findings and assessment of risks. *Annu. Rev. Public Health* **33,** 339–357.

Department of Health and Human Services (1980). "Health Effects of Toxic Pollutants," Report prepared for the U.S. Senate by the Surgeon General, Serial No. 96–15. U.S. Govt. Printing Office, Washington, D.C.

Du Muchel, W. H., and Harris, J. E. (1983). Boyes methods for combining the results of cancer studies in humans and other species. *J. Am. Stat. Assoc.* **78,** 293–308.

Farber, E. (1981). Chemical carcinogenesis. *N. Engl. J. Med.* **305,** 1379–1389.

Francis, J. D. (1984). "National Statistical Assessment of Rural Water Conditions." U.S. Environmental Protection Agency, Office of Drinking Water, Washington, D.C.

Greenberg, M. R., and Page, G. W. (1981). Planning with great uncertainty: A review and case study of the safe drinking water controversy. *Socio-Econ. Plann. Sci.* **15,** 65–74.

Guinan, D. K., Shaver, R. G., and Adams, E. F. (1978). Identification of organic compounds in effluents from industrial sources. *In* "Drinking Water Quality Through Source Protection" (R. B. Pojasek, ed.), pp. 25–37. Ann Arbor Sci. Publ., Ann Arbor, Michigan.

Hileman, B. (1984). Water quality uncertainties. *Environ. Sci. Technol.* **18,** 124A–126A.

Hoel, D. G., and Crump, K. S. (1981). Waterborne carcinogens: A scientific view. *In* "The Scientific Basis of Health and Safety Regulation" (R. W. Crandall and L. B. Love, eds.), pp. 1973–1995. Brookings Institution, Washington, D.C.

Hollaender, A., ed. (1973). "Chemical Mutagens: Principles and Methods for Their Detection," Vol. 3. Plenum, New York.

Hueper, W. C. (1960). Cancer hazards from natural and artificial water pollutants. *In* "Physiological Aspects of Water Quality." U.S. Public Health Serv., Washington, D.C.

Hueper, W. C., and Payne, W. W. (1963). Carcinogenic effects of adsorbates of raw and finished water supplies. *Am. J. Clin. Pathol.* **39,** 475–481.

Kool, H., van Kreijl, C., de Greef, E., and van Kranen, H. (1982). Presence, introduction, and removal of mutagenic activity during the preparation of drinking water in the Netherlands. *Environ. Health Perspect.* **46,** 207–214.

McDonald, M. E. (1985). Acid deposition and drinking water. *Environ. Sci. Technol.* **19,** 772–776.

National Academy of Sciences (1977). "Drinking Water and Health." Vol. 1. National Academy Press, Washington, D.C.

National Academy of Sciences (1980a). "Drinking Water and Health." Vol. 2. National Academy Press, Washington, D.C.

National Academy of Sciences (1980b). "Drinking Water and Health." Vol. 3, pp. 5–21. National Academy Press, Washington, D.C.

National Academy of Sciences (1982). "Drinking Water and Health." Vol. 4. National Academy Press, Washington, D.C.

National Research Council (1982). "Diet, Nutrition and Cancer." National Academy Press, Washington, D.C.

Ott, W. R. (1985). Total human exposure. *Environ. Sci. Technol.* **19,** 880–886.

Page, G. W. (1981). Comparison of groundwater and surface water for patterns and levels of contamination by toxic substances. *Environ. Sci. Technol.* **15,** 1475–1481.

Page, G. W. (1984). Toxic contaminants in water supplies and the implications for policy. *Environmentalist* **4,** 131–138.

Page, G. W. (1985). "Water Utility Responses to Volatile Organic Contaminants." Presentation to the Association of Collegiate Schools of Planning Conference, Atlanta, Georgia.

Page, G. W. (1987). Water and health. *In* "Protecting Public Health and the Environment" (M. R. Greenberg, ed.). Guilford Press, New York (in press).

Page, G. W., and Greenberg, M. R. (1982). Maximum contaminant levels for toxic substances in water: A statistical approach. *Water Resour. Bull.* **18,** 955–963.

Page, T., Harris, R., and Epstein, S. (1976). Drinking water and cancer mortality in Louisiana. *Science* **193,** 55–57.

Ricci, P. A., and Cirillo, M. C. (1985). Uncertainty in health risk analysis. *J. Hazard. Mater.* **10,** 433–447.

Ricci, P. A., and Molton, L. S. (1985). Regulating cancer risks, *Environ. Sci. Technol.* **19,** 473–479.

Rohlich, G. A. (1978). On drinking water and health. (pp. 47–76), *In* "Safe Drinking Water: Current and Future Problems" (C. S. Russel, ed.), pp. 47–76. Resources for the Future, Washington, D.C.

Schneiderman, M. A. (1980). The uncertain risks we run. *In* "Societal Risk Assessment, How Safe Is Safe Enough?" (R. C. Schwing and W. A. Albers, eds.). Plenum, New York.

Schwartz, D. J., Saxena, J., and Kopfler, F. C. (1979). Water distribution system, a new source of mutagens in drinking waters. *Environ. Sci. Technol.* **13,** 1138–1141.

Shackelford, W. M., and Keith, L. H. (1977). Frequency of organic compounds identified in water. **NTIS PB-265 470.**

Shy, C. M., and Struba, R. J. (1980). Epidemiologic evidence for human cancer risk associated with organics in drinking water. *In* "Water Chlorination: Chemistry, Environmental Impact, and Health Effects" (R. L. Jolley, W. A. Brungs, and R. B. Cumming, eds.), pp. 1029–1042. Ann Arbor Sci. Publ., Ann Arbor, Michigan.

Slovic, P., Fischhoff, B., and Lichtenstein, S. (1979). Rating the risks. *Environment* **21,** 14–36.

U.S. Environmental Protection Agency, Office of Toxic Substances (1975). "Preliminary Assessment of Suspected Carcinogens in Drinking Water," Report to Congress. National Technical Information Service, Springfield, Virginia.

Wallace, L. A., Zweidinger, R., Erickson, M., Cooper, M., Whitaker, D., and Pellizzari, E. D. (1982). Monitoring individual exposure: Measurements of volatile organic compounds in breathing-zone air, drinking water, and exhaled breath. *Environ. Int.* **8,** 269–282.

Wallace, L. A., Pellizzari, E., Hartwell, T., Rosenzweig, M., Erickson, M., Sparacino, C., and Zalon, H. (1984). Personal exposure to volatile organic compounds. *Environ. Res.* **35,** 293–319.

Zoeteman, B., Hrabec, J., de Greef, E., and Kool, H. (1982). Mutagenic activity associated with by-products of drinking water disinfection by chlorine, chlorine dioxide, ozone, and UV-irradiation. *Environ. Health Perspect.* **46,** 197–205.

4

Technological Approaches to Removing Toxic Contaminants

ROBERT M. CLARK
Drinking Water Research Division
U. S. Environmental Protection Agency
Cincinnati, Ohio 45268

I. INTRODUCTION

The purpose of this chapter is to present alternatives for supplying drinking water to a municipality when its normal supply is contaminated. The term alternative refers to any alternate source and includes both new supplies or treatment of the existing supply. The technological approaches described are often used when an existing supply is found to be contaminated.

Among the various alternatives that might be considered are: (1) development of new or existing water resources; (2) blending of a new and contaminated water supply to achieve safe levels; (3) treatment at the well head or each point of consumption; (4) connection to alternative existing municipal or private supplies; (5) oversized community storage facilities to compensate for loss of existing system capacity; and (6) alteration of existing groundwater flows. A few alternatives (other than treatment) are considered in the following section.

II. ALTERNATIVES TO TREATMENT

Alternatives to treatment include development of new sources or connection to existing supplies.

A. Development of New Sources

New groundwater sources that may be available include shallow wells that can be drilled up-gradient of the existing site so that the new source is unaffected by pollutants from the original site. Such an approach may also serve to retard the plume movement of contaminants down-gradient of the site if there is sufficient pumping from the new wells. If an aquifer of adequate yield is located below the contaminated aquifer and is not hydraulically connected to it, new wells can be established in this deeper aquifer. New wells also can be established a sufficient distance from the site so that, with controls to prevent additional contaminant migration, a safe water supply can be provided.

New surface water sources that may be available include streams, rivers, ponds, lakes, and reservoirs. If these possible surface supply sources have adequate yield, they may be located down-gradient of the site, provided the surface supply is not hydraulically connected to the contaminated aquifer. Alternately, these surface supply sources may be located up-gradient or a safe distance from the site.

B. Connection to Existing Supplies

If there is a community with an uncontaminated water supply in close proximity (i.e., 3 to 5 miles), connection to the existing supply may be a viable alternative.

C. Storage Facilities

If an alternative supply (or the portion of a community's supply that is not contaminated) does not have a sufficient yield to meet maximum demand, round-the-clock pumping and an oversized storage facility may provide adequate flows. However, it should be noted that such facilities are commonly used only to supply demand fluctuations, fireflows, etc., above the maximum daily demand.

D. Blending

If a large demand exists, water obtained from new sources (described above) may be mixed with existing supplies resulting in a dilution of pollutants to levels within water quality standards or criteria. Such an approach requires extensive monitoring to assure that the quality of the contaminated supply remains consistently within standards or criteria and that all areas of the distribution system receive blended water. The remainder of this chapter deals with techniques that might be used to treat contaminated sources.

III. TREATMENT OPTIONS

Depending on the contaminants present, a treatment process can be designed to remove contaminants and bring the supply to within drinking water standards. The treatment train necessary to remove the variety of contaminants that may be present can be complex and acceptable disposal of contaminants removed from groundwater or surface water must still be considered. Emphasis should be placed on treatment processes that have been cost-effective at other sites, i.e., carbon adsorption and air stripping for the removal of the most common toxic contaminants.

A. Pilot Testing

Prior to final design of an alternative water supply requiring treatment, it may be necessary to carry out some pilot testing. The normal aim of such studies would be to verify the applicability of processes previously tested on a bench scale, to refine the design criteria, and to obtain an estimate of operation and maintenance (O&M) costs. However, pilot studies may also be required to make a selection from two or more processes that performed similarly in bench scale tests or else may be required because bench scale testing cannot be done, as in the case of air stripping of volatile compounds. Except for very large plants, pilot studies should be looked upon as verifying proposed design criteria.

Where conventional treatment processes are proposed, adequate data for plant design can normally be obtained from bench scale testing of water. A conventional treatment process includes chemical addition and mixing, flocculation, sedimentation, and filtration. For nonconventional treatment processes or for conventional plants treating water of a low quality, pilot testing is often appropriate. The appropriate level of effort will depend on the processes to be piloted and the variability in quality of the water.

1. *Duration of Testing*

The effect of variability of water quality must be accounted for by operating the pilot facilities, either continuously or intermittently, as appropriate, for sufficient time to span the variations in raw water quality.

For most treatment processes treating water of uniform quality, the process will, at a given loading, quickly stabilize. For example, an air stripping process will stabilize after a few minutes, an air flotation process may require between half an hour and one hour, and a biological treatment process may require in excess of a week to stabilize.

For some physical or chemical processes, for example, carbon adsorption, it is possible to carry out accelerated testing using small, highly loaded units. Such testing can often be done in a laboratory, rather than on-site, and can be considered as bench testing.

Where groundwater is to be used, there may be a need to pump at a high rate for an extended period to obtain a reprsentative quality water prior to pilot testing. During the extended pumping it may be necessary to treat the water prior to discharging to surface waters.

2. *Operating Conditions*

Because of the constraints listed above, it is often necessary to study a limited range of operating conditions. Study of the conditions should ensure that an acceptable degree of treatment will be obtained. After construction of the full-size plant, adjusting loading rates, chemical dosages, etc., will obtain the optimum operating conditions for the plant as constructed.

3. *Installation and Configuration*

The normal aim of pilot testing is to verify and model the proposed treatment process. The sizes of the units must be large enough to realistically represent the full-size plant. Often 5% of the full-scale flow is used as a reasonable flow for a pilot plant. However, in practice, where available equipment is being used, the plant flow may have to be sized based on what is economically available. There is no need for all the units in the pilot plant to treat the same flow. Thus, it is often economical and practical to divide the flows through the plant. For example, the effluent from an air stripping tower may be split into two or more streams to allow two or more adsorption units to be run in parallel.

4. *Practical Considerations*

For many specialized processes, manufacturers have pilot plants for rent. Where a single process is being considered, the services of a manufacturer may represent the quickest and easiest way of evaluating a process.

It may be possible to rent the complete pilot facilities, either from equipment manufacturers or from specialized engineering firms. The contract with these suppliers can cover the complete testing program including sampling, analyses, and report preparation.

B. Treatment Processes

The principal or conventional water treatment unit processes will be discussed in the following order: rapid mix, coagulation and flocculation, sedimentation, filtration, chemical feed and handling, disinfection, softening, sludge handling, ion exchange, adsorption, reverse osmosis, and aeration.

C. Treatment Costs

Costs of individual processes or combination of processes will be presented. For each unit process or combination thereof, the assumptions made for the cost analysis will be given by a table of amortized capital costs plus operation and maintenance (O&M) costs. These costs have been updated from previously presented costs (Symons *et al.*, 1981). Data for two plant sizes will be presented.

The choice of a set of assumptions is not intended to reflect performance levels between processes but only to reflect costs within typical design levels. Pilot studies should be done to provide comparative performance information. Table I contains the cost assumptions used in each of the calculations.

One of the general issues that relates to cost estimating for water supply technology is that of economies of scale. As the size of the facility decreases the unit cost of the facility tends to increase. The "scale effect" is one that will apply to all technologies over the size ranges discussed.

IV. CONVENTIONAL TREATMENT

The purpose of this section is to review conventional water treatment processes as a basis for later discussion. These processes are not gener-

TABLE I

Cost Assumptions Used

Item	Level
Energy	\$0.05/kWh
Labor	\$12.10/hr
Producers Price Index (1984)	295.0
Engineering News Record Index (1984)	390.0
Interest rate	8%
Amortization rate	20 yr

ally effective in removing the toxic contaminants that are normally found in groundwater.

In conventional treatment the water to be treated is brought to the rapid mixing tank where destabilizing chemicals are added; vigorous mixing occurs for a short time, normally 1 min or less. Usually one mixing tank is used, although occasionally two tanks may be used in series when two coagulants requiring separate addition are employed. Destabilization is fast and essentially complete after this rapid mixing. The water and its destabilized particles are then introduced in the flocculation tank, where general fluid motion brings the particles into contact so that aggregates can form. This gentle mixing is usually done mechanically, although hydraulic mixing is sometimes employed using baffled tanks. Detention times of 1 hr are typical.

The rapid mixing and flocculation tanks together bring about aggregation and comprise the coagulation process. No materials are removed from the water in these tanks. In fact, materials are added in the form of coagulation chemicals. Solids are removed in subsequent settling and filtration facilities. These solid–liquid separation processes must remove the particles present in the original raw water and the chemicals added to bring about coagulation. These solids leave the treatment system in sludge from the settling tanks and in backwash water from the filters. Disposal of these water treatment plant wastes is a problem in itself. Here it is important to note that the characteristics of these wastes are functions not only of the raw water supply but also of the materials used as coagulants. The major processes that make up conventional treatment are discussed in the following sections.

A. Rapid Mix

Rapid, flash, or quick mix is the unit process used to generate a homogeneous mixture of raw water and coagulants that results in the destabilization of the colloidal particles in the raw water to enable coagulation. Typically, pumps, venturi flumes, air jets, or rotating impellers (paddles, turbines, or propellers) are used to achieve the high-energy mixing conditions required in this process. The vertical shaft turbine impeller is widely used for rapid mixing. It is composed of a vertical shaft driven by a motor with one or more curved blades. Table II contains typical design parameters for rapid mix (Amirtharajah, 1978; Hahn and Stumm, 1968; Letterman *et al.*, 1973).

B. Coagulation

Coagulation, generally followed by filtration, is by far the most widely used process to remove the substances producing turbidity in water.

TABLE II
Design Parameters for Rapid Mix

Item	Value
Mixing intensity	700–1000 sec^{-1}
Detention times	1–2 min
Power/unit volume	26.3–52.6 $kW/m^3/sec$ (1–2 $hp/ft^3/sec$)
Basin dimensions (diameter)	0.91–3.05 m (3–10 ft)

These substances normally producing turbidity consist largely of clay minerals and microscopic organisms and occur in widely varying sizes, ranging from those large enough to settle readily to those small enough to remain suspended for very long times.

Coagulation is the chemical process in which particle charge is satisfied and flocculation is the physical process that agglomerates particles (too small for gravitational settling) so that they may be successfully removed during the sedimentation process.

Coarser components, such as sand and silt, can be removed from water by simple sedimentation. Finer particles, however, will not settle in any reasonable time and must be flocculated to produce the larger particles that will settle. The long-term ability to remain suspended in water is basically a function of both size and specific gravity. The importance of size is illustrated in Table III, which shows the relative settling times of spheres of different sizes (Gemmel, 1971).

The main chemical factors that have been found to influence coagulation of natural waters are: type of coagulant, pH of raw water, concentration of humic substances (especially colloids), and alkalinity.

Commonly used coagulants include those that are iron or aluminum based, lime, and polymers. Aluminum sulfate, commonly known as alum, is effective for pH values of 5.5. to 8.0. Sodium aluminate is used in special cases or as an aid for secondary coagulation of highly colored surface waters and in lime soda softening to improve settling. Ferrous sulfate in conjunction with lime is effective in the clarification of turbid waters and other reactions that have a high pH, such as lime softening. Ferric sulfate reacts with alkalinity and is effective over a wide pH range. It removes color at a low pH and iron and manganese at a high pH.

C. Flocculation

Flocculation usually occurs by mechanical stirring, producing mass fluid motion that is commonly referred to as the orthokinetic flocculation

TABLE III
Effect of Decreasing Size of Spheres

Diameter of particle (mm)	Order of size	Total surface area[a]	Time required to settle[b]
10	Gravel	3.14 cm^2 (0.487 $in.^2$)	0.3 sec
1	Coarse sand	31.41 cm^2 (4.87 $in.^2$)	3 sec
0.1	Fine sand	314.12 cm^2 (48.7 $in.^2$)	38 sec
0.01	Silt	0.31 m^2 (3.38 ft^2)	33 min
0.001	Bacteria	3.14 m^2 (33.8 ft^2)	55 hr
0.0001	Colloidal particles	31.40 m^2 (338 ft^2)	230 days
0.00001	Colloidal particles	0.28 ha (0.7 acre)	6.3 yr
0.000001	Colloidal particles	2.83 ha (7.0 acres)	63 yr (minimum)

[a] Area for particles of indicated size produced from a particle 10 mm in diameter with a specific gravity of 2.65.
[b] Calculations based on a sphere with a specific gravity of 2.65 to settle 1 ft.

unit process. During flocculation slow-moving paddle mixers gently stir a mixture of water and coagulant to generate floc. Because of increasingly fragile nature of the floc as it grows in size, a series of flocculation chambers is usually employed rather than a single flocculation basin. The chambers are designed to enhance laminar flow conditions to prevent floc destruction in conjunction with sufficient mixing to achieve floc formation. A stepped down mixing intensity is utilized in each successive chamber. Typical mixing intensities range between 10 and 100 sec^{-1}. Flocculation time is also a governing factor in floc formation and is generally expressed along with mixing intensity as a product (Gt). Acceptable Gt factors range between 10^4 and 10^5. Mixing intensity is governed by basin size and configuration, number of paddles and their size, and power input to permit floc formation and prevent floc destruction.

D. Sedimentation

Sedimentation or clarification is the removal of particulate matter, chemical floc, and precipitates from suspension through gravity settling (Culp *et al.*, 1968). This makes water clarification a vitally important step in the treatment of surface waters for potable supply and in most cases the main factor in determining the overall cost of treatment. Poor design of the sedimentation basin will result in reduced treatment efficiency that may subsequently upset other operations (Committee Report, 1951).

Surface water containing high turbidity may require sedimentation prior to chemical treatment. Presedimentation basins are constructed in excavated ground or out of steel or concrete. Steel and concrete tanks are often equipped with a continuous mechanical sludge removal apparatus. The minimum recommended detention time for presedimentation is three hours, although, at many times of the year, this may not be adequate to remove fine suspensions. Chemical feed equipment may be provided ahead of presedimentation to provide prechlorination or partial coagulation for periods when water is too turbid to clarify by plain sedimentation (Walker, 1978).

Settling basins are usually provided for chemical coagulation or softening. These basins may be constructed of concrete or steel in a wide variety of shapes and flow mechanisms.

E. Sludge Handling

No clarifier design discussion is complete without a plan for the collection and disposal of the unwanted underflow product—sludge. Traditionally, the sludge was disposed of simply by discharge of a basin into a nearby stream, but this is now generally forbidden. Sludge treatment has become a major problem for most water utilities in the United States.

F. Filtration

Filtration is defined as the passage of fluid through a porous medium to remove matter held in suspension. In water purification, the matter to be removed includes suspended silt, clay, colloids, and microorganisms including algae, bacteria, and viruses.

Floc particle size and strength are of basic importance to the filtration process. In the water applied to the filters, floc particle size may range from 2 mm down to sizes less than 0.1 mm. In a filter containing rounded media, the pore opening sizes will range from 15 to 40% of the particle diameter. Thus, the pore openings in a bed of 0.5-mm sand will range from 0.1 to 0.2 mm in size. In coarse, angular media, such as 1.2-mm anthracite, because of greater porosity, pore openings are larger, ranging from 0.3 to 0.6 mm. It follows that the larger floc particles can be removed by simply straining at the bed surface, but that much of the flocculated matter will pass into the top 2.54 to 10.16 cm (1 to 4 in.) of the bed and lodge within it. If the media grains are small in size, or if the floc is tough, the penetration of flocculated material beneath the sand surface may be small.

Both sand and mixed media filters are normally cleaned by reversal of the flow through the bed or backwashing. Backwashing expands the media depth, allowing particles to be released into the water that is collected above the media in wash troughs. This washing process may be improved by jets of water, air injection, or mechanical agitation at the surface.

G. Softening

Softening is the removal of ions that cause hardness in water. Hardness is caused mainly by calcium and magnesium ions, and at times by iron, manganese, strontium, and aluminum ions. Hardness causes excessive soap consumption for washing operations and scale formation in pumps, boilers, and pipes. Public water supplies should not exceed 300 to 500 mg/liter of hardness; however, aesthetically, a hardness greater than 150 mg/liter is unacceptable.

There are two methods to soften water: precipitation and ion exchange. The lime–soda ash precipitation process removes most of the available hardness. Lime will precipitate calcium oxide, calcium bicarbonate, and magnesium bicarbonate, in that order. These compounds are commonly identified as carbonate hardness. Noncarbonate hardness may then be removed with soda ash. The sulfate and chloride anions associated with calcium and magnesium comprise the noncarbonate hardness. After all hardness is removed, the water is recarbonated using carbon dioxide to reduce the pH to near 7 and thereby preclude production of a corrosive product water. Lime–soda ash softening is capable of reducing total hardness to 30 to 40 mg/liter as calcium carbonate and 10 mg/liter magnesium hydroxide as calcium carbonate.

H. Cost of Conventional Treatment

Treatment costs have been developed for conventional treatment and two modifications: direct filtration, which is filtration without sedimentation, and precipitative softening. Table IV lists the unit processes assumed in each of these treatment trains and Table V contains some of the assumptions used in generating the costs. Tables VI, VII, and VIII contain the capital and operation and maintenance costs for conventional treatment, direct filtration, and precipitative softening.

TABLE IV
Unit Processes Assumed in Each Treatment Train

Direct filtration	Conventional filtration	Precipitative softening
Alum feed	Alum feed	Lime feed system
Polymer feed	Polymer feed	Chlorine feed
Chlorine feed[a]	Chlorine feed	Rapid mix
Rapid mix	Rapid mix	Upflow solids contact clarifier
Flocculation	Flocculation	Recarbonation basin
Gravity filtration	Sedimentation	CO_2 source
Hydraulic surface wash	Gravity filtration	Gravity filtration
Backwash pumping	Hydraulic surface wash	Hydraulic surface wash
Clearwell storage	Backwash pumping	Backwash pumping
Wash water surge basins	Wash water surge basin	Clearwell storage
		Wash water surge basin
		Sludge handling
		Lime recalcination

[a] Chlorine included in these unit processes.

V. CHEMICAL FEED AND HANDLING

Problems associated with chemical feed and handling are important throughout the water treatment process. The topic of chemical feeding and handling will be discussed in relationship to the individual unit processes requiring it.

One important concept in analyzing chemical handling requirements is whether the chemical under consideration is one whose application may be interruptable. Those materials used for corrosion control, taste-and-odor control, and fluoridation can have their application interrupted, and therefore less rigorous criteria apply to their storage space

TABLE V
Clarification Treatment Assumptions

Item	Dose	Assumed cost
Alum	15 mg/liter, 25 mg/liter	\$0.10/kg (\$84.70/ton)
Polymer	0.2 mg/liter	\$5.32/kg (\$2.42/lb)
Chlorine	2 mg/liter	\$0.40/kg (\$363.00/ton)
Lime	300 mg/liter	\$0.08/kg (\$78.65/ton)
Natural gas	—	\$0.017/sm^3 (\$0.0016/scf)
Diesel fuel	—	\$0.21/liter (\$0.79 gal)

TABLE VI

Capital and O&M Costs for Conventional Treatment[a]

	System treatment capacity			
	37,800 m^3/day (10 mgd)		378,000 m^3/day (100 mgd)	
Item	¢/m^3	¢/1000 gal	¢/m^3	¢/1000 gal
O&M cost	3.4	13.0	1.3	5.1
Capital cost	5.9	22.2	2.6	9.8
Total treatment cost	9.3	35.2	3.9	14.9

[a] Chemical dose: alum, 25 mg/liter; polymer, 0.2 mg/liter; chlorine, 2 mg/liter. Average operating capacity is 70%.

and handling equipment. On the other hand, substances such as coagulants and chlorine must not be interrupted under any circumstances. Adequate storage space, handling facilities, and adequate standby provision for feeding are required for them.

Chlorination is the traditional disinfection technique practiced in the United States. Chlorine is shipped, in liquid form, in pressurized steel cylinders ranging in size from 45.35 kg to 0.91 metric ton (100 lb to 1 ton). One volume of chlorine liquid yields 450 volumes of chlorine vapor. The moist gas is corrosive; therefore, all piping and dosing equipment must be nonmetal or resistant to corrosion.

Chlorine dioxide may be produced from sodium chlorite and acid, from sodium chlorite and gaseous chlorine, or from sodium hypochlorite. After production, chlorine dioxide is fed through PVC pipe using a diaphragm pump. Safety features such as chlorine gas detectors, floor

TABLE VII

Capital and O&M Costs for Direct Filtration[a]

	System treatment capacity			
	37,800 m^3/day (10 mgd)		378,000 m^3/day (100 mgd)	
Item	¢/m^3	¢/1000 gal	¢/m^3	¢/1000 gal
O&M cost	3.0	9.6	1.1	4.2
Capital cost	4.7	14.9	2.0	7.6
Total treatment cost	7.7	24.5	3.1	11.8

[a] Chemical dose: alum, 15 mg/liter; polymer, 0.2 mg/liter; chlorine, 2 mg/liter. Average operating capacity is 70%.

TABLE VIII
Capital and O&M Costs for Precipitative Softening[a]

	System treatment capacity			
	37,800 m^3/day (10 mgd)		378,000 m^3/day (100 mgd)	
Item	¢/m^3	¢/1000 gal	¢/m^3	¢/1000 gal
O&M cost	5.8	22.3	4.0	15.0
Capital cost	8.0	29.9	3.0	11.5
Total treatment cost	13.8	52.2	7.0	26.5

[a] Chemical dose: alum, 25 mg/liter; polymer, 0.2 mg/liter; chlorine, 2 mg/liter. Average operating capacity is 70%.

drains, and emergency gas masks should be available at the generation and application site. The major advantage of chlorine dioxide is for use as a residual disinfectant. It does not produce measurable quantities of by-products, such as trihalomethanes, because it does not react with many chlorine-demanding substances (possible organic precursors). Other advantages of chlorine dioxide include algae destruction, color, taste, odor, iron, and manganese removal, and residual and general disinfection properties.

Ozone is widely used in European water systems for disinfection. The ozonation system consists of four parts: (1) a gas preparation system, (2) an electrical power supply, (3) ozone-generating equipment, and (4) contacting equipment. The gas preparation system includes a filtering and drying process that removes oil, dust, and water from the air or oxygen that would interfere with and possibly damage the ozone preparation system. The air or oxygen is also cooled by refrigeration or heat exchangers before it is used. Drying is accomplished by passing the gas to be converted through a desiccant (silica gel, alumina gel, or calcium chloride).

A. Disinfection Effectiveness

Disinfection is a process designed to kill harmful organisms, and it does not ordinarily produce a sterile water. These generalizations hold for disinfection with chlorine also. Two factors are extremely important in disinfection: time of contact and concentration of the disinfecting agent. Where other factors are constant, the disinfecting action may be represented by

$$\text{Kill} = C \times t \tag{1}$$

The important point is that with long contact times a low concentration of disinfectant suffices, whereas short contact times require high concentration to accomplish equivalent kills.

B. Cost of Disinfection

In this section the costs associated with the major disinfectants are presented. The costs are calculated based on equivalent disinfectant probability.

1. *Chlorination*

O&M, capital, and total treatment costs for 37,800- and 378,000-m^3/day (10- and 100-mgd) plants are listed in Tables IX and X.

2. *Chlorine Dioxide*

The cost assumptions unique to chlorine dioxide are listed in Table XI. O&M, capital, and total treatment costs for chlorine dioxide for 37,800- and 378,000-m^3/day (10- and 100-mgd) plants operating at an average 70% capacity appear in Table XII.

3. *Ozonation*

The data in Table XIII show O&M, capital, and total treatment costs for 37,800- and 378,000-m^3/day (10- and 100-mgd) systems operating at an average 70% capacity.

VI. ION EXCHANGE

Ion exchange involves the transfer of one ion for another. For example, a cation in solution attaches to the solid exchanger (resin), which in turn releases a different cation into the solution. All exchangers exhibit selectivity; they may prefer one ion over another by a factor of 15 or more. This preference is not, however, a fixed number but varies with

TABLE IX
Chlorination Assumptions

Item	Assumption
Cost of chlorine	$0.40/kg ($363/ton)
Chlorine dose	2 mg/liter
Contact time (when used)	20 min

TABLE X

Capital and O&M Costs for Chlorination[a]

Item	System treatment capacity			
	37,800 m^3/day (10 mgd)		378,000 m^3/day (100 mgd)	
	¢/m^3	¢/1000 gal	¢/m^3	¢/1000 gal
Chlorination without contact basin:				
O&M cost	0.2	0.7	0.1	0.4
Capital cost	0.1	0.2	<0.1	0.1
Total treatment cost	0.3	0.9	>0.1	0.5
Chlorination with contact basin:				
O&M cost	0.2	0.1	0.1	0.4
Capital cost	0.2	0.7	0.1	0.5
Total treatment cost	0.4	0.8	0.2	0.9

[a] Chlorine dose, 2 mg/liter; operating at 70% of capacity on the average.

ionic strength, relative amounts of ions, the kind of exchanger, and to a lesser extent with other factors, such as temperature. Some of these factors influence preference to such an extent that the selectivity for one ion at a given concentration may even reverse if the solution is greatly diluted. The ratio of one ion to others in solution also has a significant effect on selectivity. Some exchangers have a high affinity for a particular ion, while others do not. This preference for one ion over another is the property that makes ion exchange so valuable.

Cation-Exchange Reactions

It is easy to indicate by chemical equations the reactions that occur in cation-exchange water softening and in regeneration. The exact mecha-

TABLE XI

Chlorine Dioxide Assumptions

Item	Assumption
Chlorine	\$0.40/kg (\$363/ton)
Sodium chlorite ($NaClO_2$)	\$2.66/kg (\$2420/ton)
Chlorine dioxide dose	1 mg/liter
Contact time (when used)	20 min

TABLE XII
Capital and O&M Costs for Chlorine Dioxide [a]

	System treatment capacity			
	37,800 m³/day (10 mgd)		378,000 m³/day (100 mgd)	
Item	¢/m³	¢/1000 gal	¢/m³	¢/1000 gal
Chlorine dioxide without contact chamber:				
O&M cost	0.6	2.3	0.1	1.9
Capital cost	0.1	0.4	<0.1	0.1
Total treatment cost	0.7	2.7	<0.20	2.0
Chlorine with contact chamber:				
O&M cost	0.6	2.3	0.5	1.9
Capital cost	0.3	1.2	0.1	0.6
Total treatment cost	0.9	3.5	0.6	2.5

[a] Chlorine dioxide dose, 1 mg/liter; average operating capacity is 70%.

nism of the reaction, however, is not clearly understood. Whatever the final explanation of the cation-exchange reaction may be, it is evident that its successful completion is dependent on securing adequate contact of the water with the exchange material. Other conditions being equal, adequate contact is probably the most important requisite in successful cation-exchange water softening, and to secure this contact requires experienced correlation of all such important influencing factors as hardness of water, depth of material, rate of flow, and character of the exchanger. Thus, it follows that, if continued maximum contact is to be secured, reasonably clear water must be applied to the exchanger to

TABLE XIII
Capital and O&M Costs for Ozone [a]

	System treatment capacity			
	37,800 m³/day (10 mgd)		378,000 m³/day (100 mgd)	
Item	¢/m³	¢/1000 gal	¢/m³	¢/1000 gal
O&M cost	0.2	0.8	0.1	0.5
Capital cost	0.5	1.8	0.4	1.0
Total treatment cost	0.7	2.6	0.5	1.5

[a] Ozone dose, 1 mg/liter; contact time, 10 min; average operating capacity, 70%.

avoid coating of its particles with colloidal or suspended matter from the water.

VII. REMOVAL OF ORGANICS BY ADSORPTION

Adsorption has been used throughout history as a water treatment process. Carbon-based materials have been especially useful for removing impurities from drinking water (Cheremisinoff and Morresi, 1978). The ancient Hindus filtered their water with charcoal and, in the thirteenth century, carbon materials were used in a process to purify sugar solutions. One of these applications was begun in England in the mid-nineteenth century with the treatment of drinking waters for the removal of odors and tastes. From these beginnings, water and wastewater treatment with carbon-based materials has become widespread in municipal and industrial processes, including wineries and breweries, paper and pulp, pharmaceuticals, food, petroleum, and petrochemical applications.

Certain organic compounds in the water supply are resistant to conventional treatment and many others are toxic or nuisances (odor, taste, color forming) at low concentrations. Low concentrations are not readily removed by conventional treatment methods. Activated carbon has an affinity for various organics and its use for organic contaminant removal from water supplies has been widely implemented.

The purpose of this section is to examine two particular kinds of adsorption processes. One is the use of powdered activated carbon (PAC) and the other is granular activated carbon (GAC) for treating drinking water. Both treatment techniques are useful in removing a broad spectrum of organic compounds. Granular and powdered activated carbon are particularly effective in removing pesticides and herbicides.

A. Powdered Activated Carbon

Powdered activated carbon (PAC) is finely ground loose carbon. To determine what degree of removal of dissolved organic material can be effected by adsorption, an isotherm test is usually run. The adsorption isotherm measures the substance adsorbed and its concentration in the surrounding solution at equilibrium.

B. PAC Costs

Powdered activated carbon has been suggested for removal of organic chemicals and PAC costs have been developed for such an application.

The PAC systems were sized for feeding an 11% slurry that is assumed to be stored and continuously mixed in uncovered concrete tanks that are placed below ground level, except for the top foot or so. For feed capacities of less than 320 kg/hr (700 lb/hr), 8 days of storage in two equal-sized basins are included. For greater feed rates, 2 days of storage in a single basin are included. Mixers were sized based on a G value of 600 sec^{-1}. Storage/mixing basins include equipment for PAC feed from bags in smaller installations and from trucks or railroad cars in larger installations. Table XIV contains capital and O&M costs for PAC.

Energy requirements are based on the rated horsepower of a pump motor for continuous mixing of the 11% carbon slurry at a G value of 600 sec^{-1}. PAC requirements were estimated for various configurations. Labor requirements for the mixing/storage basin are 30 min/day per basin for inspection and routine maintenance, and 16 hr/year per basin for cleaning and gearbox oil change. Slurry pumps require 1 workhour/day per pump.

C. Granular Activated Carbon

Granular activated carbon (GAC) adsorption systems used in drinking water treatment typically use stationary beds with the liquid flowing downward through the adsorbent. Under these conditions adsorbed material accumulates at the top of the bed unit until the amount adsorbed at that point reaches a maximum. Eventually the carbon reaches saturation and the contaminant level begins to rise in the effluent. This is the condition known as "breakthrough." (Editor's note: see Chapter 8, the Wausau, Wisconsin case study concerning breakthrough of GAC systems.) When the level reaches a predetermined point, regeneration

TABLE XIV
Capital and O&M Costs for PAC Treatment [a]

	System treatment capacity			
	37,800 m^3/day (10 mgd)		378,000 m^3/day (100 mgd)	
Item	¢/m^3	¢/1000 gal	¢/m^3	¢/1000 gal
O&M cost	2.0	8.3	1.9	7.7
Capital cost	0.2	0.8	0.1	0.2
Total treatment cost	2.2	9.1	2.0	7.9

[a] PAC dose is 25 mg/liter. Assumed cost for PAC is $0.66/kg ($600/ton). Average operating capacity is 70%.

of the carbon is required. The maximum amount of a contaminant that can be adsorbed on activated carbon occurs when the adsorbed material is in equilibrium with the concentration of the contaminant in solution surrounding the adsorbent.

D. Granular Activated Carbon Adsorption Costs

As discussed previously, GAC adsorption is effective for organics removal. For this analysis, two types of GAC systems will be considered (Symons *et al.*, 1981). One system uses activated carbon in separate contactors after sand filters (hereafter called "postfilter adsorbers") and the other uses GAC as a replacement for the media in existing filter beds (hereafter called "sand replacement"). Both systems will be considered with on-site thermal reactivation. Thermal reactivation is accomplished by exposing the carbon to heat, usually in a furnace. This exposure allows the carbon to regain its capacity for adsorption.

For purposes of the sand replacement analysis, a water treatment plant is assumed to consist of an integral number of 3780-m^3/day (1-mgd) filters. Design parameters assumed for the sand replacement systems are listed in Table XV, and design assumptions for postfilter adsorption systems are presented in Table XVI. Note that for sand replacement systems, a GAC loss of 10% per reactivation cycle is assumed, but a GAC loss of only 6% per reactivation cycle is assumed for postfilter adsorbers. These two assumptions are intended to reflect differences in the operation of the two systems. Sand replacement systems are labor-intensive and increase the possibility of GAC loss because the activated carbon is changed manually and frequently backwashed. In postfiltration systems, fewer possibilities exist for handling losses because the activated carbon is assumed to be changed hydraulically and is seldom backwashed between reactivation cycles.

TABLE XV

Design Parameters Assumed for GAC Sand Replacement Systems

Item	Assumption
Activated carbon cost	$1.86/kg ($0.85/lb)
Activated carbon loss per reactivation cycle	10%
Fuel cost	0.21¢/million joules ($1.80/million BTU)
Volume per filter	24 m^3 (856 ft^3)
Loss in adsorptive capacity	0%
Hearth loading	354 kg/day/m^2 (70 lb/day/ft^2)

TABLE XVI

Design Parameters Assumed for GAC Postfilter Adsorbers

Item	Assumption
Activated carbon cost	\$1.86/kg (\$0.85/lb)
Activated carbon loss per reactivation cycle	6%
Fuel cost	0.21¢/million joules (\$2.18/million BTU)
Hearth loading	354 kg/day/m^2 (70 lb/day/ft^2
Adsorber configuration:	
37,800-m^3/day (10 mgd) plant:	
No. of adsorbers	8
Diameter of adsorber	3.7 m (12 ft)
Vol./adsorber	41 m^3 (1470 ft^3)
378,000-m^3/day (100-mgd) plant:	
No. of adsorbers	28
Diameter of adsorber	6.1 m (20 ft)
Vol./adsorber	122 m^3 (4396 ft^3)
Loss in adsorptive capacity per reactivation cycle	0%

TABLE XVII

Capital and O&M Costs for GAC Adsorption [a]

	System treatment capacity			
	37,800 m^3/day (10 mgd)		378,000 m^3/day (100 mgd)	
Item	¢/m^3	¢/1000 gal	¢/m^3	¢/1000 gal
Sand replacement system:[a]				
O&M cost	1.0	3.7	0.8	3.1
Capital cost	1.6	6.9	0.7	2.5
Total treatment cost	2.6	10.6	1.5	5.6
Postfilter adsorber:[b]				
O&M cost	1.0	3.6	0.7	3.0
Capital cost	2.5	9.6	1.4	5.6
Total treatment cost	3.5	13.2	2.1	8.6

[a] Nine-minute EBCT, 3-month reactivation frequency, 10% loss per reactivation. Average operating capacity is 70%.

[b] Eighteen-minute EBCT, 6-month reactivation frequency, 6% loss per reactivation. Average operating capacity is 70%.

Table XVII contains O&M, capital, and total treatment costs for both systems operating at an average 70% capacity.

VIII. AERATION

Aeration has a long tradition in water treatment, including such applications as the addition of oxygen or carbon dioxide gases to water and the removal of dissolved hyrdogen sulfide gas (Committee Report, 1951; Langelier, 1932). During the past decade, additional applications of aeration have been developed, including air stripping for removal of volatile organic contaminants from drinking water supplies (McCarty, *et al.*, 1979). Preliminary cost analyses suggest that this process may be considerably cheaper than alternative processes including granular activated carbon or adsorbent resins for the removal of selected organic contaminants. Thus, this process merits a detailed analysis of its technical and economic feasibility for control of volatile organic contaminants in drinking water supplies.

The design principles for gas absorption and air stripping have been extensively developed in the field of chemical engineering (Perry and Chilton, 1973). In general, chemical engineers design systems to remove dissolved gases of less than 100 mg/liter. Consequently, the procedures developed for the design of absorption or stripping systems can be simplified and applied to water treatment applications.

This section presents a discussion of the basic principles of the design of air stripping systems as applied to the removal of volatile trace organics from dilute solutions. although air stripping can be conducted with a number of alternative equipment designs, this discussion will focus on countercurrent packed towers, which offer several process advantages.

The relative amounts of the volatile substances contained in air and water and their concentrations in water with respect to their saturation value are factors that control the rate at which the interchange takes place. Substances that occur in water in amounts less than or in excess of their saturation values are changed in concentration by aeration, with the saturation value being the limit of change for both conditions.

The substances involved in the reaction must be volatile. Oxygen, carbon dioxide, and hydrogen sulfide are volatile, and their concentrations in water are readily affected by the process. Many of the taste- and odor- producing compounds occurring in water as the result of algal growths or industrial wastes are not volatile at the temperatures encountered in natural waters and cannot be removed by aeration. Since higher temperatures increase the volatility of the compounds and decrease their saturation values, aeration for the removal of volatile materials is more effective in warm than in cold waters. Similarly, the removal by aeration of some gases, such as H_2S, CO_2, and NH_3, is strongly dependent on the pH of the water, although temperature and pH seem to have minimal impact on the efficiency of removal of volatile organic chemicals.

A. Design Considerations

The interchange of the substances from water to air or from air to water takes place at the air–water surface. The rate at which the interchange occurs is the result of the relative concentration of the substance at this surface and the rapidity with which new surfaces are formed and exposed.

Since molecular diffusion of gases through liquids occurs at low rates that exert little effect on the efficiency of the process, all types of aerators, to be efficient, must develop and continually change large surface areas where fresh contacts are made, and through which the interchange may take place.

The exchange can be described by the following formulas (Haney, 1954):

Gas absorption:

$$C_t = S - (S - C_0)\, 10^{-k(A/V)t} \tag{2}$$

Gas release:

$$C_t = S + (C_0 - S) 10^{-k(A/V)t} \tag{3}$$

where

C_t = concentration of gas in water
S = saturation concentration
C_0 = initial concentration of gas in water
A/V = air to water ratio in volumes per volume
k = rate constant
t = time period

These formulas and the differential equations from which they are derived indicate that:

1. At any instant, the rate of gas transfer is directly proportional to the difference between the gas saturation concentration, S, and the actual concentration, C_t, in the water.
2. The rate of gas transfer is directly proportional to the ratio of the exposed area to the volume of water, A/V.

3. The rate of gas transfer is directly proportional to the gas transfer coeefficient k that in turn is dependent on the diffusivity of the gas in question and the film resistance.

4. The total amount of gas transfer is greater as the time of aeration increases.

5. The percentage change in gas saturation deficit $S - C_0$ or surplus C_t for any given time period t is constant based on the deficit or surplus at the beginning of the time period.

6. Temperature and pressure are important factors because they influence gas solubility, S. Temperature also influences diffusivity and film resistance and hence the value of k.

The time factor and the ratio of surface area to the volume of water, A/V, are important factors that must be considered in the design of an aerator. The partial pressure or the concentration in the air of the gas to be removed must also be considered and provision made for ventilation in installations where the unit is to be housed.

Because corrosion of the structure and equipment represents a major consideration in the design of aerators, material that will resist the corrosive action of the water must be used. Concrete, aluminum, asbestos cement, copper-bearing steel, stainless steel, and creosoted lumber have been used in the construction of aerators. Bronze and cast iron are used in nozzles and a number of patented devices. Ordinary steel should be used only in those locations that are readily accessible for painting at frequent intervals.

Because slime and algal growths may be troublesome in waterfall aerators of the cascade and tray types—especially if the units are located outdoors and subject to sunlight—treatment by chlorine or copper sulfate may be necessary.

B. Physical Chemistry of Aeration

According to Dalton's law, the total pressure of an ideal gas mixture is the sum of the partial pressures of all of its components. Thus the partial pressure of a given contaminant in air is the product of the total pressure and the gas-phase mole fraction of the contaminant in the air:

$$P_a = Y_a P_t \tag{4}$$

where

Y_a = mole fraction of contaminant a in gas phase
P_t = total pressure
P_a = partial pressure of contaminant a in air

When solutions are very dilute, it is often observed that the equilibrium partial pressure of the contaminant in the air is proportional to the solution concentration of the contaminant. This phenomenon is known as Henry's law, and may be mathematically represented as

$$P_a = H_a X_a^* \tag{5}$$

where

H_a = Henry's constant of contaminant a (atm)
X_a^* = solution concentration of contaminant a

Thus, at equilibrium, the concentrations of mole fraction of contaminant k in the liquid and gas phases are related by Eqs. (4) and (5) to give

$$Y_a = X_a^*(H_a/P_a) \tag{6}$$

The larger the Henry's constant, the greater will be the equilibrium of contaminant a in the air. Thus, contaminants having larger Henry's constants are more easily removed by air stripping.

The gases and vapors of most of the volatile organic chemicals of interest follow Henry's law quite satisfactorily in the range of concentrations experienced in domestic water treatment. Table XVIII shows Henry's constants for 25 selected compounds at 20°C (McCarty *et al.*, 1979). Air stripping is relatively ineffective for those compounds with low Henry's constants.

Henry's constants are strongly influenced by temperature. The relationship of the Henry's constant to temperature can be modeled by a van't Hoff-type relation. If the enthalpy change due to dissolution of the contaminant in water is considered independent of temperature, the relation takes the form (McCarty *et al.*, 1979)

$$\sim\log_{10} H = (-\Delta H^\circ/RT) + K \tag{7}$$

where

R = universal gas constant, 1.987 kcal/kmole °K
T = absolute temperature, °K
ΔH° = change in enthalpy due to dissolution of component a in water (kcal/kmole)
K = constant

Table XIX shows values of ΔH° and K for 20 compounds of interest in water treatment. The values show that Henry's constant for most volatile hydrocarbons increases about threefold for every 10°C temperature rise.

Air stripping can successfully lower the concentration of volatile or-

TABLE XVIII
Henry's Constant for Selected Compounds

Compound	Henry's constant (at 20°C) (atm)
Vinyl chloride[a]	3.55×10^5
Oxygen	4.3×10^4
Nitrogen	8.6×10^4
Methane	3.8×10^4
Ozone	3.9×10^3
Toxaphene[a]	3.5×10^3
Carbon dioxide	1.51×10^3
Carbon tetrachloride[a]	1.29×10^3
Tetrachloroethylene[a]	1.1×10^3
Trichloroethylene[a]	5.5×10^2
Hydrogen sulfide	5.15×10^2
Chloromethane	4.8×10^2
1,1,1-Trichloroethane[a]	4.0×10^2
Toluene[a]	3.4×10^2 (25°C)
1,2,4-Trimethylbenzene[a]	3.3×10^2 (25°C)
Benzene[a]	2.4×10^2
1,4-Dichlorobenzene[a]	1.9×10^2
Chloroform[a]	1.7×10^2
1,2-Dichloroethane[a]	61
1,1,2-Trichloroethane[a]	43
Sulfide dioxide	38
Bromoform[a]	35
Ammonia	0.76
Pentachlorophenol[a]	0.12
Dieldrin[a]	0.0094

[a] Computed from water solubility data and partial pressure of pure liquid at specified temperature.

ganics in drinking water. Among the several factors that influence aeration effectiveness, perhaps the three most important are:

- The molecular properties of the contaminant(s).
- The ratio of air to water applied to the aeration.
- The ability to contact a large amount of air with a thin film of water to facilitate transfer of the contaminant into the gaseous phase.

The first factor is best represented by Henry's Law Constant, which describes the concentration of the contaminant in the gas and water

TABLE XIX
Henry's Constant Temperature Dependence

Compound	Formula	ΔH° (kcal/kmole) ($\times 10^3$)	K
Benzene	C_6H_6	3.68	8.68
Chloroform	$CHCl_3$	4.00	9.10
Carbon tetrachloride	CCl_4	4.05	10.06
Methane	CH_4	1.54	7.22
Ammonia	NH_4	3.754	6.31
Chloromethane	CH_3Cl	2.48	6.93
1,2-Dichloromethane	$C_2H_4Cl_2$	3.62	7.92
1,1,1-Trichloroethane	$C_2H_3Cl_3$	3.96	9.39
1,1-Dichlorethane	$C_2H_4Cl_2$	3.78	8.87
Trichloroethylene	C_2HCl_3	3.41	8.59
Tetrachloroethylene	C_2Cl_4	4.29	0.38
Carbon dioxide	CO_2	2.07	6.73
Hydrogen sulfide	H_2S	0.020	5.84
Chlorine	Cl_2	1.74	5.75
Chlorine dioxide	ClO_2	2.93	6.76
Sulfur dioxide	SO_2	2.40	5.68
Difluorochloromethane	CHF_2Cl	2.92	8.18
Oxygen	O_2	1.45	7.11
Nitrogen	N_2	1.12	6.85
Ozone	O_3	2.41	8.05

phases at equilibrium. Experience has shown that the efficiency of aeration varies directly with this parameter (McCarty *et al.*, 1979). For example, 1,2-dichloroethane is more difficult to strip from water than 1,1,1-trichloroethane, which in turn is more difficult to strip than 1,1-dichloroethylene. The other factors listed above are dependent on the type of aerator.

Aeration Systems

Currently (1984), an estimated 10 to 12 United States utilities are aerating to strip volatile organics from their drinking water. The three principal types of aerators being used are redwood slat aerators, packed towers, and spray towers. Two additional methods, diffused-air sparging and air-life pumping, are being evaluated for potential full-scale application.

C. Diffused-Air Aerators

Aeration equipment used in present practice employs one of two methods to control time and *A*/*V*. One method exposes water films to the air and the other introduces air in the form of small bubbles in the water. The first is generally known as the "waterfall type" and the second as the "diffused-air type."

Because waterfall aerators, as the name implies, require that the water be dropped in the process, their use results in the loss of a considerable amount of head. On the other hand, although little or no loss of head is required for the diffused-air units, they require the expenditure of energy in compressing the air and forcing it through small orifices into the water at some distance below the surface.

Difficulties in the operation of waterfall aerators can be expected from freezing during the winter in extreme climates unless they are housed. If an aeration is enclosed, careful consideration should be given to ventilation in the interest of efficiency and safety.

It has been demonstrated that good ventilation is necessary for the effective removal of gases from solution (Termaath, 1982). Based on data from pilot and full-scale operations, the media in multiple aerators should be so arranged that fresh air continuously passes through the falling water and over the media. It is impossible to secure consistent results with an aerator depending on natural ventilation alone. Better results were obtained when a breeze was blowing than when a calm prevailed.

Ventilation for an enclosed aerator is also of importance from a safety standpoint in the case of removal of carbon dioxide, ethane, and hydrogen sulfide. The first named is an asphyxiant, the second creates an explosion hazard, and the third is highly poisonous.

D. Packed-Tower Aerators

This device consists of a column, 1 to 3 m (3.3 to 9.8 ft) in diameter and 5 to 10 m (16.4 to 32.8 ft) in height, that is filled with packing material. This packing can be glass, ceramic, or plastic and is available in numerous geometrical shapes to facilitate the transfer of mass from the aqueous phase to the gas phase. The inside wall of the aeration column has several "redistributors" that force the water over the packing and thus prevent the water from simply running down the walls. If the tower has a fan at the top it is called an "induced draft" packed tower. Perhaps more common is the "forced draft" packed tower in which a blower positioned at the bottom of the tower forces air up through the packing. (Editor's note: see Chapters 8 and 12, the Wausaw, Wisconsin and South Brunswick, N.J. case studies, for a discussion of their use of packed-tower aerators.)

E. Redwood Slat Aerators

These devices are fundamentally cooling towers, but have been used to remove iron and manganese, or hydrogen sulfide, from drinking water. They also can remove a high percentage (20 to 90%) of volatile organics (Silbovitz, 1982).

F. Spray Aerators

Spray aerators are being used in Florida at a drinking water source containing trichloroethylene and related industrial solvents. In that installation, the water is pumped through four aerators in series. The removal efficiency for this operation is reported to be greater than 99% (Wood, 1982).

Early in the twentieth century, air-lift pumps were commonly used in the United States. These are simple devices with only two pipes in the well. One pipe introduces compressed air into the open bottom of the other pipe, called an eductor. In the eductor, air mixes with water. The mixture, being of ligher density than the surrounding water, will rise to the surface. The air is separated before the water is pumped into the distribution system. Because of poor pumping efficiency, typically 35%, they fell into disuse with the introduction of submersible pumps.

Several observations can be made about the aeration process.

- Because mixtures of solvents exist in contaminated water and even though the effectiveness of the process varies for each solvent, aerating to remove one specific contaminant will also reduce concentrations of the others.
- The percentage removal of a contaminant is very constant for a given set of conditions and does not vary with the influent concentration.
- The temperature of ground water is consistently around 13°C (55.4°F).
- Even when subjected to extremely cold air during the winter, the temperature of the aerated water drops but a fraction of a degree, thus these units can be operated in cold weather climates. Some housing, however, is probably prudent to protect the device from freezing in the event of a pump failure.

TABLE XX
Diffused-Air Aeration Assumptions

Item	Assumption
Basin depth	3.3 m (10 ft)
Air supply	1.52 sm^3/m^2 (5 scf/ft^2)

Questions have been raised concerning the secondary effects of aeration. These include potential air quality problems created by exhaust gases from the aerator and the potential for water quality deterioration from airborne particulates, oxidized inorganics, instability resulting in corrosion, and biological growth in the aeration device.

G. Diffused-Air Aeration Costs

Diffused-air aeration involves passing air through the process flow stream. For this analysis, it is assumed to take place in the open, reinforced concrete basins with direct-drive centrifugal compressors and porous diffusors placed at close intervals over the entire basin flow for air introduction. Process energy requirements include the operation of the air compressors 365 days per year, 24 hours per day. Maintenance materials include lubricants and replacement components for air compressors and air diffusion equipment. Estimates were developed from a review of costs associated with activated sludge aeration facilities. Labor requirements include maintenance of air compressors, air piping, valving and diffusors, and aeration basins. Table XX contains some of the key assumptions used in calculating the costs associated with diffused-air aeration.

The effectiveness of using aeration as a technique for stripping volatile chemicals depends heavily on the air/water ratios used. In turn, the cost of diffused-air aeration also depends on the air/water ratio. With the use of the design assumptions in Table XX, total treatment costs were calculated for diffused-air aeration systems with air/water ratios ranging from 1:1 to 30:1 and capacities of 37,800 and 378,000 m^3/day (10 and 100 mgd).* The systems were assumed to be operating at 70% capacity. A breakdown of costs (O&M, capital, and total) for the same systems operating at 70% capacity with a 20:1 air/water ratio is shown in Table XXI.

* These capacities are used throughout this section to reflect the differences between a small population and a large population of 75,000.

TABLE XXI

Capital and O&M Costs for a Diffused-Air Aeration System Operating at 70% Capacity with a 20:1 Air/Water Ratio

	System treatment capacity			
	37,800 m³/day (10 mgd)		378,000 m³/day (100 mgd)	
Item	¢/m³	¢/1000 gal	¢/m³	¢/1000 gal
O&M cost	2.4	9.9	1.3	5.4
Capital cost	2.2	8.5	2.2	8.7
Total treatment cost	4.6	18.4	3.5	14.1

H. Aeration Tower Costs

Estimated construction costs are for rectangular aeration towers with polyvinyl chloride (PVC) packed media. For towers smaller than 178 m^3 (6400 ft^3), units are shipped assembled and have fiberglass skins supported by a galvanized metal framework. Towers of greater volume are field-erected from factory-formed components and are similar in design and construction to industrial cooling towers. The exterior skin of corrugated asbestos–cement panels is attached to a structural steel framework. Towers are supported on reinforced concrete basins. The basin collects tank underflow and serves as a sump for the pump. (Editor's note: use Chapter 8, the Wausau, Wisconsin case study, for design and cost of an aeration tower in operation.)

The cost estimate presented here includes the tower supply pumps and tower underflow pumps. These aeration towers have electrically driven, induced-draft fans with fan stacks and drift eliminators. Process electrical energy requirements are for operation of the induced-draft fan, assuming operations 24 hours per day, 365 days per year. In some instances where pumping energy may also be required, it is estimated separately as part of the unit operation cost; but pumping head will vary from application to application. Units are assumed not to be housed, eliminating the need for building-related energy. Some localities may

TABLE XXII

Aeration Tower Assumptions

Item	Assumption
Tower height	6.1 m (20 ft)
Pumping	9.1 m (30 ft) total dynamic head
Air supply	15.92 sm³ (52.25 scf/ft²) of tower surface area

TABLE XXIII

Capital and O&M Costs for a Tower Aeration System Operating at 70% Capacity with a 500 : 1 Air/Water Ratio

Item	System treatment capacity			
	37,800 m^3/day (10 mgd)		378,000 m^3/day (100 mgd)	
	¢/m^3	¢/1000 gal	¢/m^3	¢/1000 gal
O&M cost	2.2	8.8	1.9	7.8
Capital cost	5.1	20.5	3.0	12.4
Total treatment cost	7.3	29.3	4.9	20.2

have to consider protecting the unit(s) from inclement weather, which would incur an additional cost. Table XXII contains the assumptions used in calculating tower aeration costs.

As with diffused-air aeration, the effectiveness of tower aeration depends heavily on the assumed air/water ratio. A breakdown of costs (O&M, capital, and total) for these systems operating at an average 70% capacity for an air/water ratio of 500:1 is given in Table XXIII. Some trade-offs are possible, for example, increasing the tower depth versus increasing the air/water ratio to achieve increased removal of volatile organics. These options are explored in Table XXIII.

Table XXIV can provide some insight into the important trade-offs involved in using tower aeration to remove volatile organics. For example, assume an initial design choice of a 6.6-m (20-ft) tower with an air/water ratio of 100:1. If an identical target water quality could be achieved by using a 3.3-m (10-ft) tower with an air/water ratio of 300:1, the cost would be slightly higher—1.9¢/m^3 (7.5¢/1000 gal) as opposed to 1.6¢/m^3 (5.9¢/1000 gal).

TABLE XXIV

Total Treatment Costs of Alternative Designs for a 37,800-m^3/day (10-mgd) Tower Aeration System

Tower depth		Air/water ratio					
		100 : 1		300 : 1		500 : 1	
m	ft	¢/m^3	¢/1000 gal	¢/m^3	¢/1000 gal	¢/m^3	¢/1000 gal
3.3	10	1.1	4.0	1.9	7.5	2.7	10.3
6.6	20	1.6	5.9	2.9	11.0	4.0	15.1
9.9	30	2.1	7.3	3.9	14.5	5.8	22.0
13.2	40	2.3	8.7	5.0	18.8	7.5	28.6

TABLE XXV

Cost Data Summary (VOC Removal Costs)

Contaminant	Capacity: Millions of gallons per day	Micrograms per liter	Percentage removal	Cost in dollars per thousand gallons: Tower	Aeration basis	Carbon adsorption
Trichloroethylene	0.5	100	90	0.289	0.579	0.920
		10	99	0.304	0.840	0.973
		1	99.9	0.314	1.093	1.070
		0.1	99.99	0.321	1.346	1.191
	1	100	90	0.193	0.406	0.675
		10	99	0.202	0.648	0.720
		1	99.9	0.208	0.901	0.811
		0.1	99.99	0.214	1.153	0.919
	10	100	90	0.088	0.219	0.377
		10	99	0.093	0.427	0.413
		1	99.9	0.099	0.622	0.485
		0.1	99.99	0.105	0.800	0.576
Tetrachloroethylene	0.5	100	90	0.296	0.675	0.647
		10	99	0.311	0.991	0.700
		1	99.9	0.320	1.302	0.747
		0.1	99.99	0.326	1.575	0.853
	1	100	90	0.197	0.488	0.480
		10	99	0.206	1.857	0.532
		1	99.9	0.213	1.109	0.581
		0.1	99.99	0.218	1.374	0.690
	10	100	90	0.090	0.294	0.209
		10	99	0.096	0.545	0.237
		1	99.9	0.104	0.770	0.266
		0.1	99.99	0.109	0.959	0.332
1,1,1-Trichloroethane	0.5	100	90	0.286	0.532	1.532
		10	99	0.306	0.875	1.750
		1	99.9	0.325	1.510	2.062
		0.1	99.99	0.352	2.726	2.761
	1	100	90	0.191	0.369	1.480
		10	99	0.204	0.683	1.590
		1	99.9	0.217	1.308	1.909
		0.1	99.9	0.244	2.452	2.546
	10	100	90	0.087	0.187	0.850
		10	99	0.094	0.456	1.031
		1	99.9	0.108	0.912	1.302
		0.1	99.99	0.129	1.930	1.927
Carbon tetrachloride	0.5	100	90	0.280	0.454	0.999
		10	99	0.304	0.563	1.082
		1	99.9	0.288	0.636	1.200
		0.1	99.99	0.297	0.687	1.420
	1	100	90	0.187	0.310	0.745
		10	99	0.192	0.393	0.822

Contaminant	Capacity: Millions of gallons per day	Micro-grams per liter	Percentage removal	Cost in dollars per thousand gallons: Tower	Aeration basis	Carbon adsorp-tion
		1	99.9	0.195	0.453	0.996
		0.1	99.99	0.197	0.498	1.127
	10	100	90	0.086	0.141	0.432
		10	99	0.088	0.208	0.495
		1	99.9	0.089	0.262	0.583
		0.1	99.99	0.090	0.303	0.762
cis-1,2-Dichloro-ethylene	0.5	100	90	0.301	0.771	2.663
		10	99	0.314	1.071	2.960
		1	99.9	0.322	1.358	3.342
		0.1	99.99	0.329	1.666	3.722
	1	100	90	0.200	0.580	2.290
		10	99	0.208	0.878	2.562
		1	99.9	0.214	1.164	2.926
		0.1	99.9	0.220	1.462	3.285
	10	100	90	0.092	0.371	1.840
		10	99	0.099	0.605	2.108
		1	99.9	0.105	0.809	2.467
		0.1	99.99	0.110	1.024	2.820
1,2-Dichlo-roethane	0.5	100	90	0.293	0.622	1.363
		10	99	0.302	0.794	1.553
		1	99.9	0.310	0.955	1.853
		0.1	99.99	0.315	1.117	2.461
	1	100	90	0.195	0.440	1.076
		10	99	0.201	0.602	1.248
		1	99.9	0.206	0.763	1.523
		0.1	99.99	0.209	0.923	3.159
	10	100	90	0.089	0.251	0.716
		10	99	0.092	0.390	1.060
		1	99.9	0.095	0.518	1.120
		0.1	99.99	0.100	0.639	1.660
1,1-Dichlo-roethylene	0.5	100	90	0.278	0.430	0.933
		10	99	0.281	0.475	1.021
		1	99.9	0.286	0.530	1.130
		0.1	99.99	0.288	0.563	1.318
	1	100	90	0.184	0.290	0.686
		10	99	0.188	0.325	0.764
		1	99.9	0.191	0.369	0.863
		0.1	99.9	0.192	0.393	1.036
	10	100	90	0.085	0.128	0.386
		10	99	0.086	0.153	0.448
		1	99.9	0.087	0.187	0.529
		0.1	99.99	0.088	0.208	0.678

Based on the assumptions used in this analysis, several mechanisms are available for removing volatile organics. One option for a given tower depth would be to increase the surface area of the tower, thereby increasing the amount of air induced into the water stream. Another option would be to fix the surface area of the tower (thereby fixing the amount of induced air and thus fixing the air/water ratio) and to increase tower depth.

IX. COST COMPARISONS

Because both GAC and aeration are widely accepted as technologies for removing various kinds of organics it would be useful to compare cost and performance for these technologies for a selected list of organic chemicals. Table XXV shows the costs and percentage removals for a list of volatile organic chemicals. Obviously for most of these organics, which are primarily volatile in nature, tower aeration is the most cost-effective alternative. The costs for aeration do not include off-gas control (Clark *et al.*, 1984). At this time, little information is available to provide similar cost comparisons for other less volatile organic chemicals, however, in general for higher-molecular-weight compounds, GAC is the less costly alternative.

REFERENCES

Amirtharajah, A. (1978). Design of flocculation systems, *In "Water Treatment Plant Design" (Robert L. Sanks, ed.), pp. 195–229. Ann Arbor Sci. Publ., Ann Arbor, Michigan.*

Cheremisinoff, P. N., and Morresi, A. C. (1978). Carbon adsorption applications *In "Carbon Adsorption Handbook"* (P. H. Cheremisinoff and F. Ellerbusch, eds. Ann Arbor Sci. Publ., pp. 1–5. Ann Arbor, Michigan.

Clark R. M., Eilers, R. G., and Goodrich, J. A. (1984). VOCS in drinking water: Cost of removal. *J. Environ. Eng. Div. (Am. Soc. Civ. Eng.)* **110** (6): 1146–1162.

Committee Report (1951). Capacity and loadings of suspended solids contact units. *J. Am. Water Works Assoc.* **43**, 263.

Culp, G., Hanson, S., and Richardson, G. (1968). High rate sedimentation in water treatment works. *J. Am. Water Works Assoc.* **60**, 681.

Gemmell, R. S. (1971). Mixing and sedimentation. "Water Quality and Treatment: A Handbook of Public Water Supplies," pp. 123–157. McGraw-Hill, New York.

Hahn, H.H., and Stumm, W. (1968). Kinetics of coagulation with hydrolyzed Al(111). *J. Colloid Interface Sci.* **28,** 134–144.

Haney, P. D. (1954). Theoretical principles of aeration. *J. Am. Water Works Assoc.* **46** (4), 353.

Langelier, W. F. (1932). Theory and practice of aeration. *J. Am. Water Works Assoc.* **24** (1), 62.

Letterman, R. D., Quon J. E., and Gemmell, R. S. (1973). Influence of rapid-mix parameters on flocculation. *J. Am. Water Works Assoc.* **65,** 716–722.

McCarty, P.L., Sutherland, K. H., Graydon J., and Reinhard, M. (1979). Volatile organic contaminants removal by air stripping. *Proc. AWWA Semin. Controll. Org. Drink. Water,* pp. 1–14.

Perry, R. H., and Chilton, C. H., eds. (1973). *"Chemical Engineers' Handbook,"* 5th ed. Sects. 14 and 18. McGraw-Hill, New York.

Silbovitz, A. M. (1982). Removal of TCE from a drinking water supply—A case study, Presented at National Convention, American Society of Civil Engineers, New Orleans, Louisiana.

Symons, J. M., Stevens, A. A., Clark, R. M., Geldreich, E. E., Love, O. T., and DeMarco, J. , (1981). "Treatment Techniques for Controlling Trihalomethanes in Drinking Water, EPA-600/2-81-156. U. S. Environmental Protection Agency, Cincinnati, Ohio.

Termaath, S. G. (1982). Packed tower aeration for volatile organic carbon removal. Presented at *"Seminar on Small Water Systems Technology."* U. S. Environmental Protection Agency, Cincinnati, Ohio (television video cassette available on loan).

Walker, J. D. (1978). Sedimentation. In "Water Treatment Plant Design" (R. L. Sanks, ed.), pp. 149–182. Ann Arbor Sci. Publ., Ann Arbor, Michigan.

Wood, P. R. (1982). Removal of volatile organic compounds from water by spray aeration. Presented at "Seminar on Small Water Systems Technology." U. S. Environmental Protection Agency, Cincinnati, Ohio (television video cassette available on loan).

5

Data and Organizational Requirements for Local Planning

MARTIN JAFFE
School of Urban Planning and Policy
University of Illinois at Chicago
Chicago, Illinois 60637

I. INTRODUCTION

Groundwater protection planning, because of its reliance on technical hydrogeological information, is often beyond the expertise of local planners. Guidance is therefore needed on how to best formulate data collection efforts and how to translate these efforts into management alternatives that address groundwater threats identified by the planning process. The results of this process, moreover, must be understandable by local officials and the general public, who must foot the bill for the technical and management studies. One important objective of this chapter is to give planners some basic information about the technical data required to develop local groundwater protection programs. This will enable them to better be able to work with the technical consultants who will have to be employed by the community to assist in the data collection and interpretation needed to evaluate groundwater quality threats and the responses to such threats. This information will also give the planner a good notion of why certain studies or tests are needed, and will assist the planner in better evaluating the work under way and in better assessing the likely management options that may result from this effort.

In presenting this information, this chapter draws on a number of examples of local groundwater planning programs for communities throughout the nation. It first looks at the historical basis of many of these local programs, then examines the specific data that must be collected by consultant and staff efforts. Finally, it explores how this technical information can be translated into straightforward evaluations that identify groundwater threats and the responses to such threats, in a manner that helps build public and political support for the local protection programs. This last issue is one that often is overlooked in groundwater protection planning, but one that is necessary to ensure the long-term success of local groundwater protection programs and the management alternatives that are fashioned as a result of such programs. It is therefore an issue that is addressed in detail in the last two sections of this chapter.

II. REDISCOVERING THE PAST

Although groundwater protection seems to be a relatively recent issue, a number of communities throughout the nation have comprehensively addressed this issue over a decade ago in their areawide water quality management plans. Most of these plans were funded under Section 208 of the 1972 Federal Water Pollution Control Act amendments to the Clean Water Act, and many of the 208 plans were developed by regional agencies. For the most part, these "208 plans" addressed surface water protection, but in a few regions, however, the protection of groundwater resources was also addressed.

To provide a context for discussing current local groundwater protection planning, it is useful to review these earlier 208 planning efforts (U. S. Environmental Protection Agency, 1977). Initially, in many areas of the nation characterized by rapid growth, few public sewer systems, and permeable soils, areawide water quality management plans were broadened to address groundwater as well as surface water resource management. Thus, in plans developed by the Long Island, New York, Regional Planning Board (as discussed in Chapter 6) and by the Cape Cod, Massachusetts, Regional Planning and Economic Development Commission, groundwater resources were extensively mapped and analyzed. In other areas of the nation, particularly urbanized areas dependent on groundwater supplies, 208 plans developed in the 1970s also examined groundwater resources. This was especially the case in places such as southeastern Florida and in Austin, Texas, where the carbonate aquifers providing drinking water resources also provide direct and intimate connections with surface water supplies. (Editor's note: see chapters 7 and 9). To address surface water quality in these regions, groundwater quality also had to be addressed. In the 1980s, these regions have emerged as the leaders in local groundwater protection planning, but it must be noted that this leadership role in many cases evolved after a long period of analysis and study begun in the 1970s.

These 208 programs shared many similarities, some of which are applicable to groundwater protection planning in the 1980s. One important factor that still applies to local groundwater protection planning today is the extensive use of outside technical consultants in data gathering and interpretation by the 208 agencies (where budgets often were not large enough to hire hydrologists and geologists on-staff). Recent local groundwater protection plans also have relied on outside assistance; the hydrogeological analyses required to undertake local groundwater protection planning are simply beyond the technical abilities of most local planning staffs. A second important factor, which also influences local planning today, is that the development of these 208 plans were relatively expensive undertakings. In the 1970s, federal money was available to pay for these planning programs, while today 208 funding is moribund. Local agencies in the 1980s must be willing to assume the costs of undertaking groundwater protection planning or must expand the scope of current federal funding programs (such as wastewater facilities planning grants) to address groundwater protection concerns. (Editor's note: see the summary, chapter 14, for information on new federal funding proposals.)

A third similarity shared by these early 208 plans that is not so relevant to this decade's efforts is that most of these agencies addressed groundwater protection as an element of broader concerns over growth management. Many of these regions were under heavy growth pressure and looked to water resource management as a means of controlling or guiding new development. Today, in most areas of the country, growth is no longer abhorred, but may even be encouraged to sustain local economies. Groundwater protection as an element of a local growth management program in the 1970s, however, also meant that the same public and political support that was generated for growth management also supported groundwater protection efforts.

The waning of growth management as a partial justification for groundwater protection planning is a less important factor today because of the new concern with the protection of the public health from exposure to synthetic organic chemicals. Toxic waste management issues have replaced growth management as a focus for groundwater protection planning, a process accelerated by the national media attention given to such contamination incidents as Love Canal outside Niagara Falls, New York, and in Times Beach, Missouri. Moreover, the passage of such legislation as the federal Resource Conservation and Recovery Act of 1976 (and its recent 1984 amendments) and the Comprehensive Environmental Response, Compensation, and Liability Act (the "Superfund" law) has highlighted toxic and hazardous materials management as the most crucial environmental issue facing the nation in this decade. This national attention had served to justify local groundwater protection programs that currently are being developed. If groundwater contamination is identified within a community, a political constituency can readily be fashioned to support groundwater protection planning; this new "environmental" constituency can stand in the shoes of the older "growth management" constituency.

It should also be recognized that the current groundwater protection plans being developed by local governments in this decade also differ from the earlier 208 planning programs in some important ways. The first major difference flows from the different constituencies supporting current groundwater protection planning—groundwater planning today tends to be overwhelmingly *reactive*. That is to say that most local programs are developed and supported only after a contamination incident has occurred within a community. Very few of the current programs are preventive in terms of being put into place before groundwater contamination has arisen as a critical issue in a community.

To be even more pessimistic, it might be fair to say that current groundwater protection planning has arisen from only certain kinds of contamination incidents, those that threaten *public* water supplies. As discussed in earlier chapters of this book, groundwater contamination arising from a discrete pollution source tends to be relatively localized geographically. As a contaminant plume slowly flows with groundwater from the pollution source, it tends to affect only those areas downgradient from the source. If these areas contain private wells, then only those households relying on those wells are affected. If pollution is detected (and, by the way, this is a big "if" since private wells normally do not require periodic water quality testing in most states), the pollution source can usually be readily identified and preventive and cleanup measures taken. For example, households can be provided with alternative water supplies or the polluter can assume the costs necessary to compensate the households for the contamination incident. Although the victims of the contamination incident may be quite upset and vocal about their misfortune (depending on what pollutant was contaminating their wells or threatening their health), this localized misfortune often does not translate into direct political action and support for communitywide groundwater protection planning. If, however, groundwater contamination is discovered at a public well (which, incidently, must be monitored and tested periodically under the federal Safe Drinking Water Act or analogous state legislation), and hundreds or even thousands of people are quite vocal and upset about their misfortune, rapid and strong political action usually ensues. A by-product of this action is often strong public support for groundwater protection planning and a willingness to bear the financial burdens of undertaking such plans. Even the most stingy local official realizes the economies of preventing future contamination threats when compared to the high costs of replacing public water supply wells and systems.

A second major difference between current groundwater protection planning and the "first-generation" 208 plans that addressed groundwater quality is that current plans tend to be narrowly focused and not as comprehensive in their scope as the earlier plans. This "narrow focus" typically takes one of two forms: the community adopts a plan and program that only addresses a few contamination threats within its jurisdiction, or it adopts a program that addresses a lot of different threats but only within a limited geographic area, usually a public wellfield (or possibly an area that also recharges the well field). The programs that arise from these plans can then be distinguished as *source control programs* (where various pollution sources are managed) or as *sensitive area programs* (where a number of source controls are applied only to a specific area of the community). Earlier groundwater protection plans and programs that arose from 208 planning efforts typically addressed a number of sources wherever they were located within the community or region; the newer plans are often much more limited in their scope.

The narrow focus may be the result of the reactive nature of groundwater protection planning, where only that identified threat is addressed by a local program, or it may be the result of limited local financial resources that can be allocated to undertaking these plans. Section 208 planning tended to be comprehensive because there was enough federal money available to enable a state or areawide water quality management agency to at least superficially examine a variety of water quality issues (although, some would argue, not enough money was available to carry out effective management programs based on this comprehensive examination). With the disappearance of this funding source, and a concurrent reliance on more limited federal, state, and local funding, local planners cannot be so comprehensive in their efforts. They must, instead, direct their planning program to address only those issues that are most significant and important to the community, or that represent priority groundwater quality threats. Limited economic resources mean that local groundwater protection plans must be now targeted to address specific concerns or specific geographic areas.

The reactive focus of groundwater protection planning and the need for better target data collection have implications for current efforts beyond merely restricting the scope of the planning that can be undertaken. One consequence is that those activities that individually pose the greatest threats to nearby wells may not pose the most significant threats to the community's groundwater resources; less hazardous activities that cumulatively discharge greater amounts of pollutants may, in fact, pose greater long-term risks to aquifers. For example, based on their inherent hazard, leaking underground chemical storage tanks were identified as pollution sources posing severe risks to nearby wells in Rock County, Wisconsin (Holman, 1986). The natural reactive response would have been for the county's Health Department to adopt control measures to address these localized risks. Further analysis by the health department, however, indicated that, cumulatively, the extensive use of pesticides and fertilizers in this rural county posed far greater potential long-term risks to the county's groundwater resources, even though these activities were individually less hazardous than chemical contamination by leaking underground tanks. The use of large quantities of agricultural chemicals applied over large areas of well-drained soils created infiltration risks that were determined to be a priority threat to the county's aquifers, even though groundwater monitoring indicated few pollution risks posed to individual wells from this activity. The county is currently fashioning a management program to address pesticide and fertilizer contamination threats as a preventive measure, making underground storage tank management an important, but less critical, concern for future action.

This example illustrates some of the pitfalls of adopting a reactive stance to deal with groundwater pollution. The problem of identifying the real priority threats within a community and evaluating feasible control measures to mitigate pollution risks from them is the fundamental consideration that must be addressed in groundwater protection planning. Rock County had essentially two choices available: (1) it could have used its initial risk assessment to build public and political support for a program to manage the threats posed by leaking underground storage tanks (either throughout the county, as a source control, or only in proximity to public or private wells, as a sensitive area regulation), or (2) it could have taken its analysis a step further (as it did) and addressed less hazardous activities that cumulatively posed greater potential risks to its groundwater resources. Most communities would have chosen the former alternative; Rock County, however, adopted the preventive option, thereby elevating its technical study into a groundwater protection plan.

III. LOCAL DATA NEEDS

To undertake a local groundwater protection plan, the community must collect and interpret enough technical information about its aquifer systems to guide policy-making. In many areas of the nation, a considerable amount of information has already been gathered by 208 areawide water quality management agencies when these programs were in full flower in the 1970s. Unfortunately, most of these 208 plans addressed only the protection of surface water bodies, consistent with the policy of the Clean Water Act that funded the areawide programs. But where the 208 agency had extended its mandate to include groundwater resources as well, a rich and varied data base may already exist for local governments within the region.

As a general planning strategy, it is most efficient for the community to survey existing information before embarking on the expensive and time-consuming task of generating new data on groundwater resources. The collection and interpretation of new information will undoubtedly require the use of outside technical consultants, while existing information may already have been interpreted as part of its initital analytical purpose. To guide these survey tasks, Table I lists some of the most

TABLE I

Technical Information Useful for Groundwater Planning

Category	Specific Type
Physical framework	Hydrogeologic maps showing extended boundaries of all aquifers and non-water-bearing rocks
	Topographic map showing surface water bodies and land forms
	Water table, bedrock configuration, and saturated thickness maps
	Transmissivity maps showing aquifers and boundaries
	Map showing variations in storage coefficient
	Relation of saturated thickness to transmissivity
	Hydraulic connection of streams to aquifers
Hydrologic stresses	Areas of discharge, recharge basins, recharge wells, natural areas
	Surface water diversions
	Groundwater pumpage (distribution in time and space)
	Precipitation
	Areal distribution of water quality in aquifer (ambient groundwater quality)
	Streamflow quality (distribution in time and space)
	Geochemical and hydraulic relations of rocks, natural water, and artificially introduced water or waste liquids
Model calibration	Water-level change maps and hydrographs
	Streamflow, including gain and loss measurements
	History of pumping rates and distribution of pumpage
Prediction and optimization analysis	Economic information on water supply and demand
	Legal and administrative rules
	Environmental factors
	Other social considerations

Source: U.S. Water Resources Council (1980).

critical technical information that should be a part of a local planning effort.

This list is by no means exhaustive. Finer gradations can be drawn within each category, depending on the type of groundwater contamination issue facing the community. For example, the congressional Office of Technology Assessment (OTA) identified several points about the technical information needed for hydrogeologic investigations (U.S. Congress, Office of Technology Assessment, 1984). These points are also applicable to groundwater protection planning and are therefore set forth below:

- The primary purposes for collecting hydrogeologic data are to determine the rate and direction of groundwater flow, evaluate the types of contaminants likely to be found, and determine whether the contaminants and the groundwater are likely to be moving at the same rate and direction.
- Information on the hydrogeologic environment (i.e., surface conditions—topography, vegetation, climate, and surface water hydrology; geology; and subsurface hydrology—unsaturated zone, groundwater hydrology, contaminant transport parameters, and groundwater use) is obtained primarily to describe the flow of groundwater. Evaluating flow involves the collection of data on the quantity, timing, rate, direction, and pathways of water moving from the surface through the unsaturated zone and into and through the saturated zone. Information about the hydrogeologic environment is important in understanding whether contaminants will move at the same rate as groundwater or if physical, chemical, and/or biological processes are likely to occur that will cause them to move at different rates. Analysis of the physical, chemical, and biological properties of the hydrogeologic environment, along with information on the properties of contaminants, is needed to evaluate the behavior of contaminants in the subsurface.
- Information on water quality is collected primarily to determine the nature and/or verify the extent of contamination. Water quality information also contributes to knowledge of the nature and rate of chemical and biological reactions that influence contaminant behavior.
- Information on sources of contamination is useful in predicting the types of contamination likely to be present, their locations, and their concentrations. When interpreted along with data on groundwater flow and associated contaminant behavior, source data can be used to predict the location, rate, and direction of contaminant movement. Knowledge of sources aids in determining the area to be investigated, the sites for collecting water quality samples, and sampling and analysis procedures.

- Information on properties of contaminants is important for understanding the rate and direction of contaminant movement, the location of contaminants relative to the water table and less permeable units, the persistence of the contaminants in the subsurface, and the types of techniques that can be used to detect, correct, and prevent contamination. The properties of contaminants that are most important for their detection in the subsurface relate to solubility. Hydrogeologic investigations of contaminants that are only slightly soluble (immiscible) require more information on the hydrogeologic environment and water quality than may be needed to describe contaminants that move with groundwater flow.

This overview of informational needs suggests that the collection of this data is likely to be a relatively expensive and technically demanding undertaking beyond the skills and capabilities of most local health and planning agencies. In some cases, as in Rock County, Wisconsin, communities have been fortunate enough to work out arrangements with state or federal geological or water survey agencies to select the community as a study site. This enabled the community to use state or federal technical expertise at little or no local expense. In many cases, given current budgetary constraints on both the state and federal levels of government, this expertise cannot be as easily provided as in the past (Perrine, 1985). Local communities may therefore have to reimburse a state agency for its assistance or hire a private hydrogeological consultant to assist local staff in developing this information.

Recent trends in the development of state groundwater management programs may provide some significant assistance to local governments, though, in lieu of direct technical expertise. The OTA notes that some federal agencies, such as the U.S. Geological Survey and the U.S. Environmental Protection Agency, have begun to collect information on ambient groundwater quality and that 38 states currently monitor groundwater quality or are in the process of developing monitoring systems (U.S. Congress, Office of Technology Assessment, 1984). Moreover, the OTA found that almost all states routinely monitor certain pollution sources (such as hazardous waste facilities) and collect information on water quality at public supply wells as part of their primacy obligations under the federal Safe Drinking Water Act. As this information begins to be collected and compiled by state agencies, a potentially rich data base is being fashioned that may be useful to communities wishing to develop their own groundwater protection plans and programs. These data collection efforts by the states are currently in a rapid state of evolution, a trend that can only benefit communities that must develop their own data bases for groundwater protection planning (U.S. Environmental Protection Agency, 1985).

The development of local data bases, however, is complicated by the "chicken and egg" issues of groundwater quality management—on the one hand, it is difficult to specify the type of data collection efforts that must be undertaken without first knowing something about the contamination threats facing the community; on the other hand, it is difficult to know the extent of groundwater contamination without first undertaking hydrogeologic and other types of investigations and studies. These issues only arose obliquely in the course of 208 planning in many communities, since there were usually enough funds and manpower available to address most likely surface water quality threats everywhere within the region, at least superficially. In the 1980s, with fewer funds available and the reactive character of the planning process predominating, these issues become more central to any planning process.

One way out of this dilemma is to stage data collection so that more general information is first gathered, communitywide, and, as priority issues become identified, successively more sophisticated (and expensive) investigations and studies are undertaken to address those issues in greater depth. A four-stage process might be appropriate to carry out this planning framework. First, a communitywide hydrogeologic investigation can be undertaken to identify the general characteristics of the community's groundwater resources and general geographic areas that seem particularly sensitive or susceptible to contamination. This type of data collection need not address all of the factors set forth in Table I, but should initially be more exploratory, relying heavily on existing studies and information.

Next, areas of groundwater use within the community would be identified. This process would include mapping the location of public water supply wells, and possibly concentrations of private wells tapping the aquifer if such concentrations were also within sensitive areas, as identified by the general hydrogeologic survey. These places of intensive groundwater use (or susceptibility to pollution) would be compared to generalized communitywide surveys of potential contamination sources, as the third stage of the process.

Finally, where the mapping indicates overlaps between sensitive aquifer areas and either areas of intensive groundwater use or concentrations of pollution sources, then these geographic locations could be designated as priority areas by the community. Such a designation would mean that the area would be worth more intensive and sophisticated hydrogeologic study and groundwater monitoring, in addition to other types of data-gathering efforts. If interest (and funding) by local decisionmakers can be sustained beyond this point, then the more sophisticated data-gathering efforts can be extended communitywide, and many other types of pollution sources or groundwater uses can be comprehensively addressed.

This incremental approach has many benefits. It would enable the community to best target its scarce fiscal resources to address only those issues having the most immediate effects on the public health and safety. It would also provide a more preventive focus to the planning process, rather than the reactive focus that is more typical of these types of efforts. Moreover, the generalized hydrogeologic investigation undertaken early in the process would be more consistent with the more generalized groundwater information that may already exist in state surveys and studies. These existing information sources usually are of too general a scale or lack sufficient detail to enable them to be used for sophisticated hydrogeologic investigations, but may be quite suitable for a "first cut" type of analysis preceding the more sophisticated studies undertaken later in the process (after priority areas are identified). It is therefore useful to examine how this type of staged planning process can be carried out at the local government level and to examine in greater detail the specific information that ought to be gathered at each stage.

A. Hydrogeologic Information

Three types of hydrogeologic information should be gathered to help guide local groundwater protection planning: (1) aquifer mapping (to identify and describe groundwater resources within the community); (2) the mapping of groundwater flow (to enable the community to evaluate the movement of contaminants in groundwater); and (3) the documentation of groundwater quality (to understand the nature and severity of groundwater contamination problems in the community). From an abstract perspective, these seem to be fairly straightforward issues of data collection, but actually carrying out these data-gathering tasks can be surprisingly complex and difficult, depending on the hydrological and geological characteristics of the aquifer and the community. (Editor's note: see chapter 1 for a discussion of these complexities.) The hidden degree of difficulty that may be encountered mandates the use of consultants or other experts in this stage of the planning process.

Aquifer mapping seems to be relatively simple, for example, since this type of information often exists in published studies and reports by state or federal agencies or by universities. For example, Rock County, Wisconsin, relied heavily on published hydrogeological studies from the state geological survey in developing its risk assessment process and is relying on assistance from the University of Wisconsin-Madison to evaluate appropriate management alternatives to address these groundwater quality risks. This information, however, may be too general in scope to address local conditions, or may be too fragmented (as in the case of special interest articles dealing with geochemistry or water quantity theories) to enable the community to obtain an accurate picture of its groundwater resources. In the former situation, additional local data gathering may be needed to localize the general information already available, while in the latter case, the community may have to rely on consultant expertise to piece together a description of the aquifer "forest" from the individual studies comprising a complex collection of data "trees."

To be of greatest utility, acquifer mapping should include the identification of usable quantities of groundwater in the community, the geological formations where the aquifers are found, the depth of the groundwater from the surface (or the height of the aquifer above sea level), and the characteristics of the soils and subsurface formations that overlie the aquifer. This information may be difficult or easy to obtain, depending on the data base that has already been developed for the community from published studies or reports. Groundwater quantity data, information on the geologic deposits containing groundwater, and data on the depth to groundwater may exist in published reports; such information may also have to be supplemented by examining well logs collected by a local health board or state agency having licensing authority over well drillers.

It is useful to have this information in order to identify any changes in depth or withdrawal amounts over time, to see if groundwater quantity issues ought to be addressed in the plan, and to determine the relative sensitivity of the aquifer to stress or contamination. If, for example, published information indicates karst conditions for an aquifer that is found in carbonate geological deposits, then the community would know that the groundwater resources are very susceptible to contamination (because of the relative lack of attenuation by overburden in such situations) and that, if contaminated, the aquifer pollution is likely to travel long distances or affect a substantial area (because of the high secondary permeability of these types of groundwater resources). The mapping of these types of deposits can be undertaken by examining published reports by the U.S. Geological Survey or other sources or by looking at geologic cross sections compiled from individual well logs.

The depth of the groundwater and, by implication, the characteristics of the soils or geologic formations overlying the aquifer will provide useful information on the sensitivity of the aquifer to contamination if identified and mapped. The longer it takes surface water sources to infiltrate into the aquifer, the greater the opportunities for attenuation by physical, chemical, or biological processes. The characteristics of the soils and subsurface formations within the unsaturated zone will also have a major effect on these attenuation processes, as will the depth of the groundwater. The depth of the groundwater will also provide useful information on groundwater flow, as discussed below. This information exists in well logs (if such data are gathered as part of state or local licensing or well drilling requirements) or can be generated by surveys of existing wells or by the drilling of new test wells.

It must be noted that, where well logs are not required, very little base data may exist for existing wells—the owners may know nothing of the depth of such wells, the integrity of the casing, the location of the well screens within the casing, and the types of soil and geologic deposits that the well penetrates. Without this information, the depth to groundwater (or geological) information cannot be readily gathered, nor properly interpreted; new test wells or borings may then have to be undertaken. Where this data must be gathered from field investigation, the location and construction of new test wells, the evaluation of soil and geological borings, and the need for pumping tests (to determine potential yields) must be based on the advice of the community's hydrogeologic consultant. The collection and interpretation of this information, even on a general communitywide survey scale, are likely to be expensive if published data do not already exist.

Some published information on the unsaturated zone may exist in the form of Soil Conservation Service (SCS) soil maps and interpretive manuals. This is a very useful body of information for aquifer mapping, since the SCS typically organizes its soils mapping on the county scale, which may be consistent with a communitywide hydrogeologic survey of groundwater resources. Moreover, in almost all cases, the SCS has gone a step further than merely identifying and mapping different soils types but has also looked at associated information, such as slope, depth to bedrock and to the water table, and the characteristics of the soils and their ability to support or limit certain development activities (such as roads, building foundations, or the use of on-site wastewater disposal systems). Where existing development is inconsistent with these SCS interpretations, as where higher-density development relying on septic systems is located in an area with severe limitations for such systems, potential groundwater contamination problems can be identified. Other issues can also be identified from the interpretative data accompanying the soil surveys. For example, areas with high water tables and very permeable soils may function as aquifer recharge areas, suggesting that the movement of contaminants in such areas will be downward into the aquifer and that the high permeabilities of the overburden offer little attenuation of pollutants once they are introduced into the subsurface.

Groundwater flow information is the second aspect of a community's initial hydrogeologic investigations. In unconsolidated aquifers, particularly those where geologic conditions are fairly uniform, unconfined surficial aquifers tend to follow the surface topography, the same as surface waters do. This means that groundwater flow would be "downhill" (or down-gradient, as a hydrogeologist would express it). Where unconsolidated aquifers are in less uniform deposits, as is the case in morainal settings with a series of aquifers separated by confining beds (relatively impermeable deposits) with various degrees of "leakiness," determining groundwater flow patterns may be much more difficult an undertaking. Difficulties may also arise with respect to confined aquifers, either consolidated or unconsolidated.

These types of groundwater settings may require extensive field investigation or the collection of local data if published studies or reports do not exist or are unavailable. Local data collection can build on the type of data collected to undertake aquifer mapping—the depth to groundwater must be accurately determined and the characteristics of the saturated and unsaturated environments must be evaluated. This will enable the community to determine the direction and rate of groundwater flow from areas of recharge to areas of discharge. As noted above, water levels in existing wells may provide surprisingly unreliable information if the characteristics of the wells are unknown. This may require the drilling of special test wells, called piezometers or potentiometers.

Determining groundwater flow information for confined aquifers is more complicated. The "water table" for a confined aquifer is described in terms of its potentiometric (or piezometric) surface, defining areas of different hydraulic head or flow potential. Contours can be developed to connect points of equal head the same way that contours can be drafted to connect points of equal water table depth, and groundwater can be inferred to flow perpendicularly across the contours from areas of higher head to areas of lower head. Additional complications also arise with respect to consolidated carbonate aquifers that exhibit karst characteristics (high secondary permeability, sinkholes and solution cavities, bedding planes and fractures, and interconnection with surface water resources). In these cases, field investigations (including the use of dye tracers) will probably have to be used to determine general flow patterns (although the direct flow characteristics probably cannot be identified with any accuracy because of the complexity of these aquifer formations) and some notion of the interconnection between ground and surface waters inferred.

Documenting groundwater quality is the third (and possibly the most problematic) element of the community's initial hydrogeologic investigation. Two types of studies are typically used to gather information on this issue: longitudinal studies, in which changes in groundwater quality are examined over time to determine trends, and lateral studies, in which current quality is examined at a number of locations to establish baseline data. Provided good historical data exist, longitudinal studies are probably the most useful for groundwater protection planning, although a community might have to rely on lateral studies if such past information is not available.

In either case, the scope of a groundwater quality sampling program may have to be broadened to include chemicals and other substances beyond those tested for under the federal Safe Drinking Water Act (SDWA). A number of states have voluntarily expanded their public drinking water supply testing requirements beyond the SDWA's maximum contaminant level (MCL) substances (or have reduced thresholds for potability below the recommended MCLs), and the community will have to examine its own state legislation and administrative regulations to determine if such changes apply to local sampling processes. It must be noted, however, that determining which additional substances ought to be examined can be tricky, since few data exist on the health effects of long-term exposures to some materials at low concentrations. This is particularly the case for synthetic organic chemicals, of which there are literally thousands in existence and hundreds in common use in a variety of different industries and manufacturing processes. For purposes of National Pollutant Discharge Elimination System permits, required for the discharge of contaminants into surface waters under the federal Clean Water Act, the U.S. Environmental Protection Agency has identified over 130 substances and chemicals that have been designated as "priority pollutants." To ensure consistent testing practices for both surface and groundwaters, it might be useful for these same substances to be examined as part of a groundwater quality testing program as well.

Four additional considerations must enter into the establishment of a groundwater monitoring program in order to gather information on groundwater quality (Hallberg and Hoyer, 1982). First, enough samples must be taken to establish a statistically significant data set. A monitoring program undertaken by the Illinois State Water Survey to establish a nitrate model for a single area in Winnebago County, for example, required over 1100 water samples from over 300 wells in the study area

(Wehrmann, 1983). Second, the collection of information must be undertaken at enough geographic locations to enable the community to obtain a regional overview of groundwater quality issues. Third, specific controls must be created that designate uniform data-gathering procedures (called protocols), and specific information must exist about the groundwater source, the well characteristics and construction, and well locations. It must be noted, however, that as sample size and locations increase in number, it becomes more difficult to maintain adequate control over the sampling program. Finally, the program that is developed must be economical enough to be feasibly undertaken by the community.

The cost considerations of establishing a groundwater sampling program are a particularly troublesome component of this type of data collection. As the number of substances tested for increases in scope, costs also increase since some chemicals require very sophisticated testing equipment capable of identifying substances at very low concentrations (even in the parts per billion range). The OTA notes that contaminant-specific tests for identifying and measuring organic chemicals can range as high as $1500 per sample (if mass spectrometer/gas chromatograph testing equipment is used) and that identifying specific strains of microorganisms can also be as expensive as $1000 per strain (U.S. Congress, Office of Technology Assessment, 1984). Moreover, if inadequate information exists about existing well characteristics, special monitoring wells may have to be constructed at additional cost to the community. If the special wells are used over time as part of an ongoing longitudinal study, however, their cost can be amortized over the study period and may be more feasible to undertake than the construction of test wells for a single lateral study. Narrowing the scope of the testing program to a specific geographic area, which may be susceptible to contamination or which contains a concentration of potential pollution sources, may be one way to contain costs, but at the expense of gaining information about groundwater quality issues on a communitywide scale. The hydrogeologic or technical consultant can help the community evaluate these types of economic and technical trade-offs.

B. Groundwater Use

To evaluate man-made stresses on the hydrogeologic environment and to assess the potential for disrupting groundwater supplies, it is necessary for the community to collect information on how much groundwater is being used, where it is being withdrawn, and what it is being used for. In places where there is heavy reliance on public water supply systems, and good information exists about public well fields and wells, this process can be relatively straightforward. This is particularly the case where public water supply or other large withdrawals must be reviewed and permitted by a state agency under a state water use law. (Editor's note: see Chapters 10 and 12). Existing problems will become apparent if groundwater quality problems are identified in areas that coincide with large groundwater withdrawals, especially withdrawals for public water supply purposes.

Where information does not exist on groundwater use, the collection of data on the location, capacity, and use of existing groundwater may require substantial effort. To be of greatest utility in the development of a local groundwater protection plan, information on groundwater use should also describe the aquifer being tapped, the amounts of groundwater being withdrawn, well capacity and pumping rates, and the degree of water treatment. Although some of this information may be difficult to gather for private wells (particularly if well logs are not required as a regulatory consideration), the mapping of this information may prove to be useful when overlaid on maps showing groundwater flow, sensitive areas, and potential sources of pollution within the community.

Information on groundwater use can be collected through a well survey, especially where private on-site wells are prevalent. Such a survey, for example, was undertaken by the Northeast Michigan Council of Governments (1982) as part of a groundwater protection plan for Briley Township in Montmorency County. Water use can also be estimated from housing counts (if average water use per household figures are available), from census data and housing counts (if per capita use has been established and average number of residents per dwelling unit determined), or even from sewer system effluent flows (assuming that the amount of water used is equivalent to the amount of sewage generated per household) (Jaffe and DiNovo, 1987). Water use information also is readily available from private water companies or utilities if these entities service the community.

C. Groundwater Threats

Assessing groundwater contamination threats involves examining land use activities within the community. However, traditional land use inventories and maps often do not provide sufficient detail about threats to be of much value for groundwater protection purposes. For example, although residential land uses often are mapped by local planning agencies, the information available rarely indicates whether these residences rely on public sewering or on on-site wastewater disposal systems (such as septic systems, cesspools, or tertiary package treatment plants). Information must therefore typically be collected from a variety of sources, including board of health records, sanitary district or commission studies, and capital improvements and wastewater facilities plans in order to properly evaluate the risks of groundwater contamination. These risks are likely to be highest in areas that have been identified as sensitive areas within the community, as discussed in greater detail in the next section of this chapter.

The use of hazardous or toxic materials in business, industry, and agriculture raises special problems. Typically, little is known about the materials that many industries use, which can pose groundwater threats if introduced into an aquifer. Some states, such as Illinois, have passed "worker right to know" laws to protect workers against accidental exposure to some of the more toxic materials found in the workplace, but, for the most part, these laws do not provide for notification to the community (except in the limited case of notification to a fire marshal where concentrations of materials that are readily flammable or explosive are stored on a factory site). Several communities, however, have adopted ordinances requiring that notice of hazardous materials use and storage be given to the community (Sherry and Purin, 1983). These "community right to know" laws may provide useful information about groundwater contamination risks that may arise through material spills and imprper storage practices. Fire department and municipal records should therefore be evaluated as part of a local groundwater protection program, especially where "worker" and "community" right to know disclosure is required.

Where this type of information does not exist for the community, two options are available. One approach is to classify toxic materials according to the Standard Industrial Classification (SIC) codes for the businesses using such materials (U.S. Office of Management and Budget, 1972). Many states, such as Michigan and Illinois, have adopted these industry and business lists, keyed to SIC codes, as part of their programs for regulating hazardous wastes. Spokane County, Washington, for example, keys toxic materials in its sensitive aquifer area protection zone to SIC codes for specific industries; this information, as well as "critical quantities" of the materials deemed likely to present groundwater threats, is incorporated as a Critical Materials Activity List in a section of the ordinance (Spokane County, Washington, 1983). This list is duplicated in Table II. A second approach is to undertake a hazardous or toxic materials survey of all businesses and industries within the community, requesting information on the types and quantities of mate-

TABLE II

Critical Activities in Spokane County, Washington, Aquifer Sensitive Overlay Zone Ordinance (1983)

4.16A.200 *Critical Materials Activity List.* The following types of business activity have been found to use, handle, or store critical materials. For the purposes of administering this ordinance, proposed activities fitting one of the general business descriptions provided or having one of the specified Standard Industrial Classification codes should be assumed to have critical materials on site unless the proponent provides assurance otherwise. Chemicals in addition to those listed may be found on some sites. In some cases SIC codes other than those listed may be associated with a general category.

Type of business	SIC codes	Possible critical materials	Critical quantity
Agricultural chemical	5191	Ammonia	1600 lb as NH_4NO_3
Warehouse and distribution	2873	Nitrate	370 lb as NH_4NO_3
	2874	Sulfate	3000 lb as $(NH_4)_2SO_4$
	2875	Chloride	1200 lb as KCl
	2879	Pesticides and herbicides	
Alumina rolling mills	3353	Hydrocarbon solvents	110 gal
		Methylethyl ketone	105 gal
		1,1, 1-Trichloroethane	70 gal
		Gasoline and diesel fuels	110 gal
		Chloride salts	1000 lb as NaCl
		Chromium salts	90 lb as $Na_2Cr_2O_7$
Aluminum reduction	3334	Fluoride salts	300 lb as AlF_3
	3341	Chromium salts	90 lb as $Na_2Cr_2O_7$
		Gasoline and diesel fuels	110 gal
		Fluoride and cyanide wastes	
Building materials production	2435	Pentachlorophenol	70 gal 5% soln.
	2436	Copper salts	90 lb as $CuSo_4$
	2439	Chromium salts	90 lb as $Na_2Cr_2O_7$
	2491	Phenolic resin glue	15 lb (formaldehyde)
	2492	Caustic soda	850 lb
Chemical manufacturing	2813	All types of chemicals may be on site	
	2816		
	2819		
Chemical warehousing and distribution	5161	All types of chemicals may be on site	

(Continued)

TABLE II (*Continued*)

Type of business	SIC codes	Possible critical materials	Critical quantity
Cleaning supplies, manufacturing, and distribution	2841	Isopropyl alcohol	110 gal
	2869	Chlorinated phenols	20 lb
	5087	Dibutylphthalate	3000 gal
	5161		
Drycleaning establishments	7215	Trichloroethane	2.5 gal
	7217	Tetrachloroethene	2.0 gal
		Hydrocarbon solvents	110 gal
Educational institutions	8221	All chemicals may be present in laboratory quantities	
	8222		
Electrical and electronic manufacturing	3612	Metal salts (Cr, Cu, Ni, Zn)	90 lb
	3641	Cyanide	150 gal 10% NaCN soln.
	3662	Methylene chloride	10 gal
	3674	1,1, 1-Trichloroethane	70 gal
	3677	Acetone	60 gal
	3679	Methylethyl ketone	105 gal
	3825	Formaldehyde	1 gal
	3993		
Electroplating operations	3471	Metal salts (Cr, Cu, Ni, Zn)	90 lb
		Cyanide	150 gal 10% NaCN soln.
		Sodium phosphate	300 gal 30% soln.
		Trichloroethene	2.5 gal
		Tetrachloroethene	2.0 gal
		Xylene	110 gal
		Other solvents	110 gal
Foundries	3321	Metal salts (Cr, Cu, Ni, Zn)	90 lb
	3322		
	3325	Cyanide	125 lb as NaCN
	3361	Trichloroethene	2.5 gal
	3362	Isopropyl alcohol	110 gal
	3369	Caustic soda cleaning soln.	250 gal 35% soln.
Furniture refinishing	7641	Methylene chloride	10 gal
		Acetone	60 gal

Type of business	SIC codes	Possible critical materials	Critical quantity
		Hydrocarbon solvents	110 gal
		Paint-related products	
Medical facilities	0742	Mono- and poly-	
	8062	cyclic aromatic	
	8069	hydrocarbons	
	8071	Prescription drugs	1 gal
		Biological contaminants	
Paint manufacturing and wholesale distribution	2816	Metal salts (Cr,	90 lb
	2851	Pb, Sb, Zn)	
	2865	Phthalate esters	
	5198	Methylene chloride	10 gal
		Methylethyl ketone	105 gal
		Ethylene glycol	7.5 gal
		Hydrocarbon solvents	110 gal
Paint shops	7535	Hydrocarbon solvents	110 gal
		Xylene	110 gal
		Methylene chloride	10 gal
Petroleum products production and storage; bulk distribution of petroleum products	2992	Gasoline	110 gal
	5171	Diesel fuel and heating oil	110 gal
	5172	Lubricating oils	110 gal
		Ethylene glycol	7.5 gal
		Methyl alcohol	60 gal
Photo processing	7333	Silver salts	50 lb as $AgNO_3$
	7395	Phenols	10 lb
		Cyanide	125 lb as NaCN
		Aromatic hydrocarbons	110 gal
Printing establishments	2711	Silver salts	50 lb at $AgNO_3$
	2751	Aromatic hydrocarbons	110 gal
	2752	Phenols	10 lb
	2761	Cyanides	125 lb as NaCN
		Tetrachloroethene	2.0 gal
		Hydrocarbon solvents	110 gal

(Continued)

TABLE II (*Continued*)

Type of business	SIC codes	Possible critical materials	Critical quantity
Gasoline distribution	5541	Gasoline	110 gal
		Diesel fuel	110 gal
		Lubricating oils	110 gal
		Ethylene glycol	7.5 gal
		Methyl alcohol	110 gal
Metal fabrication	3441	Metal salts (Cr, Cu, Ni, Zn)	90 lb
	3443	Caustic cleaning solutions	250 gal
	3444	Hydrochloric acid	155 gal
		Sulfuric acid	150 gal
		Hydrocarbon solvents	110 gal
		Xylene	110 gal
		Caustic soda	250 gal 35% soln.
		Sodium phosphate	300 gal 30% soln.
		Sodium hydroxide	800 lb

Source: Spokane County, Washington (1983).

rials used and the management practices (such as the type, number, age, and condition of storage containers or facilities employed in their use) (Southeast Michigan Council of Governments, 1982). (Editor's note: see Chapter 12 for a description of how South Brunswick, N.J. has conducted such a survey.)

Information may also be gathered on pesticide and herbicide applications in areas where groundwater pollution may be caused by extensive agricultural activities. The types and amounts of agricultural chemicals used, their area of application, and their time of application may be useful information for groundwater protection purposes. Groundwater contamination from these sources may not be as localized as pollution from a single source, because the chemicals may be applied over a large geographic area and may therefore pose special risks (Holman, 1986). Because of the relatively low population densities in rural areas, however, these risks may immediately affect only a limited population—usually the farm household and its livestock—if water supplies are drawn from a private on-site well (as is usually the case). It should also be noted that groundwater protection in such situations is usually simplified by voluntary actions on the part of the farmer once contamination is detected, since the farm household itself is at greatest risk. Without such mitigation measure being carried out, however, agricultural chemical use and application may pose longer-term risks to community groundwater resources, as was discussed with respect to Rock County, Wisconsin.

The final groundwater threats that must be evaluated are those generated by the municipal government itself. This is often overlooked in groundwater protection planning, but can present significant protection issues. Activities that can pose threats include the use of toxic materials by governmental agencies, such as deicing salt storage and application along roadways, the storage of petroleum products in underground storage tanks at municipal garages and service facilities, and the disposal of sewage sludge from municipal sewage treatment plants. Leachate contamination from municipal landfills is also an obvious public sector groundwater threat. These sources should be identified and mapped for the community and treated the same as private sector activities and sources that also can pose groundwater threats.

IV. IDENTIFYING PRIORITY AREAS

Once a data base on hydrogeologic, groundwater use, and groundwater quality information is compiled, the community can analyze this information to identify priority areas that warrant closer examination. These areas are worth a closer look because they are very susceptible to pollution or the effects of groundwater contamination are graver than in other geographic locations in the community. At this stage of the planning process, sensitive areas (typically recharge areas or public well fields) are delineated and pollution threats that exist within them are evaluated. Models are often used to create such risk assessments or evaluations. Once assessed, preventive actions can then be fashioned and compared. To determine which pollution threats require the most immediate response, the community may also rely on prioritization techniques to guide its planning efforts.

A. Sensitive Area Delineation

The community may decide to address either recharge areas or well fields (or both) as sensitive areas that require additional analysis. Recharge areas are places where groundwater is replenished and the flow has a strong downward component that can carry contaminants into the aquifer. Wells draw water from the surrounding area of the aquifer, called the area of influence (or area of contribution), whose boundary depends on the transmissivity, thickness, and lateral extent of the aquifer and the pumping rate of the well. Pollutants introduced into the area of influence will invariably be drawn toward the well, along with the groundwater flow.

Recharge areas can be either local or regional in scope, depending on the size and depth of the aquifer; localized flow patterns are defined by the size and location of groundwater discharge points (typically local streams or waterbodies). Both local and regional flow regimes may be present, each with their own recharge areas. In the larger regional aquifers, pollutants may be carried deeper into the aquifer and reside longer, thus contaminating greater volumes of groundwater. Existing land use (especially impermeable surface coverage), the permeability of soils, and the transmissivity of the geologic materials are important factors used to define recharge areas.

The recharge areas of consolidated aquifers often occupy a smaller portion of the entire aquifer than is the case with surficial aquifers. Recharge to consolidated aquifers may also tend to be more complex, especially where they are in contact with other aquifers (where intervening relatively impermeable deposits—confining beds—are absent or leaky, allowing aquifer interconnection). Recharge areas for consolidated aquifers can be identified by the presence of very permeable soils or unconsolidated materials overlying the aquifers; the recharge areas for surficial aquifers can be quite extensive, overlying the entire community in some instances.

Identifying the areas of influence of wells can be surprisingly complex, depending on local hydrogeologic conditions surround the well or well field. A pumping well creates a cone of depression in the aquifer as ground water is drawn toward the well point. Under idealized conditions of an aquifer of constant level and flowing through uniform material, the cone of depression will be radial with flow lines converging on the well. This simplified system can be predicted by mathematical equations, after certain technical and hydrogeological assumptions are made.

In the real world, however, identifying the cone of depression or area of influence is more complicated, since groundwater flow gradients will distort the cone of depression (compressing it down-gradient and extending it up-gradient). Moreover, few areas adjacent to a well possess uniform characteristics and various permeabilities of geological materials around the well will further distort the cone, sometimes quite drastically, from its idealized circular and radial pattern. If the areas of contribution to a well are examined through the use of modeling, by modifying the ideal flow pattern equations to reflect real conditions, then quite sophisticated computer programs are often required, along with considerable computer time to run the programs. Even then, one ends up with a simplified notion of where the boundaries of the area of contribution are (as shown, for example, by the well field sensitive area delineation problems that have arisen in the development of the Dade County, Florida, well field protection program discussed in Chapter 7).

In identifying sensitive areas around wells or well fields, it is useful for groundwater protection purposes to map this information in terms of time-related capture zones. This mapping process invariably involves the use of sophisticated computer modeling as described above, but gives the community a notion of how long it will take a contaminant introduced into the area of contribution to reach the well point (Gibb *et al.*, 1984). The map will show the same radial pattern as the cone of depression (as modified by local conditions), but the isopleths will be expressed in terms of time (typically days of travel time to the well) instead of geographic distance. The usefulness of this information is that the community would have some idea of how long remedial actions can be delayed before public health risks emerge as the contamination reaches the well. It can also serve to prioritize various groundwater contamination threats within the area of contribution, since those in closest temporal proximity to the well will ostensibly pose the gravest and most immediate risks to the public health and safety.

B. Groundwater Modeling

As discussed above with respect to well field cones of depression, models can be useful and powerful tools in some, but not all, groundwater studies. The use of a model should take the complexity of the hydrogeologic regime into consideration and should be evaluated against a study's specific objectives, resource limitations, and availability of information. Models have been used to predict the movement of contaminant plumes in groundwater, to estimate the effects of changes in recharge or groundwater withdrawal on groundwater levels, and to project changes in groundwater quality as a result of future development within a community (Mercer and Faust, 1981). Models generally work best in fairly simple and uniform hydrogeologic regimes and modeling becomes more problematic and expensive where aquifers are complex or are interconnected.

In several communities, groundwater models have been used as an important component of groundwater protection planning. In Dade, Broward, and Palm Beach counties, Florida, groundwater models have been used to establish time-related capture zones for public wells and to estimate changes in flow patterns to the wells as the result of measures that are intended to modify existing flow patterns. For example, as discussed in Chapter 7, the modeling of the area of contribution to one well field by Dade County, Florida, resulted in the discovery that an existing landfill was within the cone of depression and that leachate from the landfill posed risks to the public water supply. By constructing a "leaky" freshwater canal to intersect the area of contribution, Dade County intends to deliberately distort the cone of depression through artificial recharge, to direct groundwater within the area of influence away from the landfill site and its attendant leachate threats. These threats and preventive measures would have been impossible to evaluate without the extensive data base and the use of sophisticated modeling by the county's water resources management agency.

Groundwater models have also been used to predict changes in groundwater quality within recharge zones. Consultants to the town of Acton, Massachusetts, for example, developed a model to predict nitrate contamination (a pollutant indicative of contamination from septic system effluent) in order to establish densities for septic systems that would not violate the maximum contaminant level for nitrate under the Safe Drinking Water Act (Crystal, 1984). The septic system densities also translated into housing densities, with guidelines of one dwelling unit per 80,000 square feet of lot area in the most sensitive recharge areas (and one d.u./40,000 ft.^{2} in less sensitive areas) recommended. A "mass balance" contaminant prediction model was also developed by the Illinois State Water Survey to assess nitrate contamination from septic system discharges (Wehrmann, 1983). This model enables "down-gradient" nitrate concentrations to be predicted from "up-gradient" discharges and can be used to establish density or other guidelines for residential districts.

C. Establishing Priority Concerns

Given the limited economic and staffing resources available to undertake groundwater protection planning, it makes sense for a community to assess which groundwater threats pose the greatest risks to the community and deserve the greatest attention early in the planning process. These types of decisions are often made according to a prioritization process that takes into account the inherent hazards posed by an activity or substance, where and how such materials are used or activities undertaken, and the potential population that would be at risk as a result of groundwater contamination occurring from the introduction of the materials into groundwater supplies. It should be noted that not all of the risks may be posed by the possibility of ingesting toxic or hazardous substances in groundwater; some substances, such as gasoline, offer quite substantial risks of explosion and fire as they volatilize, in addition to the health risks that are posed by consuming these materials in drinking water.

In many cases, these determinations can be quite subjective, so comparisons are more easily drawn among alternative threats if they can be somehow quantified. The quantification process, however, must be recognized as being inherently subjective in terms of the weight given each factor to be considered; the various ranking systems developed to assess various groundwater threats, therefore, represent value decisions about the relative importance of each factor entering into a risk assessment process. A rich and varied set of alternative methods of ranking groundwater threats has arisen as a result of the federal Superfund law, by which remedial actions must be prioritized according to the hazards posed by a toxic substances disposal site or santitary or hazardous waste disposal facility (Canter and Knox, 1985). Some technical consultants will be familiar with the strengths and weaknesses of the different prioritization systems in common use and can advise the community about which is most appropriate under existing resource constraints and circumstances.

Although a variety of methods have been developed to assess the relative risks to groundwater posed by solid and hazardous waste facilities and by materials used in business and industry, all of the prioritization systems share certain common features. Generally, they all examine four factors: (1) the relative health hazards posed by the materials present and the way that they are managed; (2) the proximity of the site to a water supply well or sensitive aquifer areas; (3) the number of people likely to be affected by groundwater contamination; and (4) the susceptibility of the aquifer to be contaminated at that location (LeGrand, 1983). Each factor is evaluated according to a number of steps and ranked; those sites with the highest scores have the greatest potential risks associated with them and are given the highest priority for management.

V. ESTABLISHING A LOCAL PROTECTION PROGRAM

Local goals and objectives have to be clearly defined and concisely articulated before the community can select appropriate management techniques. One policy issue that will have to be addressed is defining exactly what is to be protected. This consideration will be largely determined by the local hydrogeologic environment, the pattern of water use in the community, and the types and priorities of groundwater threats that exist. In some cases, the formulation of this policy can be quite straightforward; if, for example, the community has no public water supply system (or one that only services a very small geographic area), then attention ought to be directed toward the management of aquifer recharge areas instead of public wells. In other cases, the establishment of these policy decisions can be more difficult. In the Acton, Massachusetts, groundwater study, for instance, the town's consultants examined a number of different boundaries to define various sensitive aquifer areas within the community and estimated the amount of land coverage (expressed as a percentage of the community's total land area) involved in selecting each sensitive area boundary alternative (Crystal, 1982). The management implications of protecting alternative sensitive areas could therefore be compared before a policy decision was reached by local officials.

A second policy concern is the amount of protection that is to be given to the resource once it is identified. This consideration hinges on the administrative and financial capacity of the community and its staff. For example, if groundwater contamination from individual residential septic systems has been identified as a priority management issue in a community, then several options exist to deal with this issue. One approach is to periodically inspect on-site systems to ensure their adequate performance, but this option depends on having enough staff available with the proper expertise to carry out these periodic inspections. A second option is to require sewering development in sensitive areas, but this alternative depends on the community having the funds available (and the willingness) to extend sewer service to such locations. A third option might be to require the use of alternative wastewater treatment and disposal technologies (such as cluster systems or package treatment plants), in lieu of sewering or on-lot wastewater disposal. This option, however, also depends on extensive inspection and monitoring to ensure adequate system performance, as well as being more expensive than the prior two options in operating and maintenance costs (U.S. Department of Housing and Urban Development, 1986). Before adopting any management alternative, the community must take a close and hard look at its current (and future) resources to determine whether the program being proposed can effectively be carried out. Without this, the community may have a false sense of complacency about its groundwater protection program and the level of protection that actually is in effect.

A third policy concern is the type of management system that is to be employed to protect the community's groundwater resources. Three choices exist: health regulations, zoning, and general police power ordinances combining both zoning and health regulations (Jaffe and DiNovo, 1987). Zoning controls are often used (and some states, such as Wisconsin, have amended their zoning enabling acts to make groundwater protection a legitimate purpose of zoning), but are limited only to new development since existing development typically would be exempt as a result of being prior nonconforming uses. Board of health regulations address both proposed and existing development, but are typically beyond the skills of most local sanitarians to administer except if the groundwater threats identified by the local planning process are those that traditionally are addressed by local health officials (such as septic system contamination). General police power ordinances adopted under a community's home rule powers or its inherent police powers (the authority to regulate to protect the public health, safety, and general welfare) may offer the greatest degree of administrative flexibility in addressing different groundwater threats by allocating management responsibility to the local agencies with the greatest experience and resources to deal with them. Certain priority threats can be made retroactive to address current priority concerns, while others can be made prospective to address future risks posed by new developments.

VI. BROADENING THE LOCAL PROTECTION PROGRAM

To broaden a local groundwater protection program beyond priority threats in priority areas, it is necessary to build broader public support for this issue within the community. Many of the earlier 208 plans had this support in the guise of strong citizen and political support for associated growth management issues, but in the 1980s support for groundwater protection as a legitimate and important local objective must be generated anew. This may best be accomplished by building in a strong public outreach and information component into the local data collection and evaluation process.

As mentioned earlier, reliance on past groundwater contamination incidents within the community to generate political support is often of limited use in building a broader constituency, since groundwater contamination tends to be a relatively localized phenomenon (except if public supply wells are affected). Even after Love Canal, for example, Niagara Falls, New York, has not undertaken a comprehensive groundwater protection program. Strong and ongoing public education efforts must be undertaken to ensure the long-term expansion of a reactionary remedial program into a comprehensive local groundwater protection program.

Several techniques can enhance local data collection and planning efforts (Jaffe and DiNovo, 1987). The creation of a steering committee of interested local citizens, representatives of environmental groups, and local officials can be established as part of the local program. Ideally, this committee should include prominent and respected citizens as well as representatives of the local government body that eventually will have to adopt any recommended ordinances, regulations, or groundwater protection measures to carry out the groundwater protection policies developed as part of the program. As part of this effort, information should be developed in a form that is useful and informative to the more general public in addition to the steering committee members. This information may take the form of brochures or posters as well as the more traditional technical report. Information can also include voluntary actions that can be carried out by the public to help protect groundwater resources, in addition to the regulatory measures that normally are considered. (Editor's note: see Chapter 6 for a case study on Long Island, N.Y. that used these techniques very effectively.)

This public outreach effort should probably also include the media, perhaps through the distribution of press releases or the staging of a local event to attract media attention. Since most people find out about important issues through radio and television, it would be a mistake to overlook these resources. It must be noted, however, that groundwater protection is a difficult issue with which to engage a listener or viewer—drilling rigs or absorption spectra of contaminants are not riveting images, nor does the recitation of multisyllabic organic chemicals cause the spirit to soar. It may be easier to use the media if the technical information is translated into health risks or some other "human interest" angle can be developed. In the long run, however, the early and continuous involvement of a public information specialist may prove as important to the development of a comprehensive local groundwater protection program as the involvement of the community's hydrogeologic consultant.

VII. CONCLUSION

Groundwater protection can be a surprisingly complex undertaking, requiring extensive data collection and evaluation. The only efficient way to approach this topic is to build on existing information to the greatest extent possible and to carefully target additional data collection and evaluation activities. Public participation and education are also important components of any planning effort, from its beginning.

The long-term success of establishing a local groundwater protection program can best be evaluated by the lack of pollution threats to public and private water supplies; in a preventive program, the absence of problems is, in some ways, a more accurate measure of success than merely identifying pollution through groundwater monitoring and responding to it by stringent regulatory measures. Perceiving groundwater contamination means that it is already too late to mitigate many of the past and current risks posed to the public health and safety. Moreover, given the slow movement and relative lack of attenuation of many contaminants in groundwater, pollution is relatively irreversible except at great cost and difficulty. Avoiding such risks initially through preventive measures therefore is the course of action that makes the greatest sense.

This chapter presents an approach to groundwater protection that can be carried out by local planners and officials concerned about this subject. Unlike many planning recommendations, it does not take a comprehensive perspective, but instead proposes initially focusing on only a few high-priority threats in a few high-priority locations. Given the fiscal and staffing constraints of many local jurisdictions, this approach seems to make the most sense. If public support for groundwater protection planning can be maintained, then the local program can always be expanded to address additional threats in additional locations. By using an incremental approach to data collection, data evaluation, and management assessment, communities can readily employ the same planning process, regardless of the scale of the inquiry or the scope of threats addressed. The guidelines presented in this chapter can, hopefully, enable communities to rationally and effectively address what is likely to be the most significant environmental issue of the next decade—the protection of groundwater resources from contamination by land use and other activities occurring on the land surface.

REFERENCES

Canter, L. W., and Knox, R. C. (1985). "Ground Water Pollution Control." Lewis Publishers, Chelsea, Michigan.

Crystal, R. (1982). "Acton Ground-Water Protection Program." Lycott Environmental Research, Inc., Southbridge, Massachusetts.

Crystal, R. (1984). Municipal ground-water quality management through land use regulations and zoning. *In* Proceedings of the NWWA Eastern Regional Conference on Ground Water Management" (D. Nielsen and L. Aller, eds.), pp. 432–455. National Water Well Assoc., Worthington, Ohio.

Gibb, J., Barcelone, M., Schock, S., and Hampton, M. (1984). "Hazardous Waste in Ogle and Winnebago Counties: Potential Risk via Groundwater Due to Past and Present Activities," State Water Surv. Contract No. 336. Department of Energy and Natural Resources, Champaign, Illinois.

Hallberg, G., and Hoyer, B. (1982). Sinkholes, hydrogeology and groundwater quality in northeast Iowa. *Open File Rep.—Iowa, Geol. Surv.* **82-3.**

Holman, D. (1986). "A Groundwater Pollution Potential Risk Index System." Rock County Health Department, Janesville, Wisconsin.

Jaffe, M., and DiNovo, F. (1987). "Local Groundwater Protection." Planner's Press, Chicago, Illinois.

LeGrand, H. E. (1983). "A Standardized System for Evaluating Waste Disposal Sites." National Water Well Assoc., Worthington, Ohio.

Mercer, J. W., and Faust, C. R. (1981). "Ground-Water Monitoring." National Water Well Assoc., Worthington, Ohio.

Northeast Michigan Council of Governments (1982). "Local Strategies for Groundwater Management Project Area: Briley Township, Montmorency County, Michigan." NEMCOG, Gaylord, Michigan.

Perrine, R. L. (1985). "The need for expertise personnel. *In* "Proceedings of a National Symposium on Institutional Capacity for Ground Water Pollution Control." University of Oklahoma, Norman.

Sherry, S., and Purin, R. (1983). "Hazardous Materials Disclosure Information Systems: A Handbook for Local Communities and Their Officials." Golden Empire Health Systems Agency, Sacramento, California.

Southeast Michigan Council of Governments (1982). "Genoa Township Policy Plan for Groundwater Protection." SEMCOG, Detroit, Michigan.

Spokane County, Washington (1983). "Aquifer Sensitive Area Overlay Zone Ordinance," Sect. 4.16A.000.

U. S. Congress, Office of Technology Assessment (1984). "Protecting the Nation's Groundwater from Contamination," 2 vols. U. S. Govt. Printing Office, Washington, D.C.

U. S. Department of Housing and Urban Development (1986). "A Reference Handbook on Small-Scale Wastewater Technology." U. S. Govt. Printing Office, Washington, D.C.

U. S. Environmental Protection Agency (1977). "Clean Water and the Land: Local Government's Role." U. S. Govt. Printing Office, Washington, D.C.

U. S. Environmental Protection Agency (1985). "Overview of State Ground-Water Program Summaries," Vol. 1. USEPA, Office of Ground-Water Protection, Washington, D.C.

U. S. Office of Management and Budget (1972). "Standard Industrial Classification Manual." U. S. Gov. Printing Office, Washington, D.C.

U. S. Water Resources Council (1980). "Essentials of Groundwater Hydrology Pertinent to Water Resources Planning." Bull. No. 16. USWRC, Washington, D.C.

Wehrmann, A. (1983). "Potential Nitrate Contamination of Groundwater in the Roscoe Area, Winnebago, County, Illinois," State Water Surv. Contract Rep. No. 325. Department of Energy and Natural Resources, Champaign, Illinois.

6

Long Island Case Study

LEE E. KOPPELMAN
Long Island Regional Planning Board
Hauppauge, Long Island, New York 11788

I. PHYSICAL SETTING

A. Location

Nassau and Suffolk Counties, occupying one-sixth of the land area of the 21-county and the City of New York Metropolitan Region, have been two of the fastest growing counties in the United States since the end of World War II. In 1960, the combined Nassau and Suffolk population of 2 million persons was one-eighth of the total regional population of 16 million. The present two-county population is 2.6 million (Long Island Regional Planning Board, 1985). This 600,000 increase represents half of the total growth in the region. This growth has had numerous impacts, but none more germane than those affecting the quality of the environment.

The counties, with their streams, lakes, rivers, ocean, bays, and sound frontages exceeding 1000 linear miles in total, are familiar natural attributes to millions of persons for resort and recreation opportunities. Long Island Sound on the north and the Atlantic Ocean on the south and east afford a decidedly unique advantage for the proper development of marine resources. The south shore is paralleled by barrier beaches that create bays between the south shore of the Island and the ocean from Long Beach on the west to the Hamptons in the Town of Southampton. Jones Island, Fire Island, and Moriches and Shinnecock Inlets connect these bays to the ocean. This portion of the Long Island peninsula is over 100 miles long and 20 miles wide at its widest point, which is near the Nassau–Suffolk boundary (see Fig. 1). The major land area extends eastward from the Queens–Brooklyn and Nassau County border for approximately 60 miles to Riverhead. East of Riverhead two forks or peninsulas continue eastward separated by the waters of Pe-

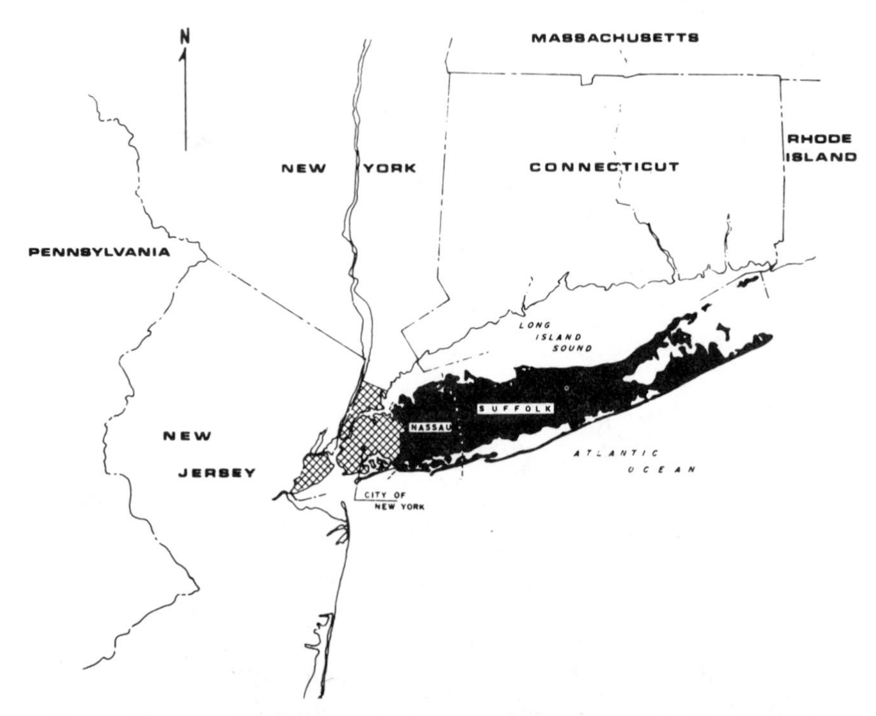

Fig. 1. Nassau and Suffolk counties are situated in the heart of the New York metropolitan region, contiguous with the Brooklyn and Queens New York City border. They are linked to New York City and the surrounding mainland by three railroad lines and three major highways; the Long Island Expressway, the Southern State Parkway, and the Northern State Parkway.

conic and Gardiners Bays. The northern fork terminates at Orient Point and is approximately 20 miles in length. The southern fork terminates at Montauk and is about 44 miles long. The land area of the two counties is approximately 1200 square miles.

B. Natural Characteristics

The topography of Long Island is uniform with a gentle to moderate downward slope from the north to the south shore. A high ridge of glacial origin running approximately east and west from the northwesterly corner of Nassau County and then running in a southeasterly direction through Nassau from the north shore reaches an elevation of about 300 feet above sea level. North of the ridge the topography is generally abrupt with an overall slope to Long Island Sound. South of the ridge is a long, gentle slope terminating in the marsh and meadow land that borders the bays on the south. Four main river watershed valleys are located in Suffolk County. These are the Nissequogue in the Town of Smithtown, Connetquot in the Town of Islip, Carmans in the Town of Brookhaven, and the Peconic, which occurs in the Towns of Riverhead, Brookhaven, and Southampton. These are not true rivers, but exposed groundwaters. In contrast to rivers created by surface drainage from uplands and melting snows, the Suffolk streams are part of the glacial aquifer that is exposed as a result of topography (Fuller, 1914).

The area is mainly composed of the unconsolidated deposits of sand, gravel, and clay laid down in more or less parallel beds on a hard bedrock surface. The rock floor is tilted downward in a southeasterly direction so that from a position of surface outcroppings in the northwest end of Long Island (Queens County) it reaches a depth of 2100 feet below sea level beneath Fire Island (Jensen and Soren, 1974). The subsoil is generally sandy and of yellow color except on the ocean side of the south shore dunes, which are of light gray sea sand. The topsoil has been particularly suited for agricultural uses. Elsewhere the ground is generally covered with scrub growth, mostly oaks and pine. North of the glacial ridge there is an abundance of flora including many of the hardwoods as well as evergreen cover.

The water supply is obtained entirely from groundwater. Natural replenishment of this supply is derived solely from precipitation, i.e., rain, snow, and sleet, which average 42 inches per year. It has been estimated that approximately 50% of the precipitation is lost to evaporation, transpiration, stream flow, and other factors, so that only about half of the precipitation reaches the water-bearing strata (Holzmacher, McLendon and Murrell, Inc., 1969).

On the basis of past experience and engineering projections, the groundwater reservoir appears to be adequate in quantity to serve an estimated population of approximately 5 million persons in the two counties.

C. Problems of Growth

Obviously this is an area where the quality of life and large segments of the economy are related to and dependent on the quality of its environment. Tourism, agriculture, seasonal homes, and residential communities in general thrive in areas of healthy and aesthetically attractive natural settings. Yet a number of major surface and groundwater pollution problems already exist. In the marine waters, these include nutrient enrichment and the closing of beaches and shellfishing areas because of bacterial contamination. These are attributed to both point and nonpoint sources of pollution (Koppelman, 1978). Some freshwater streams have dried up and others are threatened because of lowered groundwater levels due to sewering and excessive well pumpage (Kimmel, 1971). Groundwater quality has been degraded by nitrates, chlorides, and other contaminants from cesspools, fertilizers, recharge of wastewater, landfill leachate, and stormwater recharge (Porter *et al.*, 1977). Water quality—both potable and marine—is the key to Long Island's future. Planning for the orderly growth of these communities and the management of their wastes is the linchpin that will ordain the quality of that future.

In response to earlier perceived growth problems, e.g., residential sprawl, transportation deficiencies, rapidly changing community characters, increased deterioration of older downtowns and housing, and shortages of community facilities, the Boards of Supervisors of Nassau and Suffolk Counties created the Nassau–Suffolk Regional Planning Board in 1965. This agency's prime task was to prepare a comprehensive plan that would serve as a guide for all units of government in the two counties in coping with future growth and to reverse the negative aspects of past development.

The comprehensive plan was completed in July 1970. In essence, it recommends controls on the ultimate size of growth, location and form of development, and institutional changes necessary to achieve implementation. Based on environmental data extant at the time, it was apparent that the most obvious limit to growth was available potable water. It was also apparent that some degradation of these waters had already occurred. Since the projected total population of 3.3 million people by 1985 was less than 60% of the estimated yield that could be sustained, it was assumed that the plan was environmentally prudent.

The Board also recommended that additional funds be sought to conduct water quality studies to ensure that the two counties' sole source of potable water not be jeopardized. The advent of the 1972 Amendments to the Federal Water Pollution Control Act, and particularly its Section 208 planning provisions, furnished the means to start the required studies. Following Sections II and III on the groundwater system, a brief discussion of the planning programs will indicate the nature of the designation process and organization of the technical staff and identify the major substantive elements in the 208 planning program.

II. THE GROUNDWATER SYSTEM

The aquifer is chiefly composed of undifferentiated deposits of coarse to fine sand and gravels with lenses of interbedded silts, clays, and lignite. Its surface is highly irregular due to erosion by streams and glaciers. North–south trending buried valleys occur from the Borough of Queens (New York City) eastward, verifying the idea that a substantial surficial drainage system existed some time after deposition occurred. Three water layers exist—the Upper Glacial, Magothy, and Lloyd strata (Wiggin, 1957). The Magothy, which is the prime source of potable water, generally underlies the Upper Glacial aquifer with some outcropping at the surface. Magothy sediments are Cretaceous in age and of fluvial or deltaic origin. The Magothy is absent in northwest Nassau County but increases in thickness to over 1000 feet in the southern part of Suffolk County. Hydraulic conductivities vary widely, with the average horizontal conductivity estimated to be about 50 feet per day, while the average vertical conductivity is about 1.4 feet per day (Franke and Cohen, 1972). Consequently, specific capacities of wells also vary widely, from 1 to about 40 gallons per minute per foot of drawdown. Wells in the Magothy commonly yield in excess of 1000 gallons per minute. Groundwater in the aquifer may be poorly or well confined, depending on location. In areas where the overlying clays, such as the Gardiners clay, are laterally continuous and of substantial thickness, vertical movement between the water table aquifer and the Magothy is reduced.

In the central portion of Long Island, recharge to the Magothy from the water table occurs over a large area. Thus the delineation, control of land uses, and study of this region are critical in trying to assess how chemical contaminants, introduced to the water table from the land surface, will tend to move in the groundwater system.

Because of its shape, structure, and geography setting, Long Island has a rather unique groundwater flow system that can be broadly described on a regional basis. Groundwater in the water table aquifer, near the spine of the island, is subject to hydraulic gradients that tend to carry some of the water vertically downward to the deepest parts of the Magothy. To the south and north of this zone, water from the shallow deposits flows with vertical and horizontal components that result in some water moving into the middle portion of the Magothy. Farther toward the coastlines, circulation becomes shallower until, at some point, flow is essentially horizontal in both the water table and the Magothy aquifers. Beyond this area, water in the Magothy has a vertically upward component, while water in the shallow deposits flows essentially horizontally until it discharges to streams or saltwater bodies.

The lowermost aquifer is the Lloyd stratum, which lies directly above the bedrock and is generally separated from the Magothy by a thick layer of Raritan clay. The Lloyd water, obviously the oldest and most pristine, has been rarely tapped and is the most restricted of the aquifers. Figure 2 presents a generalized hydrogeological cross section of the island and simplified pictorial of the Long Island aquifer system.

III. GROUNDWATER PROBLEMS

There are many sources and causes of groundwater contamination on Long Island. They can be divided into four categories. The first two categories represent discharges of contaminants that are derived from solid and liquid wastes. The third category concerns discharges of contaminants that are not wastes, and the fourth category lists those causes of groundwater contamination that are not discharges at all. Table I summarizes the list.

TABLE I

Classification of Sources and Causes of Groundwater Contamination Used in Determining Level and Type of Control

Category I: Systems, facilities, or sources designed to discharge waste or wastewaters to the land and groundwaters.	Category II: Systems, facilities, or sources not specifically designed to discharge wastes or wastewaters to the land and groundwaters.	Category III: Systems, facilities, or sources that may discharge or cause a discharge of contaminants that are not wastes to the land and groundwaters.	Category IV: Causes of groundwater contamination that are not discharges.
Domestic on-site waste disposal systems	Sanitary sewers	Highway deicing and salt storage	Airborne pollution
Sewage treatment plant effluent	Landfills	Fertilizers and pesticides	Water well construction and abandonment
Industrial waste discharges	Animal wastes	Product storage tanks and pipelines	Salt water intrusion
Storm water basin recharge	Cemeteries	Spills and incidental discharges	
Incinerator quench water		Sand and gravel mining	
Diffusion wells			
Scavenger waste disposal			

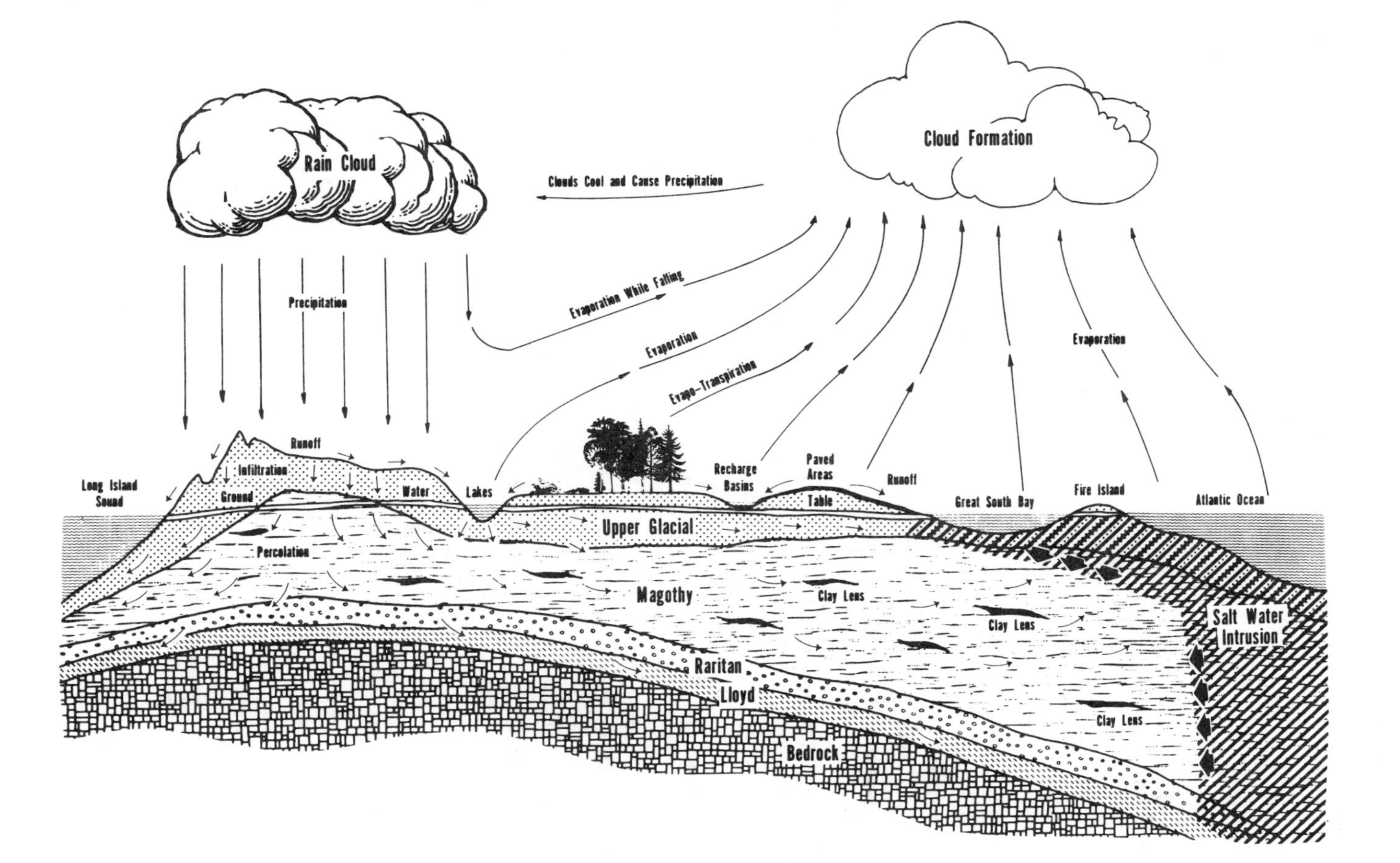

Fig. 2. The Long Island aquifer system.

A. Domestic On-Site Waste Disposal Systems

Cesspools, septic tanks, and leaching fields are a source of groundwater contamination on Long Island that has been of great concern. In on-site disposal systems, bacterial action digests the solid materials and the liquid effluent is discharged to the ground. In theory, filtration by earth materials provides additional treatment so that the liquid, when it arrives at the water table, is relatively clean. However, many constituents carried by the effluent are introduced to the groundwater system. Excessive concentrations of nitrate, organic chemicals, detergent, metals, bacteria, and viruses present the greatest threat to groundwater quality. Other constituents—previously ignored, but now recognized as a threat—are halogenated hydrocarbons. Compounds such as chloroform, carbon tetrachloride, trichloroethylene, and others are in common use in industry as degreasers and solvents or are incorporated in plastic products. These and similar compounds also regularly occur in household discharges. Many home products such as fabric and rug cleaners, workshop cleaners and solvents, and solutions to clean pipes find their way into on-site disposal systems. Septic tank cleaners are composed almost entirely of active ingredients that are frequently halogenated hydrocarbons. For example, one common cesspool cleaner contains more than 99% trichloroethylene. One gallon of this compound could raise the trichloroethylene concentrations of 29 million gallons of water to the state-recommended maximum of 0.05 parts per million (Koppelman *et al.*, 1978). Other compounds containing a variety of hydrocarbons have also contaminated groundwaters. The Suffolk County Department of Health Services (1981) has found more than 800 houses whose wells were contaminated.

B. Sewage Treatment Plant Effluents

These are minor threats since most of the effluent is discharged directly to the sea. However, almost 9 million gallons per day are discharged into the ground. Although this amount is only slightly more than 1% of the daily fresh water from precipitation recharge, it nevertheless may have a significant effect when concentrated at a few sites.

C. Sanitary Sewers

Approximately 120 million gallons per day of raw sewage flows through thousands of miles of sewers in the bicounty area. Sewers frequently leak, and depending on the type of sewer and its altitude relative to the water table, groundwater can infiltrate or sewage can exfiltrate. The contamination that takes place in the latter case is normal domestic sewage, plus those constitutents in industrial effluent discharged to sewers.

D. Industrial Waste Discharge

Approximately 21 million gallons per day of permitted industrial wastes are discharged to the ground in a few industrial areas. There are additional commercial and industrial discharges not covered by the permit process. These include car washes, coin-operated laundries, and industries discharging wastewater with constituents not covered by permitting regulations.

E. Storm Water Basins

Research has determined that half the annual precipitation finds its way to the groundwater reservoir as recharge. This averages roughly 1 million gallons per day per square mile in a 760-square mile recharge area. As the western portion of the region has become increasingly urbanized, however, permeable soil areas have been replaced by impermeable roofs and paved areas. The water cannot seep into these surfaces, so it accumulates and runs off.

As a water conservation alternative to offset reductions in groundwater recharge and to eliminate the need for expensive trunk sewers leading to the sea, a system of small storm sewers draining to unlined recharge basins was initiated in Nassau County in 1935. There are now more than 2000 recharge basins on Long Island (Seaburn and Aronson, 1973). They range from less than 1 to more than 30 acres in size, but most are about 1 acre. They average 10 to 20 feet in depth.

Recharge basins have been considered to be highly beneficial to the overall water conservation program since they account for approximately 20% of all recharge to the underlying aquifers (Aronson and Seaburn, 1974). Although the basins restore potentially lost recharge, they are also sources of contamination. Inflow into the basins is a combination of precipitation plus constituents that are dissolved and suspended by the water as it falls to the earth and runs over the ground. Typical sources of contaminants are fertilizers, pesticides, deicing salts, organic debris, grease and road oil, rubber, asphaltic materials, hydrocarbons, animal feces, and food wastes. Many of the contaminants are not biodegradable and persist in groundwater.

F. Landfills

Long Island disposes of its solid wastes by burial in landfills, with incineration of combustible refuse at some of its sites. There are more than 40 major active and abandoned solid waste land disposal sites extant (Koppelman *et al.*, 1978). The current annual solid waste output exceeds 2,300,000 tons. An average of between four to five pounds of solid waste per capita per day is generated from residential sources. Industrial, commercial, agricultural, demolition, and incinerator wastes have also been landfilled along with residential refuse.

One of the most important problems associated with landfills is the generation of leachate. The amount of leachate produced depends on the landfill's overall extent, volume, and absorptive capacity and the amount of recharge water infiltrating it. In theory, 100 acres of landfill can generate 40 million gallons of leachate per year under Long Island climatic conditions (Holzmacher, McLendon and Murrell, Inc., 1975).

The older landfills typically range in size from 20 to 60 acres. Most are in mined-out sand and gravel pits and are unlined. Several of the newer landfills are lined.

These landfills receive a wide range of materials, including paper products, food wastes, septic tank sludge, construction debris, tires, autos, leaves, plastics, glass, chemicals, textiles, cans, oils and hydrocarbons, street and building sweepings, dead animals, and wastewater and water treatment sludges. The type and concentration of contaminants in leachate are of great importance in determining potential effects on ground and surface water quality. Table II lists pollutants expected from municipal waste sites. Significant pollutants in landfill leachate are biochemical oxygen demand (BOD), chemical oxygen demand (COD), iron, chloride, ammonia, heavy metals, and organic chemicals.

G. Highway De-icing

The application of salt to highway surfaces in freezing weather can cause groundwater contamination. The salt melts snow and ice, and the resulting solution of brine, combined with other pavement contamination, runs off the impermeable road surface and most of it either seeps directly into the ground or is diverted to a storm water recharge basin. In some places along the shores of Long Island, street runoff is diverted to sewers that discharge to the sea. However, the bulk of the runoff enters the ground, where it seeps down to the water table and migrates through the groundwater system.

Contamination also occurs around inadequately protected salt storage

TABLE II

Leachate Characteristics Based on 20 Samples from Municipal Solid Wastes (Constituents Given in Parts per Million, where Applicable)

Constituent	Median value	Ranges of all values
Alkalinity ($CaCO_3$)	3,050	0 –20,850
Biochemical oxygen demand (BOD) (5 days)	5,700	81 –33,360
Calcium (Ca)	438	60 – 7,200
Chemical oxygen demand (COD)	8,100	40 –89,520
Copper (Cu)	0.5	0 – 9.9
Chloride (Cl)	700	4.7 – 2,500
Hardness ($CaCO_3$)	2,750	0 –22,800
Iron (total) (Fe)	94	0 – 2,820
Lead (Pb)	0.75	<0.1 – 2.0
Magnesium (Mg)	230	17 –15,600
Manganese (Mn)	0.22	0.06– 125
Nitrogen (NH_4)	218	0 – 1,106
Potassium (K)	371	28 – 3,770
Sodium (Na)	767	0 – 7,700
Sulfate (SO_4)	47	1 – 1,558
Total dissolved solids (TDS)	8,955	584 –44,900
Total suspended solids (TSS)	220	10 –26,500
Total phosphate (PO_4)	10.1	0 – 130
Zinc (Zn)	3.5	0 – 370
pH	5.8	3.7 – 8.5

Source: Holzmacher, McLendon and Murrell, Inc. (1975).

piles. Rain falls on the pile, dissolves the salt, and runs into the ground. Increasingly salt piles are being stored under cover on paved impermeable surfaces to prevent dissolution.

H. Fertilizers and Pesticides

Most fertilizers and pesticides used on Long Island are for nonagricultural uses, including home lawns, golf courses, and nurseries, notwithstanding the fact that Suffolk's agricultural industry is the leader in dollar value in the entire State of New York. Increasing evidence from groundwater quality testing programs supports the association between fertilizer use and nitrate in groundwater (Porter *et al.*, 1978). In fact, lawn fertilizers applied (or more accurately, misapplied) by homeowners may be as significant a source of nitrogen contamination of groundwaters as discharges from septic tanks.

Generally, pesticides have not posed a serious threat to groundwater resources, despite the heavy use of the chemicals on intensely farmed agricultural land. Metallic-based pesticides and hydrocarbons were used until quite recently. The "first generation" pesticides are very persistent and substantial residues may still be found in the soil. However, the residues are strongly retained by soil forces and do not usually leach to groundwater. In the past 15 years, new generation pesticides have been widely used. Some of these, including certain herbicides that are now used very extensively, are more soluble in water and have higher acute toxicities. Most, however, are broken down rapidly in soil or water and, therefore, are not regarded as a groundwater hazard. The one major exception has been the use of Temik (aldicarb) in the Town of Southold on the north fork of Suffolk County. The compound has exceeded safe limits in the groundwater in that area (Zaki *et al.*, 1982). Temik was formulated to dissolve in soil moisture and be taken up by plant roots as a systemic insecticide. It appears to work well in denser soils. Unfortunately, the very porous soils of the north fork allowed too rapid an infiltration with resulting contamination.

I. Product Storage Tanks and Pipelines

A number of products are stored in surface and subsurface tanks and are transmitted in pipelines. Among the most frequently stored fluids are liquid petroleum products, gases, both liquefied and gaseous under pressure, industrial chemicals, and drinking water. Among these products, industrial chemicals are not commonly transmitted long distances through pipelines, and water poses little threat to groundwater. The other products, however, have all contributed to groundwater contamination.

A major threat is posed by liquid petroleum products stored in tanks and transmitted through pipelines. Three types of petroleum products and their means of storage are common on Long Island. Gasoline service stations store various grades of gasoline in two or more subsurface storage tanks, with capacities of 2000 to 12,000 gallons. Many individual homes and businesses store heating oil indoors, below ground, or at the surface. Oil depots store one or more grades of fuel oil in surface tanks of various sizes. For several reasons, service stations are of most concern as a threat to groundwater. First, they are spread over the entire two-county area. The greatest concentration of service stations occurs in the most densely populated areas, which are also the areas of greatest groundwater pumpage. There are estimated to be more than 3000 fuel storage and dispensing sites on the Island (Long Island, Gas Retailers Association, personal communication, January, 1977).

The volume of storage is enormous. More than 1 billion gallons per year of gasoline and diesel fuel are consumed. Most of this fuel is initially delivered from the import docks by tank trucks, and almost all of it is stored underground for some period of time.

Most service stations have steel tanks, although fiberglass tanks are slowly gaining in popularity. The steel tank, usually of welded construction, is highly susceptible to corrosion, especially at the welds. To inhibit corrosion, steel tanks are often coated on the outside with tar. With care, the coating may be temporarily effective, but many tar-coated tanks develop leaks at the point where the tar was scratched away during installation. Gasoline additives often accelerate tank corrosion from the inside. Fiberglass tanks have effectively alleviated the corrosion problem, but they can crack when exposed to cold temperatures and excessive surface loads.

Home and industrial heating and power generation use an even greater volume of products. More than 1.5 billion gallons of No. 2 oil are used to heat homes and nonresidential buildings. Over 2 billion gallons are consumed annually, mainly for energy generation. Three-fourths of the homes on Long Island are heated by oil. Home storage tanks are 275, 550, or 1050 gallons in capacity. Basements can accommodate one or more 275-gallon tanks, but larger, underground tanks are more common. Welded steel tanks are the most popular.

Underground steel tanks at homes and industries are as susceptible to leaks as tanks beneath service stations. However, their contamination potential is less for several reasons:

1. Stored volumes are smaller so a single loss of product would have less of an impact on groundwater.
2. The loss of a few hundred gallons of fuel oil should be more obvious to a homeowner than the loss of a similar amount by a gasoline retailer.
3. Because the tanks are smaller, the bottom of the home fuel oil tank will likely be at a shallower depth and, therefore, less likely to be installed below the water table where corrosion could be accelerated.

The distribution and number of depots that store petroleum products are significant. Several cases of leakage into groundwater have already occurred (R. Early, Melville Fire District, personal communication, 1977; Maloney, Suffolk County Department of Environmental Control, personal communication, 1973). Most depots store fuel oil for home and business use, although many also store gasoline, kerosene, and diesel fuel.

Most industrial chemicals are stored in tanks on or above the ground. It is doubtful if the location of many of these hazardous substance storage facilities are known to regulatory agencies because most are small and unobtrusive, except for flammable products that should be known to fire departments. There are occasional spills that contaminate groundwaters.

J. Incinerator Quench Water

A typical 600-ton per day municipal incinerator requires about 10,000 gallons of water each day to dissipate heat from fuel gases and gas residue. Some of this quench water is retained in the ash, but most of the unevaporated water, which may contain hazardous constituents, is siphoned away for discharge. Of the eleven active municipal incinerator sites on Long Island, four dispose of quench water directly to the ground; the others discharge to surface waters. Contaminants of concern are similar to those found in landfill leachate. The quality of quench water from one incinerator in the study area did not meet sewage treatment effluent discharge standards for pH, suspended solids, and turbidity. At another site, several constituents were in excess of the concentrations permitted under the State Pollution Discharge Effluent Standards (SPDES) regulations (Roy F. Weston, Inc., 1976). Owners of incinerators must now apply for SPDES permits, a recent requirement because the potential for contamination from this source was not previously regulated.

K. Airborne Pollution

Both the atmosphere over Long Island and precipitation contain appreciable amounts of potential groundwater contaminants. For instance, emissions from motor vehicles on Long Island may amount fo 4000 tons per day of carbon monoxide, 800 tons of hydrocarbons, and 740 tons of nitrogen oxides. Other sources of air contamination and potential incorporation of chemicals into precipitation include incinerators, power generators, aircraft, railroads, industrial emissions, and open burning of refuse.

The result of recent studies shows an average sulfate content in rain water of 3.60 mg per liter, total nitrogen of 1.36 mg per liter, sodium of 1.89 mg per liter, and chloride of 3.22 mg per liter (Frizzola and Baier, 1975).

L. Sand and Gravel Mining

The extent of sand and gravel mining on Long Island is uncertain. Although figures for current operations are available, an inestimable acreage of small pits exists that is mined for a short period and then abandoned.

For the most part, large operations are located in the hilly areas where mining can proceed laterally into the hillside. Vertical pitmining is prevalent only where a sufficient depth can be attained before the water table is reached.

The practice of mining does not in itself present a major groundwater pollution threat. The greater threat is from pollutants that have been introduced into the pits. Abandoned pits serve as attractive but illegal disposal sites for domestic, industrial, and scavenger wastes. Careless handling of materials at mining sites, for example, where road salts or petroleum fuels are stored, can cause these materials to enter the groundwater.

The principal cause of groundwater pollution at the pits has been the use of brackish and salty water to wash the fines from the mined material. Salt water pumped from bays is spread on the land surface or stored in ponds where it can leach to the underlying fresh groundwater.

M. Scavenger Waste

Scavenger waste pits are open, unlined seepage ponds for disposal of liquid and sludge waste from septic tanks, cesspools, and municipal and industrial sources. Each of the thirteen towns has a designated location for scavenger waste; most are located near landfill sites or sewage treatment plants. Suffolk County has eight major waste pit areas discharging directly to groundwater. However, a large percentage of the sludge generated by the municipal treatment plans is trucked to the five available scavenger waste treatment plants. Nassau County septic tank waste goes directly to treatment plants.

N. Diffusion Wells

State regulations require that all water used for cooling must be pumped through a completely closed system and then returned to the source aquifer through one or more diffusion wells or equivalent structures (DeLuca *et al.*, 1975). Depending on the cooling system, the water is raised 10° to 40°F above the natural groundwater temperature.

Groundwater temperatures to a depth of about 200 feet are usually 3° to 6°F warmer than the average annual air temperature. Below that depth, the natural gradient is approximately 1°F increase for each 60 to 125 feet in depth. Natural groundwater temperatures on Long Island range from 50° to 70°F.

Thermal loading has already been observed in a number of industrial areas where large quantities of hot water are returned to the ground (New York State Department of Environmental Conservation, personal communication, 1977). Up to now the return of cooling water to the ground as a source of recharge has been considered an environmentally beneficial practice and not a source of pollution. Consequently, institutional activity has been directed toward encouraging the use of diffusion wells rather than their control. In my judgment, far more water quality data are called for to determine whether this policy is justified.

O. Water Wells

Water wells themselves are not normally sources of contamination, except in the cases of improper construction, well failure, and, in some instances, use of the well. Some of the most common conditions that have resulted in groundwater contamination are given in Table III.

There are more than 80,000 private residential wells on Long Island in addition to the more than 2000 public supply, industrial and commercial, and agricultural wells (New York State Department of Environmental Conservation, personal communication, 1977).

Domestic wells are usually of small capacity (45 gallons per minute or less, and 6 inches or less in diameter). Typically they are installed as shallow as water table conditions and local regulations permit. Agricultural wells, on the other hand, are usually of high capacity (500 to 1500

TABLE III

Examples of Conditions under Which Water Wells Can Cause Groundwater Contamination

Imperfect construction	Inadequate surface protection
	Poor or no grouting
	Well finished at or below land surface
	Open annulus around casing
Illegal construction	Poor location
	Split screen where prohibited by situation
	Improper abandonment of well
	Improper backfill of test holes
Well failure	Casing corrosion
	Casing electrolysis (chemical)
	Stray currents in ground
	Accidental holing of casing during construction or maintenance
Well use	Direct recharge of contaminant
	Movement of contaminants caused by pumping
	Salt water intrusion

gallons per minute), but are also constructed as cheaply as possible and as shallow as possible. Since cost is often of prime importance, little is done to secure these wells from contamination. Both domestic and agricultural wells are rarely, if ever, cemented in place or sealed at the land surface. Consequently, surface and groundwater percolation outside the casing to the water table may take place. Such percolation may introduce pesticides, fertilizers, septic effluent, or other contaminants to the groundwater.

Industrial and commercial wells are normally constructed more carefully, although the water is used principally for cooling. Public supply wells are the best constructed, in some cases with double casings combined with cement grout. Modern construction and proper inspection during installation generally minimize the chance of contamination due to improper design or structural failure. The gravel packing and backfill are sterilized with chlorine compounds and, upon completion, the entire well is disinfected. Casings and wells are inspected for structural integrity, and the volume of gravel or grout used is carefully checked to avoid the creation of voids or openings that might allow vertical migration of contaminated water. In many cases, however, the annular space between the outside wall of the drilled hole and the inside casing are backfilled with the natural material that was removed from the hole and grouted only near the surface. Migration of contaminants along this annular space between the well casing and natural sediments may be an important source of contamination for deeper wells on Long Island.

Another way in which a well can become a vehicle for contamination is through improper abandonment. If the well is not filled with impermeable or low-permeable materials, the open casing can provide easy access for contamination, which can enter at the surface or through a break in the casing. When a well is unused, flooding from surface water and even deliberate dumping of various, sometimes toxic, substances into the well can occur.

P. Spills and Incidental Discharges

Carelessness and poor housekeeping account for a large percentage of the spills and incidental discharges of contaminants that occur on Long Island. Their occurrence is random and frequently not recorded. In most cases, failure to confine the spilled material, and even cleanup methods themselves, increases the probability of groundwater contamination. For instance, when accidents occur, especially when hazardous flammable products are involved, the fire department normally washes down the road. This procedure moves the flammable substance from the road surface to unpaved surfaces, sewers, or catch basins, where it then may seep into the ground. Some of the principal causes of spills experienced on Long Island are:

Industry	Mishandling of delivered or stored chemicals on-site.
	Spills or accidents involving trucks containing industrial chemicals
	Miscellaneous discharges to the ground of waste products that cannot be disposed of in other ways.
	Surface discharge of blowdown water from stacks and boilers that contains chemical additives that inhibit corrosion and scaling, but are otherwise harmful
Sewage	Major breaks in sewer lines that allow for rapid incidental discharges to the land surface or below the surface.
	Handling accidents or lagoon breaks and overflows at sewage treatment plants or scavenger waste plants.
	Dumping by scavenger waste trucks in deserted areas because it is inconvenient to haul to a designated area.
	Sewage discharge in violation of the Sanitary Code.
Agriculture	Mishandling of chemical fertilizers and pesticides on the farm.
	Disposal of chemical fertilizers and pesticides on the farm.
Petroleum Products	Spills during loading/unloading of truck tankers.
	Incidental discharges that result at the completion of the filling of service station tanks.
	Spillage during home heating deliveries when the tank is overfilled or when the hose is rolled up.
	Mishandling and disposal of crankcase oil.
	Dumping of sludge on-site after tanks are cleaned.
	Spills from carelessly attended motor vehicles in service stations.

Q. Animal Wastes

In the two-county area, there are approximately 260,000 cats, 425,000 dogs, 30,000 horses, 120,000 chickens, 1800 cattle, 750,000 market ducks, and assorted turkeys, swine, and sheep, let alone the wild animal and bird population (U.S. Soil Conservation Service, 1977). Domesticated animals alone are estimated to deposit more than 300 tons daily of excrement. Beyond question, the animal population is sufficiently diverse and numerous to add to the BOD, COD, bacterial, nitrogen, and phosphorus content of groundwater.

R. Cemeteries

There are more than 7 square miles of cemeteries on Long Island. The potential contamination to groundwater is unknown. At the very least, the lawn maintenance practices do add to the nitrogen loadings on the aquifer.

IV. INSTITUTIONAL ROLES

Long Island, being totally dependent on groundwater aquifers for potable supply, in contrast to the city of New York with its reservoir system or other parts of the state that have access to surface fresh waters, has received a great amount of governmental attention from all levels on the issues and problems affecting this sole source aquifer. The United States Coast and Geological Survey (USGS) has maintained a very significant research staff on the Island for decades and has contributed major hydrogeological findings on the system that has enabled local government to better understand groundwater regimes, both here and throughout the country where similar conditions exist (Kimmel and Harbaugh, 1976). The USGS, ever-scrupulous in maintaining their scientific integrity and objectivity, has deliberately avoided direct participation in policy formulation. Nevertheless, their work has served as the base on which local and state water policies have been formulated.

For more than a decade the Environmental Protection Agency (EPA) has played a major fiscal and policy role. Initially, prior to the advent of the 1972 Amendments to the Federal Water Pollution Control Act (FWPCA), the federal role was mainly in the major funding for sewer construction programs in the two counties. Subsequent to the passage of the 1972 Act, and particularly in regard to Section 208, the emphasis shifted from contract administration to planning and policy formulation. In one sense, virtually all of the major achievements on groundwater protection can be traced directly to the support Long Island received from EPA.

The state role has in similar fashion increased in recent years. Prior to the 1972 Act, the state mainly contributed to groundwater protection through the New York State Department of Health and its surrogate on Long Island, i.e., the Nassau–Suffolk County Departments of Health. Water monitoring, sanitary facility permits, inspection procedures, etc., were the main contribution. However, the prime concern was directed to issues of public health. Namely, the groundwater issues addressed had to have a strong, if not direct, relationship to pathology. If the consuming public was not in danger of disease, then the Health Departments felt that their jurisdiction was limited. The State Department of Health played a second role in response to an amendment passed by the New York State Legislature to the Conservation Law (Article 5, Part V-A). This law authorized the Health Department to issue contracts for the preparation of comprehensive public water supply studies. Suffolk County received a $340,000 contract that was coupled to a county expenditure of $800,000 for a Test Well Program. Consultants were retained and the first freshwater resource plan was prepared and released in 1969

(Holzmacher, McLendon and Murrell, Inc., 1969). This pioneering effort, published in three volumes, detailed the present use of water, including projections of use to the year 2020, an inventory of the groundwater resources, and plan recommendations for the development of the resources to meet the future needs. One of the major outputs of the study was the detailed analysis of "safe yield." The study also identified the impact on surface waters from the increased drawdown of the aquifer over time. Its major shortcomings were the limited examination of water quality issues and the lack of land use alternatives in contrast to engineering solutions. Of course, the omissions were more due to the "state of the art" in the 1960s rather than any deliberate bias on the part of the participants.

A second state effort was launched in the early 1960s with the creation of the Temporary State Commission for Water Resources Planning. The group consisted of elected officials, appointed health department sanitary engineers, consultants from academia, industry representatives, and one regional planner. At the time, the major obvious groundwater problem on Long Island was the presence of hard detergents in the system (Perlmutter and Guerrera, 1970). The surface manifestation of foaming from the nonbiodegradable phosphates was an aesthetic eyesore that gave visual proof of the increasing degradation of the groundwater aquifer. When it was suggested that the State Commission recommend the passage of state legislation banning the use and sale of hard detergents in the state, the proponent of the proposal was shortly afterward presented with a scroll from the Commission for "Outstanding Service to the Public Interest in New York State" and not invited back. Thus, this state role came to naught. Suffolk County itself enacted a local law—the first in the nation—banning the sale and use of hard detergents in Suffolk County in 1971 (Baier, 1976, Proceedings, 1971).

In 1970 the New York State Legislature abolished its Department of Conservation and replaced it with a Department of Environmental Conservation (NYSDEC). Instead of a limited focus with major interest in hunting and fishing, the new agency was greatly expanded with broad regulatory and permitting procedures, particularly in the area of water resources—encompassing fresh surface, saline, and groundwaters. The voters passed a $1.6 billion bond issue for the environment largely based on the goal of clean waters. The DEC opened a regional office on Long Island and has played an important role in the research, laboratory testing, planning, and regulation of the Long Island groundwaters.

At the local level, the two county health departments, as mentioned earlier, played the key roles in terms of local water policies and enforcement. Yet all of these agencies, activities, and interests failed to produce a fully comprehensive and coordinated approach to the management of Long Island's groundwaters. It was not until the passage of the FWPCA Amendments of 1972 that the opportunity arose to coalesce the fragmented, piecemeal, multijurisdictional, and competitive efforts into a unified framework.

A. Groundwater Planning

The passage of the Federal Water Pollution Control Act Amendments of 1972 heralded a new era in environmental management (P. L. 92-500). National interest and purpose were stated in the goal to achieve wherever attainable "water quality which provides for the protection and propagation of fish, shellfish, and wildlife and provide for recreation in and on the water" by 1983. The Act not only set forth a timetable for action but also provided some new approaches to solving the problems of our nation's polluted waters. For the first time there was a clear recognition that improved quality and the prevention of further pollution would also require changes in land use and management of growth policies, in addition to the prevailing pratice of building sewage treatment works. Nonpoint sources of contamination that result from poor construction and agricultural activities, highway runoffs, widespread discarding of the plastic residues of modern society, and the lack of control over animal wastes must now be considered. The Act is a landmark in three other aspects. It provides for a comprehensive areawide planned approach, requires strong citizen participation in the planning process, and mandates a commitment from state and local governments to implement the results of the planning effort.

These three elements—planning, citizen participation, and guaranteed implementation—make the 208 program unique among all prior examples of planning legislation. The Act clearly meant to move planning beyond the advisory role that so often in the past was blithely ignored by the elected officials. It also represented a new departure in that the clear emphasis was placed on "areawide" or regional planning. Wastewater management prior to the passage of Section 208 was dealt with municipality by municipality. Furthermore, the solutions were mainly based on engineering or "end-of-pipe" techniques, i.e., sewer installations and treatment plans. Now land use planning, or nonstructural solutions, was mandated on a par with engineered structural ones.

The regional focus was a clear indication that the authors of the Act recognized that groundwater management transcends artificial municipal boundaries. Watersheds must be the basic building block. Effective planning must be based on the natural boundaries of the groundwater regime with management responsibilities assigned according to the specific needs rather than the converse. In short, jurisdictional assignments—local, county, and state—should reflect the hydrogeological system, rather than past attempts that tried to tailor wastewater management to fit within the limits of a municipality's boundaries. An expanded discussion of the regional approach follows in Section V.

Unfortunately, the program was delayed until 1975 by presidential impoundment of the funds authorized and appropriated by Congress. This undoubtedly set back the attainment of the 1983 goal for several years at a minimum. In fact, it is premature at this time to predict with any certainty that the goals ever will be attained. Until a successful plan is fully implemented somewhere in the nation, some doubt exists about the feasibility of such an ambitious undertaking. Many nagging questions persist. Is there a sufficient body of scientific data and knowledge? If it is not all currently available, can it be developed? Does the technology exist to do the job? Can the public afford the cost? Will the public support the effort? On this last point there is already evidence that Congressional support is showing signs of disenchantment. Conflicts over environmental enhancement and its impacts on economic development are strident. Additional constraints exist within the administration of the Act itself. For example, although generously funded in contrast with other environmental planning legislation, limitations had to be imposed on the scope of services performed by each designated agency. Region II of EPA (New York, New Jersey, Puerto Rico, and the Virgin Islands) was allocated $15 million for planning programs. The Long Island Regional Planning Board requested $17.2 million. The Board subsequently was awarded $5.2 million. While this was more than EPA's allotment formula based on population and land areas allowed, it still was far short of the amount considered necessary to mount a maximum effort. A further fiscal limitation resulted from the stringent time eligibility for full funding. All designated agencies that received contract approvals from EPA prior to July 1, 1975,, were entitled to 100% funding. After that date, state and regional participants were required to contribute 25% of the cost. Needless to say, this did not stimulate widespread interest from local governments. Since the states were obligated to do the planning for those portions of their territory not covered by regional studies, a new dimension was introduced to further limit the chances for success. Many states were not equipped to conduct these studies from a technical and fiscal point of view. Among those states having the resources, many lacked the means to implement their plans, as land use options—a significant aspect of waste treatment management—are generally under the control of local governments as a result of enabling legislation.

Other shortcomings have become apparent through practices that tend to impair the quality of individual planning efforts. Although the Act required plan development within 24 months, provision was made for project design prior to this for up to 12 months. EPA determined to dispense with the preplanning opportunity. In a sense this fiat is the antithesis of the planning concept itself. Agencies were forced to "learn on the job," thus creating plans by work assignment amendments. The Act also prohibited any work elements that might be deemed "research-oriented" rather than accepted state of the art. This mandate was probably meant to ensure the feasibility of meeting the Congressional timetable. Strict interpretation of this clause in actuality tends to inhibit this objective. A detailed example of this last point is discussed in the section on citizen participation.

Lest this litany of pitfalls, pratfalls, and problems be deemed a dire prediction, one should instead consider the program as a challenge and opportunity. Granted, the first generation of plans may show scars and evidence of misfit between expectation and achievement. Nevertheless, optimism should be the hallmark. This was the first national effort to cope with a major environmental crisis on a multidisciplinary and interdisciplinary basis, all under the rubric of comprehensive regional planning. Any success from the first programs should strengthen future efforts and also provide the methodological base for solving other environmental problems. Thus far the Long Island experience has proven to be most positive. Significant levels of implementation have already occurred between the release of the plan in 1979 and 1986. This includes new regulatory controls, the creation of a New York State Joint Commission for Long Island Groundwaters, pine barrens legislation, and proposed federal legislation to implement the watershed recommendations in the 208 Plan. A more complete discussion of the extremely beneficial value of the 208 program in accomplishing groundwater protection and enhancement on Long Island follows in Section VII.

B. Designation

The Act provides that regional planning boards with a history of accomplishment could be designated by the governor of the state as eligible to file a proposal request to EPA. Where such agencies do not exist, the state would assume the responsibility. In the New York Metropolitan Region there was no shortage of interested parties. The Tri-State Regional Planning Commission expressed interest in being the lead agency for all 30 counties in the region. Since a commitment to implement the plans was also a requirement, and Tri-State's popularity in some sectors of the region was at a low ebb, EPA did not encourage Tri-State's participation. In Nassau and Suffolk Counties, several of the operating agencies, e.g., Health and Environmental Control, also expressed interest, as did the Long Island Regional Planning Board. The two county executives agreed that the Planning Board should be the applicant. At the same time some members of the New York State Department of Environmental Conservation indicated that their office would have to do all of the planning for nondesignated regions of the state, and therefore perhaps they should do the entire job. The strong home-rule prerogatives on Long Island mitigated against this suggestion. Be that as it may, all agencies that wished to be eligible for 100% funding had to receive designation before January 1, 1975, and receive contract approval from EPA prior to July 1, 1975. It was already late November 1974. A work program was hastily assembled by the planning staff and a request was made to the governor. County Executive Caso of Nassau County was asked by the Board to personally support the request. He had just run with the governor as a candidate for lieutenant governor in the November elections. On December 28th, three days before the expiration of his term of office, Governor Malcom Wilson gave the first designation in the State of New York to the Long Island Regional Planning Board. As a footnote to the designation process, some of the state technicians pointed out that the law also required that a public hearing be held prior to designation. The governor observed that earlier in the year EPA held a meeting in Manhattan to discuss the Act and since Long Island officials were present, that in effect met the requirements. End of case!

C. Technical Staffing

The program represented a joint effort of the Board and the various operating agencies in Nassau and Suffolk Counties. A Technical Advisory Committee (TAC) was formed with seven voting members. The Nassau Departments of Health, Public Works, and Planning were matched by representatives from the Suffolk Water Authority and the Departments of Health and Environmental Control. The executive director of the Planning Board, who was the project director for the program, also served as the Chairman of the TAC.

In addition, the New York State Department of Environmental Conservation, Region II of EPA, the Long Island office of USGS, and the Interstate Sanitation Commission also participated as nonvoting resource persons. The TAC was composed totally of nonpolitically appointed professional technocrats. It was organized in this fashion for several reasons. First, it was considered essential by the Board to keep this planning effort absolutely free from partisan politics. Second was the recognition that the agencies with the ultimate responsibilities for carrying out portions of the plan should have the opportunity to participate in its formulation. Third, these participants represented the cumulative local expertise and knowledge base at the onset of the program.

A fourth factor that adds credence to the original reasons is that the original intent to balance planners with engineers—reflective of the dual intent of the program to find structural and nonstructural solutions—is subtly amplified by the range of structural alternatives and approaches supported by the respective engineering members.

Preparation of a detailed work program was the initial and principal work conducted by the TAC. At first, a laundry list was compiled of each agency's pet projects. Much culling, addition, deletion and alteration occurred as the items were compared with the requirements and/or limitations of the Act. The document finally submitted to EPA and the state for contract approval was reduced from $17 million to $5.2 million during this iterative design stage and quite faithfully adhered to the requirements in the Act.

A secondary concern during this preplanning period was expressed in regard to the work to be conducted by consultants where specialized talents and equipment were needed. Some of the agencies felt quite comfortable with firms that had been or were working for them. The TAC followed the procedure mandated by the Long Island Regional Planning Board. Sole source contracts were to be awarded only to those governmental and/or academic institutions that had a unique talent. In all other cases the TAC was expected to prepare "Requests for Proposals" (RFPs). The RFPs would then be advertised nationally and the consultants would be chosen on merit. This procedure was meant to eliminate political favoritism and to hopefully secure the best talents in the country. Although this technique was time-consuming and often onerous, it has proven meritorious. Some anticipated provincial malcontent was expressed locally, but this soon disappeared. The consultants were selected by the voting members of the TAC based on the substance of the proposals. Pricing information was kept sealed until each selection was made and then the winner's envelope was opened. Fortunately, little negotiation was required to bring the contracts into harmony with the TAC's estimates of reasonable cost.

D. Work Elements

The general goals of the program to achieve the water quality criteria established by federal and state laws required an extensive knowledge of: the existing and proposed land uses and demography of the region; the quantity, quality, and hydrology of the ground and surface waters, including interrelationships between both; the types, sources, amounts, and impacts of contaminants entering the waters; the alternative technologies best suited for any specific area or set of problems; and the legal, fiscal, and institutional laws, agencies, and regulations needed to implement the completed plan. This last aspect will be discussed in Section VII.

The comprehensive land use plan released by the Long Island Regional Planning Board in 1970 still serves as the basic guideline for growth in the two counties until the year 2000. The existence of the land use plan provided strong advantage to the likelihood of producing a workable waste treatment plan for the two counties. With this plan in existence, it was much easier to develop and test two major treatment approaches. Assuming that the land use plan would be realized, it then became a task of estimating the contamination loadings that would result from the various land uses and recommending treatment facilities, e.g., sewers, treatment plants, etc. Conversely, the land use plan can be modified and population densities altered to avoid the need for structural solutions. Both models were examined by the Board.

The first study of water quantity for Suffolk County was conducted in 1957 (Wiggin, 1957). An exhaustive examination, completed almost a decade later, estimated that the county would have a "safe yield" (this is solely a quantity term) for a population of approximately 3.5 million people (Holzmacher, McLendon and Murrell, Inc., 1969). These studies did not indicate changes in the aquifers that would result from various wastewater treatment proposals. Fortunately the United States Geological Survey has conducted groundwater studies on Long Island for several decades and had developed an analog model that could identify hydrologic conditions resulting from various wastewater management alternatives. (The Survey was one of the sole source consultants retained by the Board.)

Water quality is a corollary concern and exponentially more complex to analyze. Although field tests have been carried out for decades to monitor water quantity (and these have been far from all-embracing), little has been done in comparison for water quality, beyond periodic public water supply testing for specific bacteria, and even less for private wells. Testing for trace organics and virus is still in the embryonic stages of practice. The application of sophisticated test equipment such as mass spectroscopy and gas chromatography is costly, difficult to use, and until recently difficult to verify. Nevertheless, such studies were an integral and vital portion of the Long Island study. Water samples were taken on a representative basis and analyzed for virus, trace organics, and heavy metals. The results indicated a prevalent occurrence of numerous organics in the glacial or upper aquifer. Some of these potentially carcinogenic compounds were found in the deeper or Magothy drinking supply wells at Bethpage, Long Island (Nassau County Department of Health, 1977). These wells have been closed. The wastewater study did not definitely respond to these concerns for the entire potable supply. However, as a result of preliminary findings, the Board recommended that a permanent monitoring program for all public water supplies be instituted. The two County Health Departments have responded by expanding their monitoring efforts.

To have simulation capabilities for water quality evaluation, Dr. George Pinder and his associates of Princeton University were retained to adopt and apply existing digital models (one, two, and multidimensional) to permit determination of the impacts of various groundwater levels on saltwater intrusion and further groundwater pollution (Pinder, 1973; Pinder and Page, 1976; Tetra Tech, 1976). The modeling runs were paralleled with quality evaluations based on empirical data and related literature.

The utility of the models bears direct relationship to the quality of the input into them. Point sources of contamination, e.g., sewage treatment plants, monitored wells, etc., have been routinely identified and tested. In the past this was sufficient because wastewater studies were directed toward engineered solutions to these sources of contamination. Current planning efforts are also concerned with nonpoint sources of pollution, e.g., storm water runoff, sedimentation, uncontrolled deposits of animal wastes, etc. Several responses to these ubiquitous, pervasive, and unknown sources, forms, and amounts of pollutants were included in the work program. The Cooperative Extension Service and the Suffolk County Soil and Water Conservation District conducted field studies and literature searches to identify and evaluate the volumes of pesticides, fertilizers, and other home and agricultural poisons used in the two counties. They also identified other nonpoint pollution problems associated with various animal populations, runoff, and sedimentation.

All the information gathered from the quantity and quality sampling, analyses, and evaluation studies served as inputs to the final phase of the technical studies, namely, the determination of engineering options in relation to water quality objectives. This phase of work paralleled the land use or nonstructural options—it was based on the same data—and afforded the broadest possible array of policy choices as to method, location, timing, and cost of solutions (Nassau-Suffolk Regional Planning Board, 1977).

The alternatives and the consequences of each were considered by the TAC for the purpose of selecting the optimal combination of solutions—structural and nonstructural—that in the aggregate formed the Comprehensive Areawide Waste Treatment Management Plan for Nassau and Suffolk counties.

V. PLAN RECOMMENDATIONS

Many approaches can be utilized to manage Long Island's groundwaters. In the past, the response to groundwater problems was to construct facilities, e.g., sewage collection and treatment facilities. Less traditional approaches include guides and limits to growth and the initiation of programs and regulatory means either to prevent problems from occurring or to develop solutions that do not require engineered facilities. The Plan included both structural and nonstructural solutions.

A. Structural

Sewers have and continue to be a significant element in wastewater management. However, there was a clear desire on the part of both the TAC and CAC to avoid the need to create additional (and extremely costly) sewer districts. Instead, we recommended the testing of several new concepts, including marsh-pond treatment and improved septic tank systems that denitrify wastes.

B. Nonstructural

Heavier concentration was given in the Plan to management options that would address the control of potential sources of pollution. These included:

1. options to prevent the establishment of sources
2. options to better manage existing sources
3. options to eliminate existing sources

Land use controls can be used to prevent new sources of pollution from arising. An activity may be prohibited, or it may be permitted to develop in a controlled manner consistent with environmental objectives.

Good management practices can minimize pollution discharges. For example, in areas of low soil permeability, septic systems may malfunction as a result of septage not being periodically pumped from the tank. Hence an appropriate management policy may be to require pumping according to a formally required procedure. Likewise, wastes from domestic animals on streets and highways may constitute a major source of contamination in storm drainage water. A nonstructural option may be to prohibit littering or to require owners of dogs not to allow their dogs to defecate where water pollution may result. Regular sweeping of streets would also limit the amounts of pollutants transported in storm water. Another management practice that may be particularly relevant to Long Island is to encourage the use of fertilizers and pesticides at levels not exceeding the requirements of cultivation. Thus, householders could be encouraged to adopt "low-maintenance" methods of cultivating lawns as opposed to growing species of grasses that demand high levels of water and chemicals for their maintenance.

If the use of inorganic fertilizers by householders were to be banned outright, the ban would constitute an example of the total elimination of a potential source of contamination. For example, the prohibitions of chronically toxic and persistent pesticides such as DDT have begun to reduce the levels of these chemicals in the environment (Harr, 1972; Baier, 1976). Similarly, certain organic materials should be controlled. The substitution of less harmful materials in place of such organic chemicals may provide a partial nonstructural solution.

C. Legal/Institutional

Another set of possible management options deals with the formulation of legal and institutional programs. These types of approaches might include the following:

1. strengthening of existing laws or regulations
2. enacting new laws or regulations
3. restructuring of existing county- and town-level agencies
4. establishing new agencies
5. elimination of existing agencies
6. redefinition of responsibilities

The thrust of these types of management options is to ensure that proper regulatory power exists, that regulation and operation of facilities do not conflict, and that an adequate system of monitoring and control is available.

An institutional approach recognizes that improvements can be made in the administration and implementation of water quality management programs. It recognizes the need for improved communication between operating agencies and for a clear definition of responsibilities.

D. Hydrogeological

The Long Island 208 program identified management needs and the relevant structural and nonstructural alternatives that may be used to ameliorate them. A large number of wastewater management alternatives were defined, all having as their general objective the development of a comprehensive management plan. During the course of the planning process alternatives were subjected to constant evaluation in terms of engineering and economic feasibility, probable environmental effects, and legal or regulatory requirements, and where any alternative or set of alternatives was found to violate environmental, economic, engineering, or legal constraints, it was eliminated or modified.

It was well known that the groundwaters of the bicounty region were strongly interdependent, as far as quality and quantity were concerned, but that waste management alternatives could conveniently be defined in terms of protecting one or the other. However, great difficulties arose when attempts were made to assign the various solutions to the specific units of local government. Although the emphasis was on regional planning, there nevertheless was a constant reminder that a good deal of the implementation would have to be carried out by the individual (and independent) towns and villages. Yet every attempt to relate solutions to the constraints of municipal boundaries failed. We were dealing with a Promethean problem. Either the "legs" or the "head" had to be amputated to make the solution fit.

This led to the division of the land area into a number of hydrogeologic zones distinguished by differences in groundwater flow regime and degree of development (see Fig. 3). It was then possible to present the areawide recommendations under eight main headings. In this fashion the recommendations were coterminous and in harmony with the realities of the complex groundwater regime of Long Island. Of course, this left the issue of interjurisdictional matters to be resolved. However, this is doable since jurisdictional boundaries are somewhat arbitrary, whereas hydrogeologic zones are not.

So much said for the technical phase of the planning effort. All will be for naught if the recommendations are not implemented. The general history of planning accomplishment throughout the nation has been quite limited. The planners proposed and the politicians and elective officials have disposed. A large measure of fault lies with the planners. They often considered their work to be finished with the delivery of the

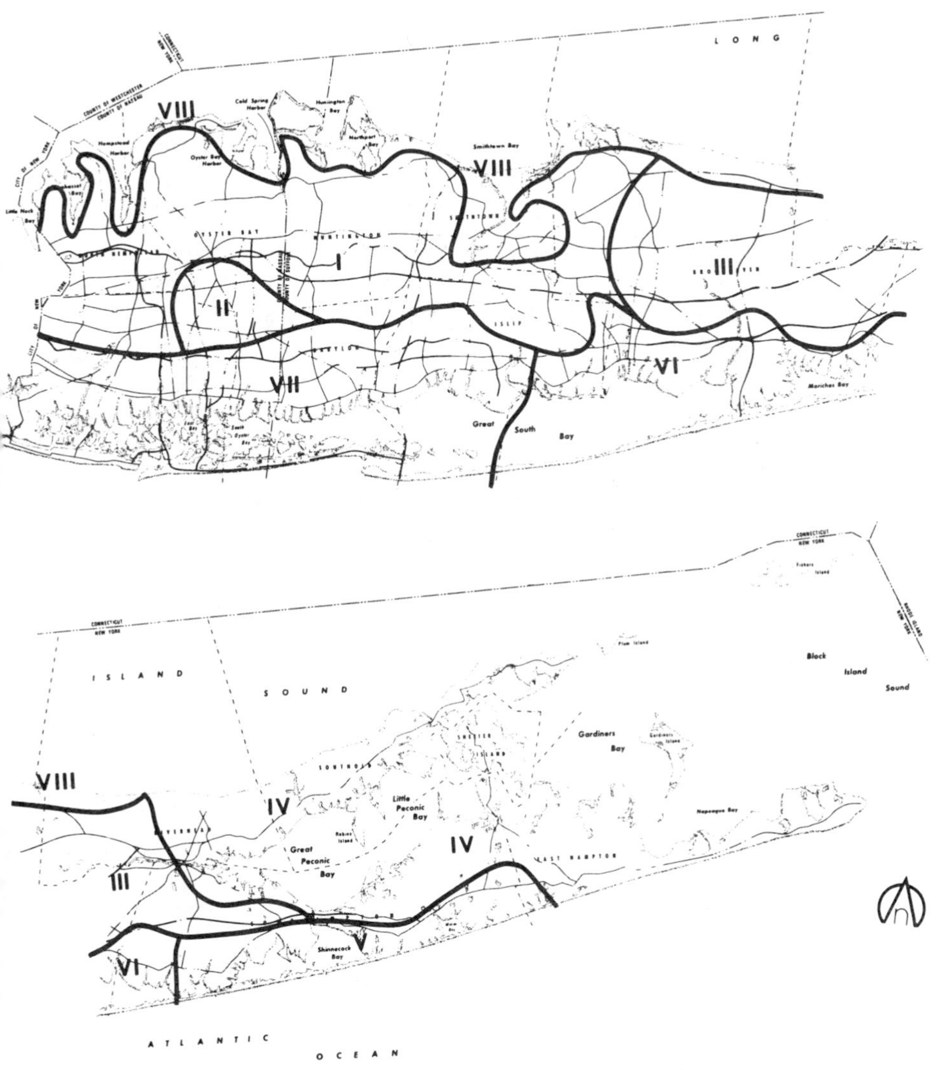

Fig. 3. Hydrogeologic zones of Long Island.

plan documents to the elected decisionmakers who more often than not had little training or awareness of what the planners were talking about. A greater deficiency was the lack of any broad-based citizen constituency. Elected officials often try to be responsive to the general public. Too often the public is less aware of planning proposals and the impacts that will affect them than are their elected representatives. Ergo, no support. This program is different. Citizen participation was mandated in the Act. This did not automatically guarantee widespread support but at least it did ensure citizen awareness and perhaps even some citizen input during the planning process. Section VI summarizes the history of citizen participation.

VI. CITIZEN PARTICIPATION

Governmental decision making in the United States is generally based on the notion that in a federated republic the elected representatives shall act as surrogates for the citizens. The ballot box serves as the arbiter or how well the "voice of the people" is heard. Even communities with strong attachments to the colonial home-rule town meeting format honor the concept of citizen participation in theory—not in practice. Of course, special interest groups have had access to the process by the use of lobbying or, in more limited instances, by seeking recourse through litigation. Either of these means is substantially limited to those individuals or organizations with sufficient financial support to wage a successful campaign.

The Great Society programs of the Johnson administration, in attempting to resolve the problems of poverty and foster a further redistribution of economic resources originally instituted by the New Deal administration of Franklin Delano Roosevelt, formalized the concepts of participatory democracy by including language in the Congressional acts that mandated citizen participation and, in some instances, citizen control. This was a sharp departure from the technocratic governmental initiation, design, and implementation of social programs.

A recital of the decade's short history of these efforts is not one of noticeable success, either in the achievement of the high noble purposes of the "War on Poverty" or in the development of any widespread citizen input. Instead it did serve as a means of governmental access for a small number of publicly or politically oriented individuals who were locked out of the traditional political routes to power. In fact, these limited successes in broadening citizen participation were important factors in the subsequent decline of governmental support. As soon as the participants elected by the community groups became visible, they were considered to be threats to the established political regimes. The Economic Opportunity Councils were seen as parallel political forces that were competing for constituencies and resources. Thus, it was not surprising that local governments were not overly enthusiastic in their support. Some accommodations have occurred. This has usually happened to the extent that the councils have become institutionalized within existing political systems, and therefore serve to reinforce the status quo.

Despite the early creakings and groanings of disenchanted citizens, frustrated technicians, and concerned politicians, the Congress has been mindful of the benefits to democratic myths if workable citizen participation could be achieved. The confidence crises engendered by Watergate and the Vietnam War have given impetus to the growth of consumer, environmental, and social planning advocacy organizations. The common thread linking these diverse groups in common allegiance is their adverse posture toward government. The common credo is simple. Each of the loose constituencies that may be affected by governmental actions or inactions relative to their interests argue that they are not able to secure relief because the government represents the "establishment" rather than the citizenry. A solution to such unresponsiveness and unrepresentation is to use public interest lawyers as advocates for these segments of the population in direct litigation against governmental programs and actions.

An obvious alternative to such confrontation politics is to render government more responsive by greater citizen involvement. This choice appears to be the conscious commitment of Congress, specifically in domestic legislation relating to community planning, environmental, and social issues. For example, comprehensive health planning undertaken by health service agencies must be under the direction and control of councils made up of "consumers and providers," with a minimum of 51% of the membership in the consumer category. Since most government personnel are defined as providers by virtue of public programs in social services, health, community planning, and general health, virtual assurance is provided that at least on paper the private citizens will prevail.

The federal guidelines prepared for the Coastal Zone Management Act of 1972 provided a framework for state coastal zone planning. Local government participation and public input from the widest range of interests were stressed. Public input was encouraged during all phases of the program. During the same session identical requirements were included in the Water Pollution Control Act of 1972. This recital of our experience in trying to meet the spirit and letter of the law in fostering the most open and rigorous public participation is integral to the success of the Long Island groundwater protection efforts. Organizational strategies, personality conflicts, role consciousness, and an appraisal of the effectiveness of the efforts are discussed below.

A. Organizational Strategies

Since planning is essentially a governmental function, it must provide a broad and desirable rationale for attracting and holding the support of political leaders, appointed officials, and the general citizenry. The appeal must be broad to cover the multitude of aspirations of such diverse groups. Obviously, the process itself, aside from the technical aspects, is basically a political activity.

The Long Island Regional Planning Board at the outset faced the issue of how best to meet the requirements for citizen participation. We examined the literature and our own experiences and soon concluded a new approach was needed. The earlier examples of limited or stacked citizen representation could not provide the broad inputs required for a program as complex, as costly, and as unpopular as we expected wastewater management planning to be. Our only guidelines were that the effort had to be as open and representative as possible. This meant that some of the approaches, such as the appointment of select and elite "citizen" representatives or the hiring of a consultant to "organize the public," were to be avoided. Instead, the Board advertised widely in the press and radio that this new program was in the organizational stage and that anyone interested in the issues of maintaining the quality of the potable and swimming waters of Long Island should attend an informational session (*Newsday*, 1975). In addition, over 300 letters were sent to anyone who had ever contacted the Board for information about any matter related to water conservation, pollution, and/or wastewater management. In particular, we strove to identify those respondents whose views were antagonistic to prior governmental programs. All local governments were similarly notified about the impending project. At the conclusion of the open meeting attended by over 100 people, approximately 60 persons volunteered to participate.

At the organizational meeting the staff briefly outlined the purposes, duration, and anticipated outputs of the program. The citizens were informed that the Board would prefer to have the chairman of the Nassau–Suffolk Regional Marine Resources Council serve as the nonvoting chairman for the Citizens Advisory Council (CAC). This was to ensure impartiality in the conduct of the meetings and to avoid jurisdictional maneuvering over the position. The citizens were also informed that the rest was entirely up to them. After a brief discussion they concluded that they would prefer a structured organization rather than an open-ended one. They agreed to meet every second week and to form subcommittees as necessary with additional meetings to occur as required. Within one month's time the formal structure was accomplished. Ten categories of representation were selected. They included the academic community, agriculture, business, and industry, civic associations, disadvantaged citizens, environmental/conservation groups, government, labor, the Long Island Water Conference, and recreation interests. Committees were set up in each of these categories from the representatives who agreed to convene a caucus for each grouping on a regional basis, for the purpose of having each group vote for two voting members and alternates for each category. This was accomplished within one month's time. In concept the CAC had the freedom to establish its agenda, to prepare and publish a newsletter on 208 activities, and to react and contribute to the technical work conducted by the Technical Advisory Committee (TAC) as the program was carried out. The CAC, to foster the widest possible public information program, also agreed to inform their constituents on a regular basis regarding the progress of the planning effort. In theory, the CAC should have been able to provide the widest possible dissemination of findings and to have received feedback from the general community. However, as important as citizen participation is, it must always be kept in mind that the concept of universal civic participation should not be overromanticized nor be expected to be achieved. Anyone with experience with any type of voluntary organization soon realizes that a relatively small number participate in the decision-making role beyond the selection of officers. This is but one of the problems inherent in the very nature of participatory democracy. As the activites of the CAC began to mesh with those of the TAC, other problems soon became apparent.

B. General Problems

The CAC, being representative of the broad variety of interests and differing levels of technical competence, gave rise to a set of circumstances that frustrated citizens and technicians alike. For example, to assure a working partnership between the TAC and the CAC, it was agreed by the TAC to furnish all technical materials prepared by consultants or staff to the CAC as the material was developed. Theoretically, this was to achieve two purposes: to bring the knowledge level of the CAC abreast of the TAC and also to receive commentary and input from the citizens' point of view.

It was soon discerned that there was no happy medium in the manner of presentation. Initially, the technicians presented their findings in common language that was resented by those citizens with technical backgrounds who felt they were being talked down to. In response, further presentations were given identical to those presented to the TAC. This produced objections from the nontechnicians on the CAC. There was no reasonable middle approach. There also arose a set of problems having to do with role consciousness. Many of the citizen participants joined the CAC because of their very deep and abiding concern with water conservation and wastewater management. They perceived themselves to be knowledgeable and, in fact, were in most instances. Therefore, there was also some chagrin and frustration with the TAC's scientific approach in trying to build the case one brick at a time. Some of the citizens felt that some of the desired solutions were obvious. Other expressions of impatience were evidence of the fear that the technicians would merely produce a "plumber's approach" to wastewater management. They did not quite trust the governmental technicians to be truly comprehensive and to consider the broadest array of alternatives. In fairness, the TAC had as many misgivings about citizen participation as the citizens had about them.

As the program moved forward another more subtle problem began to appear. The very matter of working together tended to produce an element of co-optation. The TAC perhaps overreacted in its desire to indicate that nothing was being held back, and strong signs of growing trust between the two groups began to emerge. Let me stress that this is a two-way street. It certainly is more pleasant to work within "one happy family," yet the potential danger in this pattern of growing harmony was that one of the prime roles of the CAC, namely, to be objective critics, could have weakened. Fortunately, the possibility of co-optation was resisted by some of the more active citizens.

The process of trying to meld so many personalities into a productive and positive process should not be construed negatively despite the recital of problems. On balance, the problems were minimal in contrast to the merits and benefits that occurred. One of the more noteworthy examples follows.

C. A Success Story

Shortly after the inception of the 208 program on Long Island, the CAC raised the question of why viral studies were not included within the approved scope of services. They were informed by the director that in fact the initial application to EPA did include such work prior to the formal establishment of the program. This segment was not approved since viral studies were considered to be of a research rather than applied status because of the relative embryonic state of the art, and thus were ineligible according to the guidelines of the Act that prohibit research projects.

The concerns voiced by the citizens, who in part joined the program (with a resonable amount of distrust of politicians and technical bureaucrats, as well as a well-entrenched bias against current engineering solutions to wastewater problems) to promote recharge as a viable alternative to ocean disposal of sewage effluents, maintained that it would be essential to understand and assess the potential constraints against recharge. They did not accept the EPA ruling that viral sampling and analysis were properly categorized as research and therefore beyond the scope of the Long Island program. They demanded that discussions be held by the CAC and the representatives from EPA to demonstrate their support of the expurgated viral section of the original application. At the same meeting a subcommittee was formed to meet with the two county health commissioners in the hope of securing professional concurrence and support. These activities took place during June–July, 1975.

By July 7, 1975, the subcommittee reported to the CAC that strong support was guaranteed. They further recommended that the CAC request the scope of services be amended to include virus studies and that funds be earmarked for this purpose. The resolution adopted by them was forwarded to the TAC, which responded favorably within two weeks time, subject to approval by EPA. However, it was stressed that the chances for approval appeared slim.

The CAC then invited Dr. James Vaughn, a noted virologist from the Brookhaven National Laboratory, to address the group. He accepted, and in his presentation he discussed the state of the art—referring to work in other parts of the country and further elaborating on the analytical processes he used at the Laboratory to concentrate the viruses. At the conclusion of his talk the CAC, by resolution, directed their chairman to request of the project director that a formal application be submitted to EPA in accord with their wishes.

As a result of their vigorous pursuit of the issue, the EPA agreed to conduct limited viral investigations and that additional work would be given fair consideration if deemed essential. This response was received by August 4, 1975. By the following week Dr. Vaughn submitted a proposal to the TAC with a cost estimate of $60,000. After a number of working sessions between Dr. Vaughn and the TAC, with continuous feedback and response from the CAC, and subsequently similar meetings with the TAC, CAC, and EPA, a formal application was submitted to EPA on March 12, 1976. An affirmative response was rather quickly received from EPA on April 27, 1976.

It is absolutely clear that the motivating force in securing approval for this integral portion of the 208 study must be attributed to the strident efforts of the CAC. Subsequent to this experience, additional modifications were made to the program at the initiative of the CAC. These included investigations into the presence of heavy metals and trace organics in the aquifer of Long Island. Beyond any question the citizens dared to transgress by challenging federal rulings and succeeded in doing the undoable. This accomplishment alone merited the difficult marriage between technocrats and the general public.

D. Conclusions

Planning, as carried out in the context of this discussion, involves mediation between diverse groups and individuals who seek to influence land use and infrastructure decisions. Planning technicians conduct such mediation within the governmental framework and among the public-at-large. Thus planning must be understood to be fundamentally a political activity. It is political in that it is a governmental process presumably set up to formulate and execute policy on land use activities. Administratively, most planning agencies are part of the executive branch of their respective jurisdictional levels within government and therefore directly linked to the political power structure. It is also political in the sense that the interplay between different departments and levels within government, and with the private citizens who participate, requires mediation and compromise—the very essence of politics.

This reality is not necessarily negative. The crass aspects of partisan politics do not have to be the controlling factor. To the contrary, partisanship should be strenuously avoided. Politics herein is held to be the conduct of the public business in nonpartisan fashion or, perhaps more accurately stated, multipartisan fashion. However it is viewed, planning will be more successful if it is conducted *with* the public involved, rather than *for* the people. The correctness of including public participation in the planning process, as demonstrated by the experiences on Long Island, was seen in several ways. Staff and citizenry had more access to the interests of the general public, which was invaluable for goal formation and project design. This was of value in selecting the most suitable allocation of staff and fiscal means. Although co-optation of the public by the planners was not the aim, continued interactions indicated that the plan drawn by the consultants and agency staff was not being imposed by outsiders or from untouchable bureaucrats. It did promote greater identification and a "pride of authorship" on the part of the public with the plan. It also helped to mobilize resources and support by fostering mutual trust and understanding, particularly in instances where trade-offs had to be made. In addition, continuous evaluation of the plan and its implementation was encouraged by all who had a role in the plan's formulation.

Hence, public participation does render the planning process more responsive, more democratic, and often more comprehensive. And this leads to the last major issue of this chapter, namely, implementation. Even the best of plans coupled to the best of intentions is for naught if not carried out.

VII. IMPLEMENTATION

The Act and the regulations drafted by EPA thereto required a commitment from the 208 agencies to indicate how they would strive for the implementation of the chosen plan. This inferred that technical, fiscal, and political objectives could be achieved and that major constraints would somehow be ameliorated. Perhaps the easiest set of concerns to deal with are technical.

Sanitary engineering and waste treatment technology are both well-developed practices. The current state of the art can provide technical solutions for most pollution problems, especially those of point source origin. The only conflicts that arose in the program were jurisdictional rather than technical. It was assumed by the TAC that the New York State Department of Environmental Conservation would develop and furnish waste allocation loadings as input to the study. This in fact was a state work item in law. These loadings set the limits for the amounts and types of contaminants that may enter a surface water. The state representatives requested that the 208 agency develop the criteria and furnish them to the state. A complete reversal of responsibility! Their contentions were that the modeling and engineering consultants to the agency had to produce the methods and analyses for modeling the various surface waters and therefore would be more prepared to make the calculations. Since there was some logic to their argument—and more to the point, evidence that the state was not ready to do the work—the TAC had to assume this extra assignment.

Another constraint was related to the relative completeness of the technical data. As mentioned earlier, more must be known about trace organics and viruses. The two-year 208 program therefore had to base its recommendations on partial results. Thus when implementation was considered, it was in the context of a continuous planning effort. As the data base improves, modifications of the initial plan may become necessary. For the moment though, let us assume that a workable plan did evolve from an engineering point of view. Each subplan had to be evaluated as to its probable environmental impacts on marine freshwater ecosystems and on groundwater. This review was conducted by an independent consultant. Fortunately, conflicts were minor and the TAC amended the plan to be in conformity with the environmental limitations.

Each of the subplans and the final plan was subjected to cost–benefit analyses. There was considerable doubt as to the significance of such findings. Federal agencies responsible for construction projects are quite prone to justify management decisions on the results of cost–benefit studies. From my experience, these exercises are analogous to grantsmanship. One starts with the bottom line and works backward. Be that as it may, the real question beyond total cost is who has to provide the revenue? The crux of the implementation issue is cost. Technical debates can be resolved, political squabbles put to rest, and general consensus achieved over the merits of a particular public policy—but the deciding factor invariably is "who pays for what."

Historically, waste treatment programs centered around engineering solutions. Thus, it was not surprising that the federal response in terms of intergovernmental relations had been to sponsor major grants for construction. Sewage construction projects eligible for EPA support were able to receive grants up to 87.5% of the total project cost. This is a very necessary and strong support if the communities are to meet the needs for a water pollution-free society. However, in view of the impetus set forth in the Act that encourages nonpoint source and planning control measures, it is perhaps timely for Congress to consider adequate funding to achieve these ends as well. For example, the Southwest Sewer District in the Towns of Islip and Babylon, N.Y., built a network of collection systems, treatment facilities, and an effluent disposal outfall that eventually cost $1.1 billion. Major segments received 87.5% funding from EPA. Even the 12.5% balance financed by the users placed a heavy burden within the district. If this is to be the future pattern, aid will have to be greatly magnified. An obvious alternative would be to limit the density of growth and thereby avoid the need for communal sewage systems. This objective can be partially realized through zoning controls and successful enforcement. In addition, extensive watershed management will be required, including the reservation of significant areas of open space. These types of public programs are virtually dependent on local fiscal resources. I submit that the purchase of development rights and/or the fee simple for open space lands, or for the maintenance of low-density communities, should be construed as a modified form of construction project. Wherever land development has not already occurred, this option has to yield a greater return on the public investment. The Southwest Sewer District, which was designed to serve a population of 200,000, averages out to $7500 per person (Koppelman, 1980). The purchase of development rights in Suffolk County is currently averaging $5000 per acre (Koppelman, 1985). Since one acre can accommodate five to seven persons per acre without the necessity for sewers, it would be cheaper to buy the land and give it away for low-density use than to have to build sewers. In short, it is fiscally irresponsible to continue to allow past urban sprawl patterns to further the pollution cycle. I recognize the pitfalls inherent in expecting bureaucratic entities—especially federal ones—to overcome their own inertia. Perhaps fiscal policy improvement can occur if waste treatment programs were written in a format similar to that of the Community Development Act, namely, to assign block grants to communities with the general requirement to reduce water pollution. Let the local governments decide the most efficacious way of doing this. The states and localities would then have a greater incentive to exercise more stringent land controls. Much more could be said about fiscal reform, broader revenue bases, etc. Let us instead move on. The remaining implementation of subjects addressed in the 208 program dealt with legal and administrative regulations and the nature of institutional mechanisms necessary to administer the 208 plan to exert controls in order to limit further pollution.

To this end the TAC reviewed and evaluated land use controls and land management authorities at the various levels of government. The TAC developed several model ordinances and/or other regulatory devices to facilitate land use strategies regarding water pollution abatement (Koppelman *et al.*, 1984). Fortunately, the area met the institutional requirements for management and regulatory agencies to carry out the program. County planning commissions can manage any land use controls called for in the plan. The Nassau and Suffolk County Departments of Public Works are the operating agencies responsible for the construction, operation, and maintenance of sewer districts and programs. The two County Health Departments execute various monitoring and regulatory programs. In addition, the New York State Department of Environmental Conservation has the major state responsibility for pollution abatement. In the Tri-State New York Region, the Interstate Sanitary Commission acts in a similar role to monitor and regulate pollution matters of interstate concern. Seemingly, all the management and regulatory agencies necessary to manage and implement a 208 plan already existed.

And so we on Long Island are perhaps in a unique position. We do not suffer from a lack of institutional entities. This is encouraging as it does tend to simplify the implementation process. After all, a host of vested interests supported by a variety of constitutencies is already extant. Yet, at the completion of the 208 plan a crucial question had to be addressed. Namely, would the elected officials support the plan and its voluminous set of recommendations? The answer to this question lies in a simple policy truth.

If, in carrying out the program, the technicians translate the analyses and recommendations into understandable language, and if the news media strongly inform the public and editorially support the plan, and if the fiscal resources are adequate, then the planners will have transformed good planning into good politics. And that's the name of the game!

The Campaign for the Plan

In a sense the completion of the plan in terms of the day-to-day work, including the preparation of the final two-volume summary books for publication, which normally would signal the end of the program as far as the TAC was concerned, was in reality the beginning of the next and perhaps most vital element in the planning process—the quest for official and general public acceptance of the findings and commitments for implementation.

The first step was a critical examination of the findings by the CAC. The citizen participants were given full freedom to comment and were guaranteed that their unedited critique would be published as an integral chapter in the final report. This proved to be a sound approach since the work of the TAC was written in typical nonnormative professional style. The chapter by the CAC became, in fact, not only an endorsement of the overall plan, but the opening salvo in the fight for its adoption.

Technically speaking, NYSDEC was supposed to evaluate the plan and then recommend to the governor its adoption, amendment, or nonadoption. Despite the fact that their representatives participated on the TAC throughout the process, they could not seem to get their act together. The governor's approval was essential to secure EPA approval. In fact, the procedure was reversed. After several months of delay during which time the plan was receiving broad coverage and attention, EPA indicated its support publicly, which was followed by an endorsement from the governor. In a most anticlimactical fashion, NYSDEC indicated its approval almost one-and-a-half years later—an action that became gratuitous and ineffectual.

Three major sets of participants were instrumental in achieving the plan's acceptance: the TAC, the CAC, and Long Island's daily paper *Newsday*.

The project director and members of the TAC prepared a summary presentation of the major findings of the plan, which was taken to each of the municipalities on Long Island and then performed before the two counties' legislative bodies. Parallel to this effort, members of the CAC prepared their own speakers bureau complete with an annotated slide show that they presented before civic, business, and environmental organizations. *Newsday*, which had periodically covered the planning effort throughout the development period with news update stories and occasional support editorials, worked closely with the TAC and simultaneously with the announced completion of the plan published a multipage, multistory panorama of the plan over a two-week period (*Newsday*, 1978). They then released a public service reprint of the entire set of stories (over 100,000 prints), which both the TAC and CAC had available for free distribution. They also gave strong editorial support to the plan. It soon became clear that the protection of Long Island's groundwaters was of top concern and priority and that the public was almost totally behind the effort. Thus, political support was anticipated and received. Virtually every unit of government on the island—villages, towns, and the two counties–backed the plan.

The responses from the federal and state governments were no less supportive. One of the major findings of the study was that contrary to earlier assumptions that point sources of contamination were the major problem, it was the nonpoint sources that constituted the larger and more difficult set of issues. EPA of their own volition, offered the TAC and unsolicited grant to participate as one of approximately two dozen selected areas in the United States to undertake a detailed examination of nonpoint pollution. This was part of the National Urban Runoff studies (Koppelman and Tannenbaum, 1982).

A second area of concern expressed in the plan findings was the realization that a significant portion of the implementation follow-through would have to come from the local units of government, which, in many instances, did not have the technical and professional staffs necessary to translate the findings and recommendations of the plan into municipal action. EPA concurred with this concern and gave the TAC a second grant to enable the TAC to prepare a common language implementation handbook on "Better Management Practices" (Koppelman *et al.*, 1984). This was carried out with the view of making it transferrable to other similar groundwater areas of the country.

The state response was also forthcoming. The plan called for a number of changes and creations in both local and state laws. Frankly speaking, the members of the TAC expected local support, especially since the two County Departments of Health had independent rule-making powers and the recommendations for legal changes came from their staff members on the TAC. The immediate response from the state was most welcome indeed. The New York State Legislature created a bipartisan New York State Joint Legislative Commission for Long Island Groundwaters. This commission was headed by a Nassau County assemblywoman (Democrat) and a Suffolk County senator (Republican) and was staffed by two co-chairpeople who had served on the CAC. Since the assembly is controlled by the Democrats and the senate by the Republicans, we could not have asked for a more viable arrangement. The efficacy of this arrangement has already borne solid fruits. The first effort was the passage of a strong law banning all "sanitary" landfills on Long Island by 1990 (State of New York Law, 1983). To help with the implementation of this law the governor signed additional legislation (State of New York Law, 1985) calling for the creation of a special commission to recommend a suitable disposal site for incinerator ash.

In addition, the Legislative Commission is also sponsoring funding legislation to enable the expanded continuance of groundwater monitoring and watershed planning.

The two County Health Departments have also followed the plan recommendations and have passed some of the strongest laws in the nation over the control of subdivisions and industrial locations. Suffolk County's recent adoption of Article 7 of the Health Code (Suffolk County Sanitary Code, 1985) controls all toxic and hazardous chemical uses within the "water-sensitive" areas of the county.

In short, what was mistakenly assumed by the participants to be a planning effort of limited duration has become an ongoing, continuous governmental function. The TAC is a permanent fixture. The State Legislative Commission appears to be at least of long duration. The interest of the citizens continues unabated on all issues of groundwater and watershed protection. And the interest of local governments has echoed the wishes of the voters. For example, the eastern towns of Suffolk County, where the majority of undeveloped lands occur, have all followed the recommendations of the plan and have upzoned from 2- to 5-acre residential categories; they have removed vacant industrial zoning within deep recharge areas and are following the better management practices called for in the plan. The county of Suffolk, early in 1985, created the Suffolk County Pine Barrens Commission, which is staffed by the Planning Department. This agency reviews all actions within the major pristine deep recharge area on the Island—the Suffolk County pine barrens—in order to control any action that might be injurious to the entire watershed region. Members of the commission include several of the key participants on the original CAC.

Thus every zoning change, every subdivision, every industrial location, and every governmental action, e.g., new roads, buildings, etc., are now scrutinized as to their compatibility with the 208 plan. One common criticism of regional planning has been that this review does take place, but that it is not effective in changing municipal actions that are clearly not in line with the "spirit" of the plan. For example, the plan may designate certain recharge areas to be protected, yet the regional planning agency may be unable to change municipal approval of development proposals deemed inappropriate for certain areas. In short, the plan has been institutionalized at all levels as the guide and standard to be followed and complied with. This is not too surprising. After all, Long Island has no other choice. Its groundwater aquifers are sole sources in the full meaning of the term. So goeth the water—so goeth Long Island.

REFERENCES

Aronson, D., and Seaburn, G. (1974). Preliminary appraisal of the operating efficiency of recharge basins on Long Island, N.Y. *Geol. Surv. Water-Supply Pap. (U.S.)* **20001-D.**

Baier, J. H. (1976). "Detergent Concentrations in Suffolk County Groundwater." Suffolk County Department of Environmental Control, Hauppauge, New York.

DeLuca, F. A., Hoffman, J. F., and Lubke, E. R. (1965). Chloride Concentration and Temperature of the Waters of Nassau County, Long Island, N.Y. *N. Y. State Water Resources Commission Report* **55.**

DeLuca, F. A. *et al.* (1975). Chloride Concentration and Temperature of the Waters of Nassau County, Long Island, N.Y. *Geol. Surv. Bull.* (*U.S.*) **55.**

Franke, O. L., and Cohen, P. (1972). Regional rates of ground water movement on Long Island, N.Y. *Geol. Surv. Prof. Pap.* (*U.S.*) **800-C,** C271–C277.

Frizzola, J., and Baier, J. (1975). Contaminants in rain water and their relation to water quality. *Water Sewage Works* **122,** Nos. 8 and 9.

Fuller, M. L. (1914). The geology of Long Island. *Geol Surv. Prof. Pap.* (*U.S.*) **82.**

Harr, A. C. (1972). Major and minor constituents and pesticides in water from selected sources in Nassau and Suffolk Counties, Long Island, N.Y. *Geol. Surv. Open-File Rep.* (*U.S.*)

Holzmacher, McLendon and Murrell, Inc. (1969). "Report Comprehensive Public Water Supply Study," Vols. 1, 2, and 3. Suffolk County, New York.

Holzmacher, McLendon and Murrell, Inc. (1975). "Study of Leachate at Landfill Sites," Vol. 1. Suffolk County Department of Environmental Control.

Jensen, J. M., and Soren, J. (1974). Hydrogeology of Suffolk County, Long Island, N.Y. *Geol. Surv. Hydrol. Invest. Atlas (U.S.)* **HA-501.**

Kimmel, G. E. (1971). Water level surfaces in the aquifers of western Long Island, N.Y., in 1959 and 1970. *Geol. Surv. Prof. Pap. (U.S.)* **750-B,** 224–228.

Kimmel, G. E., and Harbaugh, A. W. (1976). Analog-model analysis of effect of wastewater management on the ground water reservoir in Nassau and Suffolk Counties, N.Y. Report 1: Proposed and current sewerage. *Geol. Surv. Open-File Rep. (U.S.)* **76–441.**

Koppelman, L. E. (1978). "The Long Island Comprehensive Waste Treatment Management Plan," Vols. 1 and 2. Hauppauge, New York.

Koppelman, L. E. (1980). Presentation to Suffolk County Planning Commission in which the estimated cost of the Southwest Sewer District of $1.5 billion was apportioned on an average basis to the population served.

Koppelman, L. E. (1985). Presentation to Suffolk County Farm Select Committee for the Preservation of Suffolk County Farmlands.

Koppelman, L. E., and Tannenbaum, E. (1982). "The Long Island Segment of the Nationwide Urban Runoff Program." Long Island Regional Planning Board, Hauppauge, New York.

Koppelman, L. E., Tannenbaum, E., and Swick, C. (1984). "Nonpoint Source Management Handbook." Long Island Regional Planning Board, Hauppauge, New York.

Long Island Regional Planning Board (1978). "Landfill Survey." LIRPB, Hauppauge, New York.

Long Island Regional Planning Board (1985). "Population Survey." LIRPB, Hauppauge, New York.

Nassau County Department of Health (1977). "Summary of Organic Sampling—Nassau County Municipal Water Supplies." Mineola, New York (unpublished status report).

Nassau-Suffolk Regional Planning Board (1977). "Section B of the Areawide Waste Treatment Management Plan." N-SRPB, Hauppauge, New York.

Newsday (1975). Announcement. Citizen input for water study, (Jan. 21).

Newsday (1978). L.I. develops major plan for waste management (special reprint). Reprinted from *Newsday,* April 28-May 3.

Perlmutter, N. M., and Guerrera, A. A. (1970). Detergents and associated contaminants in ground water at three public supply well fields in southwestern Suffolk County, Long Island, N.Y. *Geol. Surv. Water-Supply Pap. (U.S.)* **2001-B.**

Pinder, G. F. (1973). A Galerkin-finite element simulation of groundwater contamination of Long Island, N.Y. *Water Resourc. Res.* **9,** 1657.

Pinder, G. F., and Page, R. F. (1976). "Finite Element Simulation of Salt Water Intrusion of the South Fork of Long Island." International Conference on Finite Elements in Water Resources, Princeton University, New Jersey.

Porter, K. S., Zach, D., and Baskin, L. (1977). "Nitrogen: Sources and Potential Impact." Cornell University/Cooperative Extension Service, Ithaca, New York.

Porter, K., Bouldin, D. R., Schoemaker, C. A., Baskin, L., Zach, D., Pacenka, S., and Fricke, D. (1978). "Nitrogen on Long Island, Sources and Fate." Nassau-Suffolk Regional Planning Board, Hauppauge, New York.

Proceedings of the County Legislature of Suffolk County, N.Y. (1971). Local Law No. 21-1970 relating to a prohibition against the sale of certain detergents in Suffolk County, as amended (p. 23).

Roy F. Weston, Inc. (1976). "Nassau-Suffolk 208 Domestic and Industrial Point Source Inventory and Evaluation." Roslyn, New York.

Seaburn, G. E., and Aronson, D. A. (1973). Catalog of recharge basins on Long Island, N.Y. in 1969. *Bull.—N. Y. State Dep. Environ. Conserv.* **70.**

State of New York Law (1983). Landfill Law, Chapter 299.

State of New York Law (1985). Ashfill Law, Chapters 358 and 359.

Suffolk County Department of Health Services (1981). "Status Report on Aldicarb Contamination of Groundwater as of September 1981." Hauppauge, New York.

Suffolk County Sanitary Code (1985). Water Pollution Control, Article 7.

Tetra Tech (1976). "User Guide for the Estuary Water Quality Models," Rep. TC-662 Long Island, New York.

U.S. Soil Conservation Service (1977). "Animal Waste Control Alternatives, Nassau and Suffolk Counties, N.Y." U.S. Soil Conserv. Serv., Washington, D.C.

Wiggin, T. H. (1957). "Report on a Comprehensive Plan for the Development and Distribution of the Available Water Supply of Suffolk County, Long Island, N.Y.," Report to the Suffolk County Water Authority.

Zaki, M. H., Moran, D., and Harris, D. (1982) Pesticides in groundwater: The aldicarb story in Suffolk County, N.Y. *Am. J. Public Health.*

7

Dade County, Florida, Case Study

REGINALD R. WALTERS
Planning Department
Metropolitan Dade County, Florida
Miami, Florida 33128-1972

I. INTRODUCTION

Southeastern Florida is viewed, by residents and visitors alike, to be a water wonderland. Lakes and wetlands abound. The Everglades occupies one-third of Dade County. Yet southeastern Florida communities are becoming hard-pressed to meet their ever-increasing demands for potable water. One of the reasons is that these Florida communities are building directly on top of their drinking water reservoir—the Biscayne Aquifer. This poses significant water quality problems, and given the emerging knowledge regarding the contamination of groundwater with synthetic organic chemicals and the potential health effects of this contamination, the challenge to local governments is becoming obvious whereas the solutions are not. The subject of this chapter is the program that Dade County, Florida, has instituted to protect the quality of its sole source potable water supply, the Biscayne Aquifer.

A. Problem

The water quality problem presently being addressed is that of contamination of groundwater by Synthetic Organic Chemicals (SOCs). SOCs are man-made chemicals containing carbon, many of which are known or suspected to be toxic at very low concentrations. Many of the SOCs are volatile, meaning that components of these organic chemicals convert easily from the liquid to gas phase. This volatile subgroup of SOCs are referred to as VOCs.

The issue of contamination with synthetic chemicals was first raised by the U.S. Environment Protection Agency during the mid-1970s when they analyzed drinking water supplies of a number of cities around the

United States. Dade County's North District water supply, treated at the county's treatment plants in Hialeah, was one of the municipal supplies tested. It revealed the presence of low concentrations of a wide variety of SOCs in both the raw and treated water (Symons *et al.*, 1975).

In light of this finding, Dade County expedited the construction of a new regional water supply wellfield, in a nonurban location, that had been in the planning stages for many years. A 3-square-mile site for such a wellfield had been acquired from the state of Florida in 1945. Demand projections identified the need for development of an additional supply by the mid-1980s, and design and construction proceeded accordingly. The new wellfield, completed in 1983, contains 15 wells that can supply 150 million gallons per day (mgd) of water when the pumps are operating at low speed and 220 mgd when operating at high speed. The cost of this wellfield and its raw water transmission pipe to the Hialeah treatment plants was just under $40 million. It was originally intended that the new Northwest Wellfield would be used to supplement the old 135-mgd wellfields that are located in the heavily urbanized cities of Hialeah and Miami Springs. However, with the detection of organic chemicals in the water produced by the old wellfields, the water was deemed to be unacceptable, and its use was substantially curtailed as the new wellfield came on line in 1983. Because the water obtained from both the old and the new wellfields is treated at the same treatment plants, the old wellfields will be used only for supplemental supply, to be blended with the high-quality water from the Northwest Wellfield. The present challenge is to ensure that the new wellfield does not suffer the same fate as the old ones.

B. Location and Area

Dade County is located at the southeastern extremity of the Florida peninsula (Fig. 1). The area of the county is approximately 2000 square miles, roughly the size of the State of Delaware. Its population in 1985 was estimated to be approximately 1,770,000 with just under half of that total residing within various incorporated municipalities and half residing in unincorporated areas of the county. The principal cities are Miami, Miami Beach, Hialeah, Coral Gables, North Miami, and North Miami Beach.

While Dade County contains a seemingly ample supply of land and water resources to accommodate its rapid growth, a number of constraints make much of the area unavailable for these uses. Of the county's 2000-square-mile land area, half of it lies within the Everglades National Park, Big Cypress National Freshwater Preserve, and the State

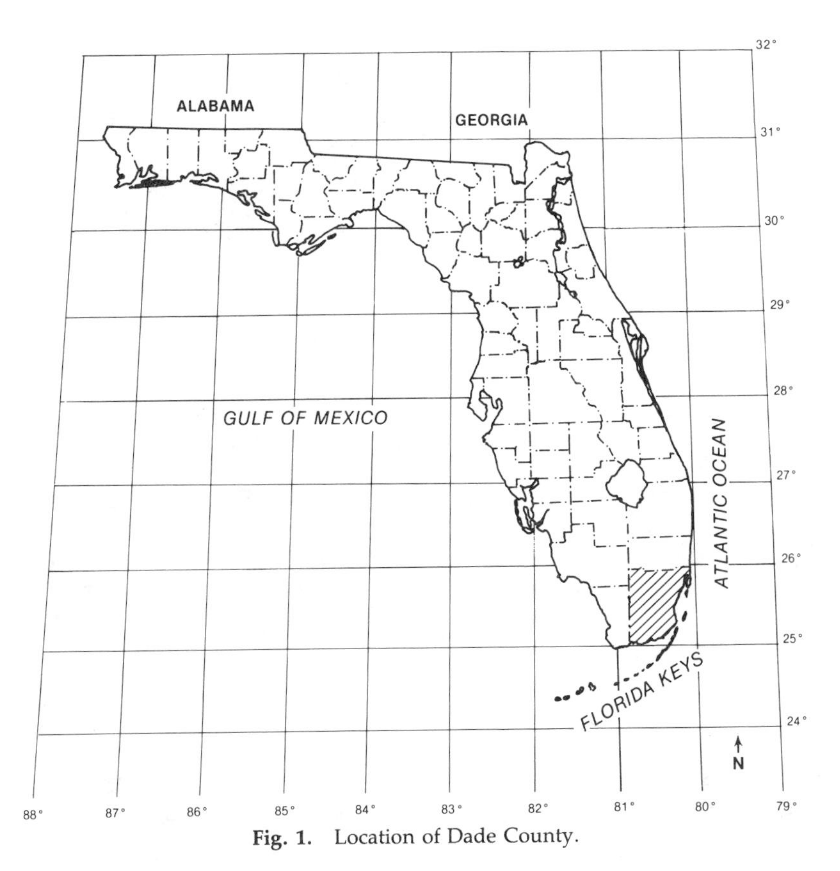

Fig. 1. Location of Dade County.

of Florida's Everglades Conservation Areas (Fig. 2). Of the remaining 1000 square miles, approximately half is already urbanized or used for agricultural production. Most of the remaining open land in Dade County consists of various types of wetlands, and while these areas experience periodic inundation, the freshwater resources in many of them are not amenable to development of large municipal water supplies because of proximity to salt water or because of ecological needs or other constraints.

C. Climate

The climate of Dade County is subtropical marine with freezing temperatures being a rarity. Temperatures average in the mid-80s (Fahrenheit) during the summer and in the middle to high 60s during the winter.

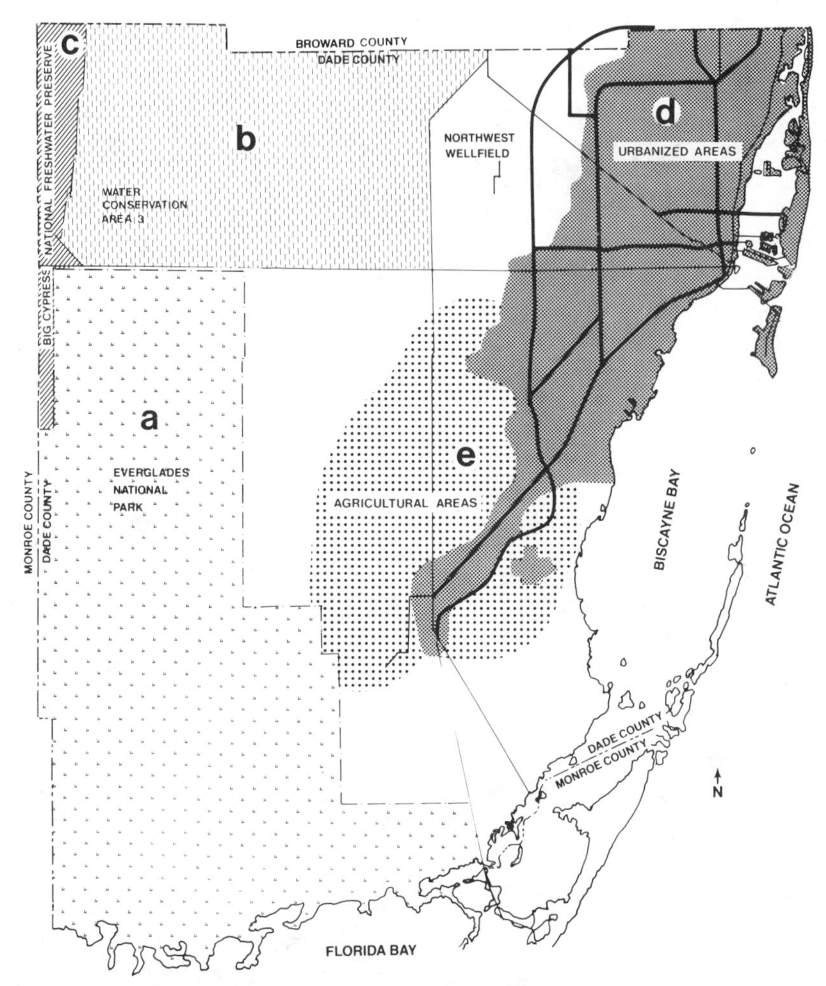

Fig. 2. Protected and developed areas in Dade County: (a) Everglades National Park; (b) Everglades Water Conservation Areas; (c) Big Cypress National Freshwater Preserve; (d) urban and suburban areas; (e) agriculture areas.

The county receives an annual average rainfall of 60 inches but great variations occur from year to year, season to season, and across short distances. About three-fourths of the rainfall occurs during the distinct summer wet season that usually begins in middle to late May and continues through October. During the winter and spring months, the area typically receives only 10 to 15 inches of rain, and during many winters much less than that. During the warm, dry spring months, water in storage is typically depleted and the necessity for water use restraint is becoming more frequent.

One of the area's water supply problems is that its flat terrain does not provide for efficient storage of water on the land surface, but just a few feet below is the very productive Biscayne Aquifer.

D. Geology and Soils

The Biscayne Aquifer is a highly permeable water table aquifer encompassing some 3300 square miles in southeastern Florida. It extends over 100 miles from southern Palm Beach County south to Florida Bay at the southern tip of the peninsula (Fig. 3). This shallow aquifer provides the

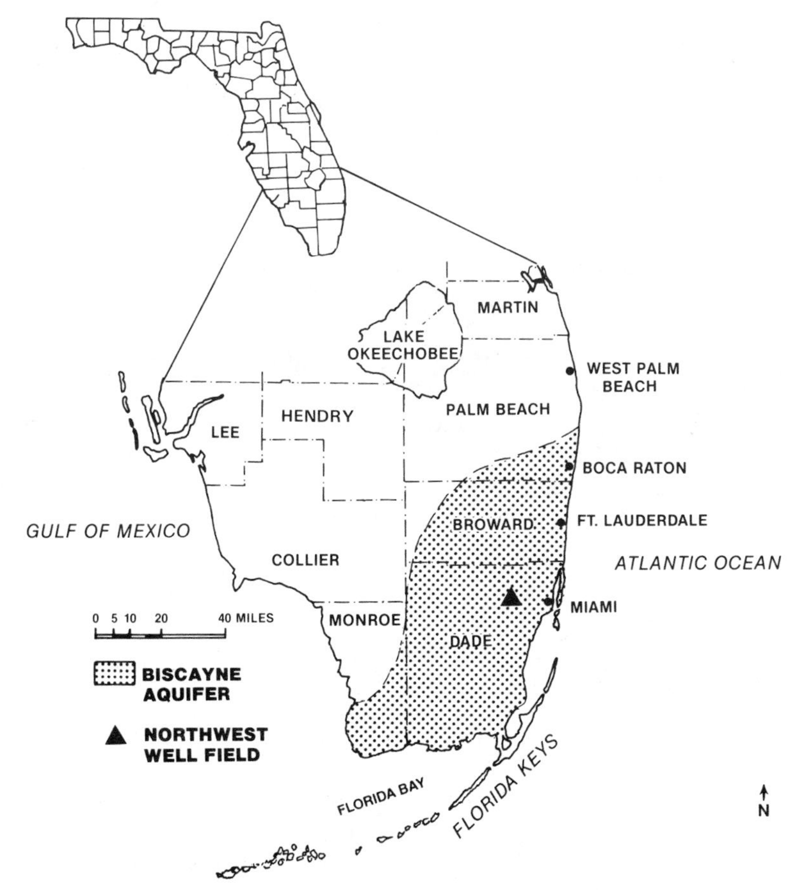

Fig. 3. Areal extent of the Biscayne Aquifer.

sole source of freshwater supply for all of Dade and Broward counties and much of the Florida Keys via pipeline, and it underlies most of Everglades National Park.

The Biscayne is a wedge-shaped aquifer consisting of limestone, sandstone, and sand. Its thickness reaches 120 feet under Miami at the coastline and thins to the west until it tapers to an edge only 3 feet thick at the western boundary of Dade County, which is some 40 miles inland (Fig. 4). At the western edge of the urbanizing area, some 15 miles inland, the aquifer extends to a depth of around 60 feet. The Biscayne Aquifer is extremely productive and it is one of the most permeable aquifers in the world. Transmissivity ranges from about 4 mgd/ft in north Dade to 12 mgd/ft in south Dade (Klein and Hull, 1978). It is very productive at all depths but the deeper layers tend to be most productive. The high transmissivities combined with large water withdrawals result in cone-shaped depressions in the water table below the wells. These changes in the water table are known as cones of depression. In Dade County the cones of depression are shallow but very extensive.

The Biscayne is an unconfined, water table aquifer. Lying on top of it is a very thin veneer of immature soils only a few inches to a few feet deep. In the southwestern parts of the county, the soil is little more than weathered limestone that is actually the exposed surface of the aquifer. Greater Miami is virtually built on top of its drinking water reservoir and much of its infrastructure is built directly in it.

Because the aquifer has so little protective cover, it is highly vulnerable to degradation by man's activities. The community has become acutely aware of this during recent years. In the past five years, two municipal water supply wellfields in north Dade, with a combined capacity of 65 mgd, had to be shut down because of contamination by industrial chemicals, and another wellfield in south Dade has periodically experienced elevated nitrate levels from agricultural chemicals.

E. Regional Context

The Biscayne Aquifer is recharged primarily by local rainfall. Local recharge is supplemented by inflow from the regional hydrologic system, known as the Kissimmee–Lake Okeechobee–Everglades basin, of which the Biscayne Aquifer is an important unit (Fig. 5). This drainage system originates in the center of the Florida peninsula over 200 miles north of Dade County near the City of Orlando and flows south through the center of the lower peninsula. Lake Okeechobee covers about 700 square miles and serves as the primary reservoir for the south Florida region. Extending south from the southern rim of Lake Okeechobee to

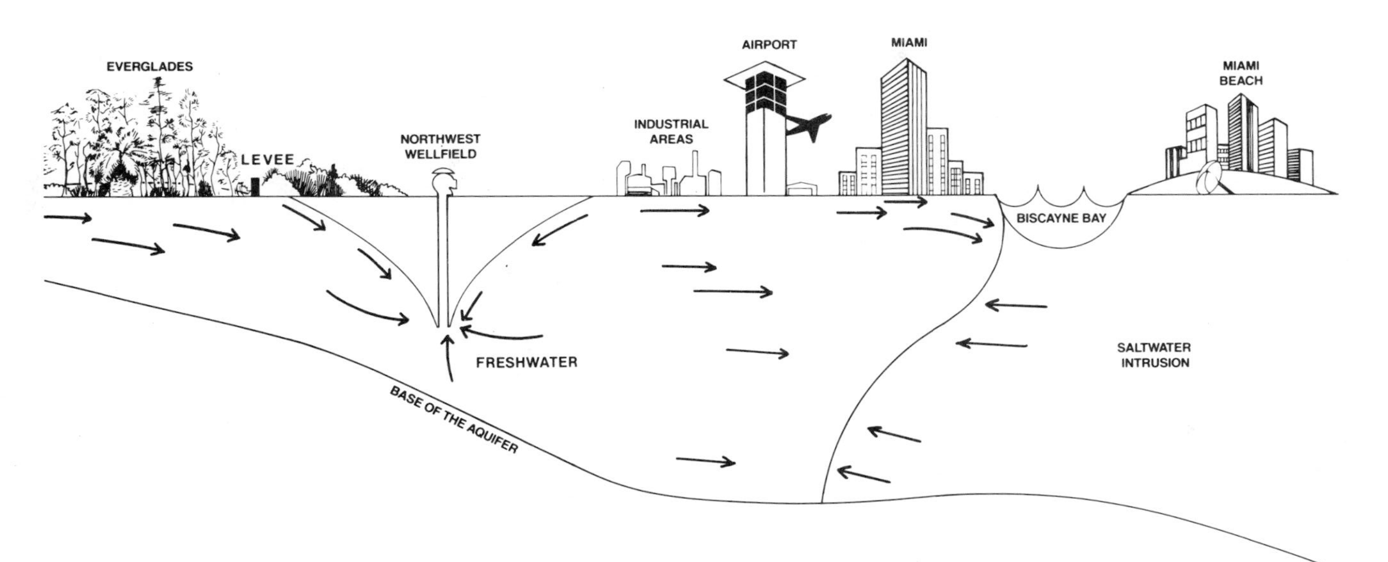

Fig. 4. Generalized cross section of the Biscayne Aquifer.

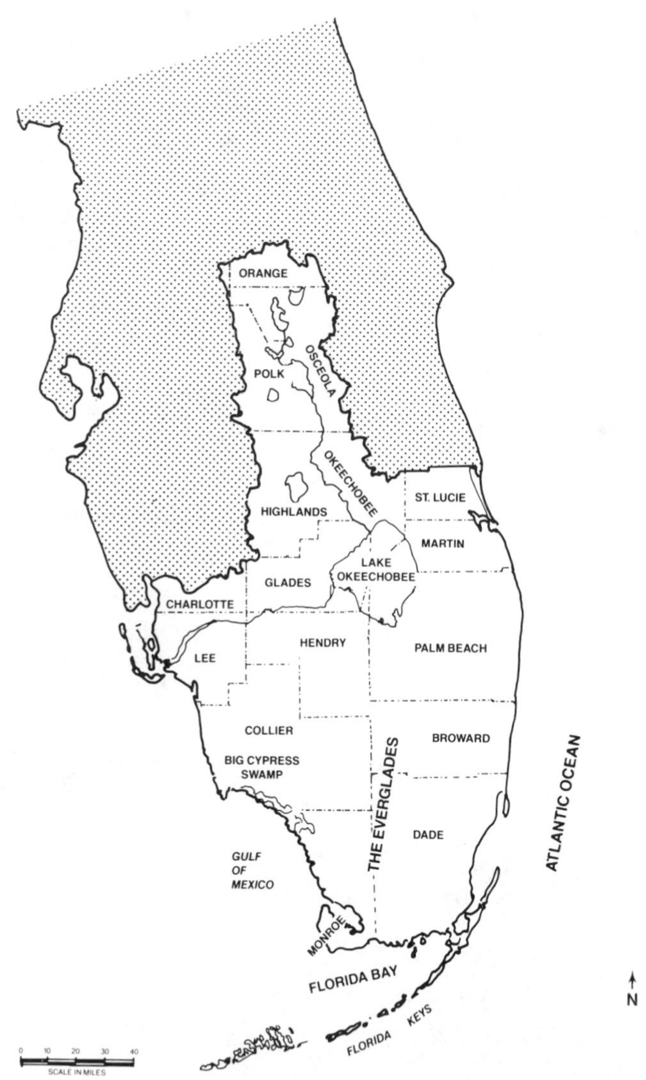

Fig. 5. Kissimmee–Lake Okeechobee–Everglades drainage basin.

the brackish glades and saline swamps along Florida Bay is the Florida Everglades. Once a continuous, shallow "river of sawgrass," the Everglades is now controlled by a vast network of dikes, canals, and pumping stations operated by the South Florida Water Management District, an agency of the State of Florida.

F. Northwest Wellfield Area

The area around the Northwest Wellfield remains undeveloped largely because of poor environmental conditions. The area is relatively low-lying, is poorly drained, and is subject to flooding during the summer wet season and severe rainfall events. The presence of 4 feet of peat and muck soils, unsuitable for building foundations, poses another constraint that must be overcome by muck removal and replacement with limestone fill prior to development.

The predominant land use in the vicinity of the Northwest Wellfield is quarrying of limestone for use as fill material throughout Florida and for the manufacture of cement products. The fringe of the urbanizing area was over 3 miles east of the wells at the time of their construction. Zoning in the area also remained predominantly in nonurban classifications.

G. Past Problems

The sustained recharge of the Biscayne Aquifer with high-quality water in the interior Everglades maintains a coastward or eastward flow of groundwater in Dade County. This abundant and easily developed water resource, coupled with the area's benign climate, facilitated rapid settlement. At the same time, the abundance of water posed a major impediment to development of the region. The answer to that problem was the extensive system of canals and levees constructed over the years to control flooding. While this system enabled the agricultural and urban development of the region, it also resulted in Dade's first major water quality problem—that of saltwater intrusion. The saltwater problem was solved, at considerable cost, with the installation of salinity control gates near the mouths of most of the canals, the abandonment of coastal wellfields, and construction of new wellfields farther inland.

The second major water quality problem faced by the community was the pollution of the canals and groundwater by domestic wastewater. Dade County's rapid growth was served in large part by a multitude of septic tanks and small wastewater treatment plants that discharged effluent into the ground and canals. The hazardous condition that this created was solved by the regional sewering of the area, secondary treatment of wastewater, and disposal through ocean outfalls and deep well injection.

The problem of SOCs is the latest in the series of local water quality problems. While the EPA found concentrations of VOCs in the range of a few parts per billion (ppb) and parts per trillion, it added up to a basis for concern.

Since the initial findings in the mid-1970s, both the EPA and Dade County's Department of Environmental Resources Management (DERM) have conducted numerous studies of raw and treated water throughout the county. The total concentration of VOCs found at some of the urban wellfields has been reported by Dade County DERM (1984) to be in the range of 20 to 30 ppb (Fig. 6). Vinyl chloride has been found in fairly high concentrations of around 4 ppb at two wellfields, but vinyl chloride is not heavily used in Dade County. It is believed that vinyl chloride forms in the aquifer by chemical and biological transformation from other industrial chemicals that are used in large amounts. One noteworthy analysis conducted during the Wellfield Protection Study revealed a close relationship between the presence of industrial development in land areas above public supply well cones of influence and the presence of VOCs in the water that those wellfields produce. The comparison is shown in Fig. 6. Total concentration of VOCs is shown on the left side of this bar graph and the right side indicates the percentage of the land area overlaying a cone of influence devoted to industrial development. While much of this industrial development occurred before the county enacted the relatively stringent standards that govern construction today, this evaluation did confirm the association between industrial development and degraded groundwater.

H. Governmental Setting

Partially offsetting Dade County's vulnerable environmental situation is its somewhat enviable position of possessing a centralized form of countywide government—a modified two-tiered form of government that is unique in the United States. The upper tier is the county government that provides broad "regional" or countywide functions such as metropolitan planning, welfare, health, and transit services. The 27 municipalities in the county represent the lower tier of government and provide a varying array of municipal services including local zoning administration within their jurisdictional boundaries.

Countywide programs now include a comprehensive planning program under which urban service boundaries are established, pollution control and other environmental regulation, and provision of water supply and wastewater services. The county's Comprehensive Development Master Plan (CDMP) provides the broad parameters within which countywide development may occur. It denotes the extent and general classes of land use that may occur throughout the county and it establishes population growth management policies for subareas of the county. This provides the policy framework for planning and program-

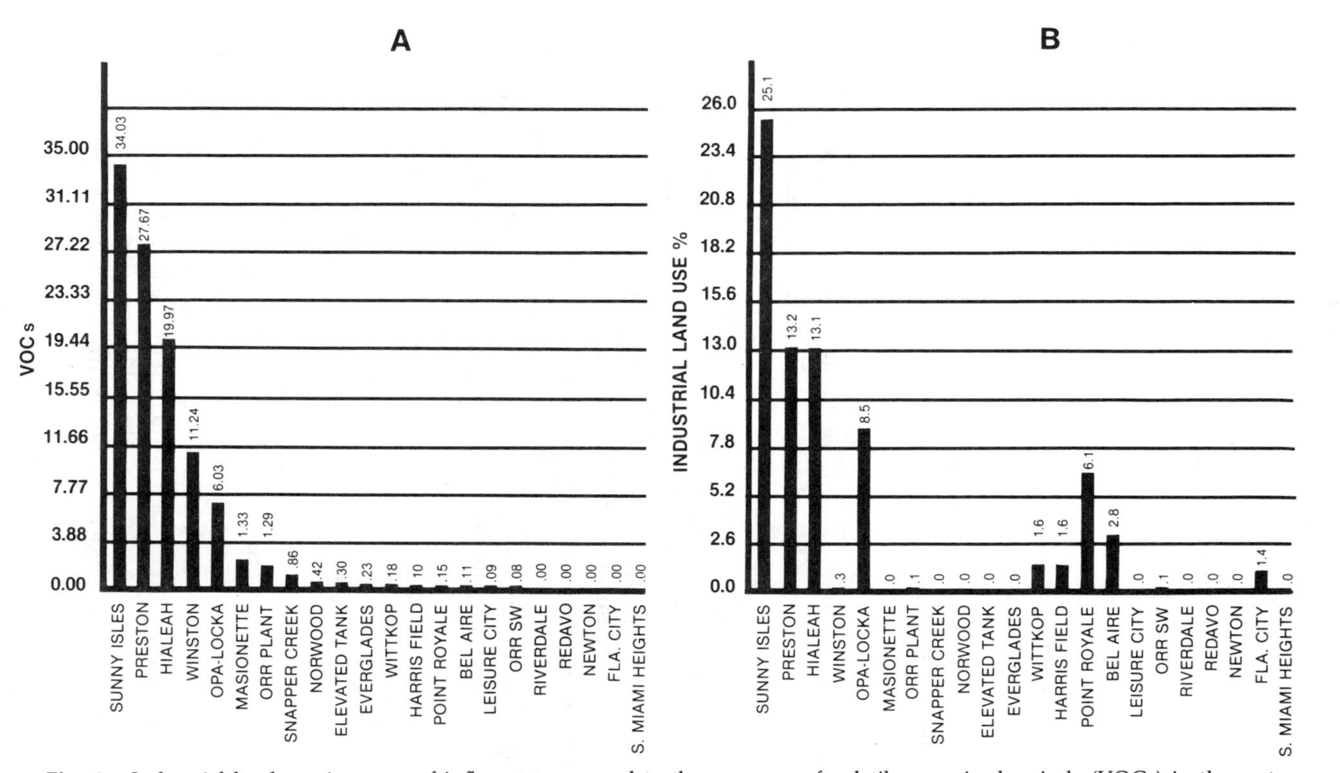

Fig. 6. Industrial land use in cones of influence compared to the presence of volatile organic chemicals (VOCs) in the water produced. Graph A indicates the total concentration of all VOCs in the water produced by each wellfield. Graph B indicates the percentage of the cone of influence area devoted to industrial uses.

ming countywide capital improvements such as highways, transit, water supply, and sewerage, as well as a framework for neighborhood-level planning and zoning administration in the unincorporated areas. While the county government has direct zoning authority only in the unincorporated areas, it does possess the authority to establish minimum standards for zoning in municipalities and has done so to impose height limitations in areas around airports and to require sensitive treatment of new development along the shoreline of Biscayne Bay. On the other hand, the county's environmental regulations directly apply throughout the county, both inside and outside of the incorporated municipalities. This has given the county an advantage over many other communities in its effort to manage its growth and protect its environment.

Another very important agency involved in providing water supply in south Florida is the South Florida Water Management District, which is an agency of the State of Florida. Florida's Water Management Districts are regional agencies that manage water resources and regulate certain uses of water throughout multicounty districts based on drainage area boundaries. The South Florida District encompasses the 18,000-square-mile area of the Kissimmee–Lake Okeechobee–Everglades Basin, including the Big Cypress Swamp. The South Florida Water Management District is critically important to Dade County because it operates all of the water storage areas and the major canals and water control structures in the region. Dade County relies frequently on conveyance of water from Lake Okeechobee and the district's Water Conservation Areas in the Everglades to help provide adequate recharge to wellfield areas and to the aquifer's coastal saltwater interface during the dry spring months until the wet tropical weather begins in late May.

II. WELLFIELD PROTECTION STUDY

In response to the community's increased concern for protection of its water supply, Dade County conducted a planning evaluation of the policies and regulations that govern water management and land use around its new Northwest Wellfield. The Wellfield Protection Study, as it is called, was initiated at the instruction of the Board of County Commissioners in July 1983 and culminated with the adoption, by the board, of the Northwest Wellfield Protection Plan in November 1985.

A. Study Background

In March of 1981, Dade County's Board of County Commissioners (1981) enacted a Wellfield Protection Ordinance imposing special restrictions on land development within certain distances from all public water supply wellfields in the county. The intent of the Wellfield Protection Ordinance was to curb the degradation of groundwater quality in urbanized wellfield areas and to preserve the unspoiled groundwater quality in nonurbanized wellfield areas. The Ordinance established a series of concentric ring-shaped regulatory areas, called Wellfield Protection Zones, around each wellfield. The boundaries of the protection zones are based on groundwater travel times. Boundaries were set at 10-day, 30-day, 100-day, and 210-day groundwater travel distances from the wells. The worst-case condition was based on the concept that contamination outside the 210-day travel distance would not reach a well without undergoing some attenuation, and 210 days was the longest period of no rainfall recorded in the area. The regulations are most stringent for the protection zones in close proximity to the wells and are more liberal in the outer zones. The countywide environmental regulations govern many of the same activities as the wellfield protection ordinance but are generally more lenient.

The wellfield protection ordinance was formulated for urban environments where development already surrounds the wellfields. Given overwhelming precedents of established development, it would have been impractical to attempt to drastically reverse or eliminate the established trend of development. However, the regulation was intended as a first step to prevent further degradation of urban wellfields.

Among other provisions, the original Wellfield Protection Ordinance restricted residential density in the immediate proximity to the wells through the imposition of limits on discharges to sewerage systems and on permissible density of septic tanks; it established stricter performance standards for sewer exfiltration; it limited allowable methods of storm water disposal in close proximity to the wells; and it prohibited new land uses that would involve the use or handling of hazardous materials. Table I summarizes provisions of the 1981 Wellfield Protection Ordinance.

With this ordinance in place in early 1981, pressure grew to amend the county's Comprehensive Development Master Plan to open more of northwest Dade for industrial expansion. Advocates of renewed economic expansion efforts successfully argued that the Wellfield Protection Ordinance and other environmental regulations would ensure the necessary groundwater protection. In July of that year the Board of County Commissioners concurred and amended the Master Plan to classify an additional 2400 acres of land as eligible for immediate industrial rezoning. By this action the commission established a long-range policy to allow industrial development to occur immediately outside of the regulated 210-day groundwater travel distance from the wellfield.

TABLE I

Significant Provisions of Original Cone of Influence Ordinance (1981–1983)

Regulated activity	Wellfield Protection Subzone (distance in feet or groundwater travel time from wells)					
	Less than 100 feet	100 ft–10 days	10 days–30 days	30 days–100 days	100 days–210 days	Countywide regulation (beyond 210 days)
New uses involving handling of hazardous material	Prohibited	Prohibited	Prohibited	Prohibited	Prohibited	Permitted
Residences/ septic tanks	Prohibited	1 DU[a]/2.5 acres	1 DU/acre	1.7 DU/acre	2.4 DU/acre	2.9 DU/acre
Nonres. use/ septic tanks	Prohibited	1400 ft^2/acre	3500 ft^2/acre	6000 ft^2/acre	8500 ft^2/acre	15,000 ft^2/acre
Residences/ sewers	Prohibited	2.4 DU/acre	4.6 DU/acre	No limit	No limit	No limit
Nonres. use/ sewers	Prohibited	8500 ft^2/acre	16,000 ft^2/acre	No limit	No limit	No limit

[a] DU = dwelling unit.

Two years later, during the 1983 review of the Comprehensive Plan, the commission responded to concern expressed by numerous civic and environmental organizations about the land use policy that had been established 2 years earlier. To temper the effect of the 1981 Plan amendment, the commission instructed the Planning Department and the Department of Environmental Resources Management to evaluate the adequacy of the existing wellfield protection regulations and land use policies and to report back on the need for revisions and the possible need for a temporary building and zoning moratorium in the area.

After close consultation, the two departments agreed that the original regulation was inadequate to prevent degradation of the "pristine" Northwest Wellfield and that regulatory protection should be extended to all land under which wellfield pumpage creates a cone of influence. (The comparative quality of water withdrawn from the Hialeah–Miami Springs, Preston, and Northwest wellfields is presented in Table II.) Therefore a short duration building and zoning moratorium was recommended to be imposed while the full extent of the cone of influence could be determined and an additional ordinance prepared to protect it.

TABLE II
Comparison of Water Quality at Three Wellfields

		Concentration (ppb)		
Compound	Standard (ppb)	Preston Wellfield	Hialeah–Miami Spgs. Wellfields	Northwest Wellfield
Vinyl chloride	1.00	3.79	4.25	0.00
Vinylidene chloride		0.51	0.88	0.00
trans-1, 2-dichloroethene		8.71	0.36	0.00
1,1-Dichloroethane		3.06	1.80	0.00
cis-1, 2-Dichloroethene		12.12	5.19	0.00
1,1, 1-Trichloroethane	200.00	—	—	—
1, 2-Dichloroethane	3.00	0.00	0.00	0.00
Tetrachloromethane	3.00	0.00	—	—
Trichloroethylene	3.00	2.97	0.22	0.00
Tetrachloroethylene	3.00	0.00	0.00	0.00
Chlorobenzene		0.99	3.78	0.00
O, M, P-Chlorotoluene		1.12	0.62	0.00
M, P-Dichlorobenzene		0.99	0.99	0.00
O-Dichlorobenzene		0.76	1.27	0.00
Benzene	1.00	—	—	—
Ethylene dibromode	0.02	—	—	—
Trihalamethanes	100.00	22.50	20.00	28.80

Source: Dade County Department of Environmental Resources Management (1983).

A new ordinance extending protection to the land surface above the full extent of the cone of influence was adopted by the Dade County Board of County Commissioners (1984a) on September 30, 1983, just 2 months after the commission's request. While the original 210-day regulatory boundary extended less than 2 miles from the wells, the new boundary extended 4 miles and encompassed approximately 85 square miles of area, including some established and industrializing areas. The boundary line depicting the estimated full extent of the cone of influence is the outer line shown on Fig. 7. The inner line is the 210-day travel distance. The September 1983 revisions to the wellfield protection regulation extended the prohibition on use or handling of hazardous materials to activities located above the full extent of the Northwest Wellfield's cone of influence. The use of septic tanks was also prohibited for all new land uses within the area defined by the full cone, and new heavy industrial zoning was prohibited. In adopting the new regulation, the commission instructed the two departments to further evaluate the regulations and land use policies, this time with extensive citizen involvement to review the staff's assumptions and to determine whether the new rules were adequate or excessive. Two advisory committees were established to provide this input.

B. Project Organization

The Wellfield Protection Study was staffed by members of the two departments; the Planning Department had overall staff responsibility for administering and coordinating project activities. DERM had specific responsibility for performing the technical environmental work including managing the activity of the groundwater modeling consultant. Other county departments were called on when needed, as were state and federal agencies.

One of the committees that was established was an advisory Technical Committee, which consisted solely of representatives of governmental and academic agencies that possess technical expertise in water resources and public health. Appointees were individuals who were at or near the policy-making level of their agency. The primary purpose of this committee was to assist staff in securing and interpreting environmental/health data and in evaluating pollution control and water treatment technology.

Also, a Policy Advisory Committee was established to provide a broad spectrum of community opinion. This committee was composed of representatives of 27 private and governmental organizations affected by water management or land use regulation. Interests represented on the

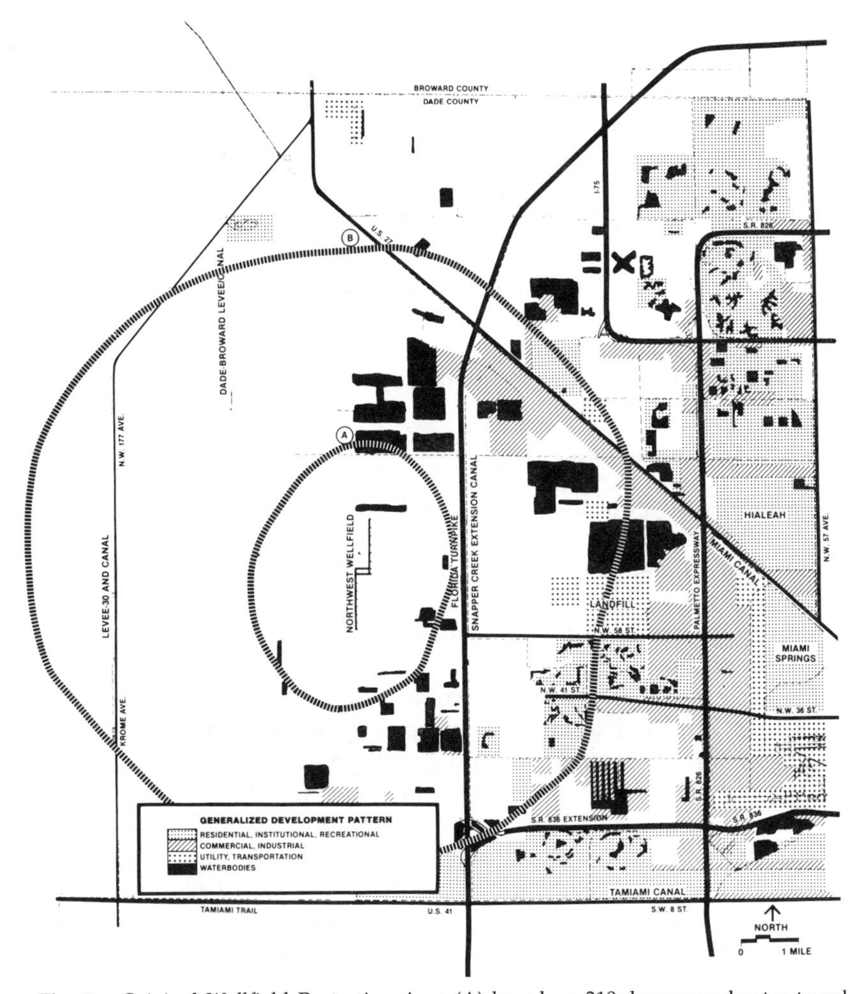

Fig. 7. Original Wellfield Protection Area (A) based on 210-day groundwater travel distance and full cone of influence (B) estimated by "second-generation" groundwater flow model assuming 220-mgd pumpage from Northwest Wellfield and 75-mgd pumpage from wellfields in Hialeah and Miami Springs, and extreme drought conditions.

committee ranged from environmental organizations, including the Audubon Society, Sierra Club, and Friends of the Everglades, to the economic development interests, such as the Greater Miami Chamber of Commerce, Industrial Association, and Industrial Development Authority of Dade County. Where technical input was required by the Policy

Committee, Technical Committee members or other experts were called in as needed.

The project was initiated with an intensive period of education and fact-finding. The Policy Committee was provided background and testimony regarding the area's geology, hydrology, and water management, cone of influence dynamics, groundwater modeling, water quality, risk assessment, water treatment, pollution control practices, and land use controls. While some committee members were impatient and tried continually to press ahead, other committee members endlessly solicited more information. The result was that the project that was originally scheduled to take 12 months took twice that long to complete. Notwithstanding the diversified interests represented on the committee, all but one voted in favor of the resultant plan.

C. Groundwater Flow Model

From the outset of the project the Technical Committee became heavily involved in an effort to upgrade the county's computerized groundwater flow model. The county's original computations of cone of influence drawdown had been made using a basic Theis equation (Dade County Department of Environmental Resources Management, 1980). Since 1980, however, three "generations" of computerized programs were developed to enable the county to perform more sophisticated analyses. The first model was prepared by the South Florida Water Management District. A "second generation" model was prepared for joint use by Dade, Broward, and Palm Beach counties by the local office of a national consulting firm (Camp Dresser and McKee, Inc., 1982). That model was constructed at a regional scale using average regional conditions as assumptions. While it served as an excellent tool for the initial planning and regulatory purposes, there was broad agreement that a wellfield-specific model should be employed to enable the full extent of the cone of influence under the wellfield to be determined with maximum accuracy. Accordingly, the consulting firm was retained to prepare a "third generation" area-specific model for the Northwest Wellfield. Because point-specific data for most variables are not available, values were generalized by interpolation across groups of points. Calibration runs were made to adjust some of the values, and the model was verified by comparison with the most recent available field data (Camp Dresser and McKee, Inc., 1985).

The model first predicted water table elevations that would result under the assumed meteorologic and pumping conditions. Groundwater travel times were then computed based on the resulting gradients of the groundwater surface. A subroutine was also incorporated to simulate dispersion of pollutants in the water. This tended to result in pollutant travel times that were 20% shorter than the hydraulic travel time estimates that did not account for pollutant dispersion.

In defining the boundaries of cones of influence for regulatory purposes, a variety of criteria were used. On the up-gradient or west side of the cone of influence, where groundwater elevations are naturally higher and slope toward the well site, the cone of influence boundary was defined as the location where well pumpage caused groundwater elevations to be depressed $\frac{1}{4}$ foot. That is illustrated on the left side of Fig. 8A. On down-gradient sides of the wellfield, as illustrated on the right side of the figure, the cone boundary was defined as the location where the water table was predicted to be at its highest elevation. Where two cones of influence of different wellfields intersect, as occurs between the Northwest Wellfield and the older wellfields in Hialeah and Miami Springs, the high point in the water table or "Water Divide" was used to define the boundary between the two cones (Fig. 8B). Seasonal changes in the size of the cone of influence and in the corresponding velocities and direction of groundwater flow were also estimated. Figure 8C shows these differences in a schematic fashion. The cone of influence is smallest during the summer wet season when recharge is the greatest, and it expands during the winter and spring dry season. It is largest, by far, during severe drought when recharge is deficient.

Traditionally, the groundwater flow model had been used by Dade County only to predict cone of influence boundaries and travel times and distances under assumed conditions. These predictions were, in turn, used to establish regulatory boundaries. However, the Northwest Wellfield situation has proven to be more complex. The updated modeling confirmed that the cone of influence encompasses a large county landfill site that is known to be contaminating groundwater (Fig. 9). If this situation is allowed to persist, the traveltime computations indicate that it would take approximately 15 years for contaminated groundwater to reach the wells. The Wellfield Protection Study proposes that two measures be taken to prevent this from happening:

1. The construction of additional aeration facilities, air stripping towers, at the Hialeah water treatment plants. This will remove VOCs and allow temporary reactivation of the older wellfields in Hialeah and Miami Springs. The associated reduction in pumpage from the Northwest Wellfield and increased pumpage at Hialeah and Miami Springs are expected to create a water divide between the two wellfields at a location west of the area anticipated to become contaminated by the time the air

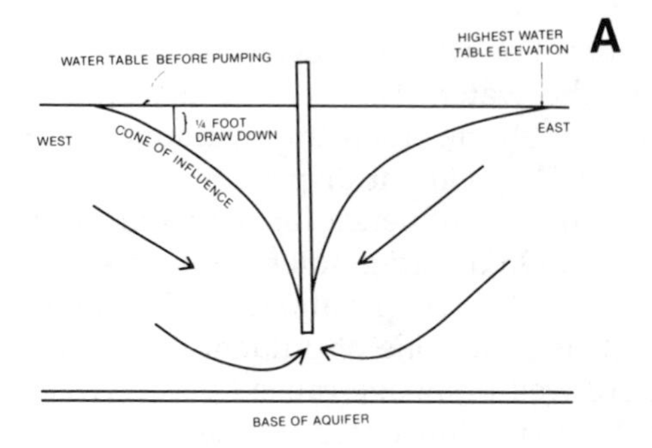

Fig. 8. Cone of influence thresholds. (A) One-quarter-foot drawdown (up-gradient) and point of highest water table elevation (down-gradient); (B) water divide between two wellfields; (C) seasonal variations in size of cones of influence.

stripping facilities are operational. The air stripping facilities should take approximately 3 years to construct.

2. The construction of major improvements to the canal system to recharge the east side of the wellfield with water from the Everglades. In essence, this would create a partial hydrologic barrier between the

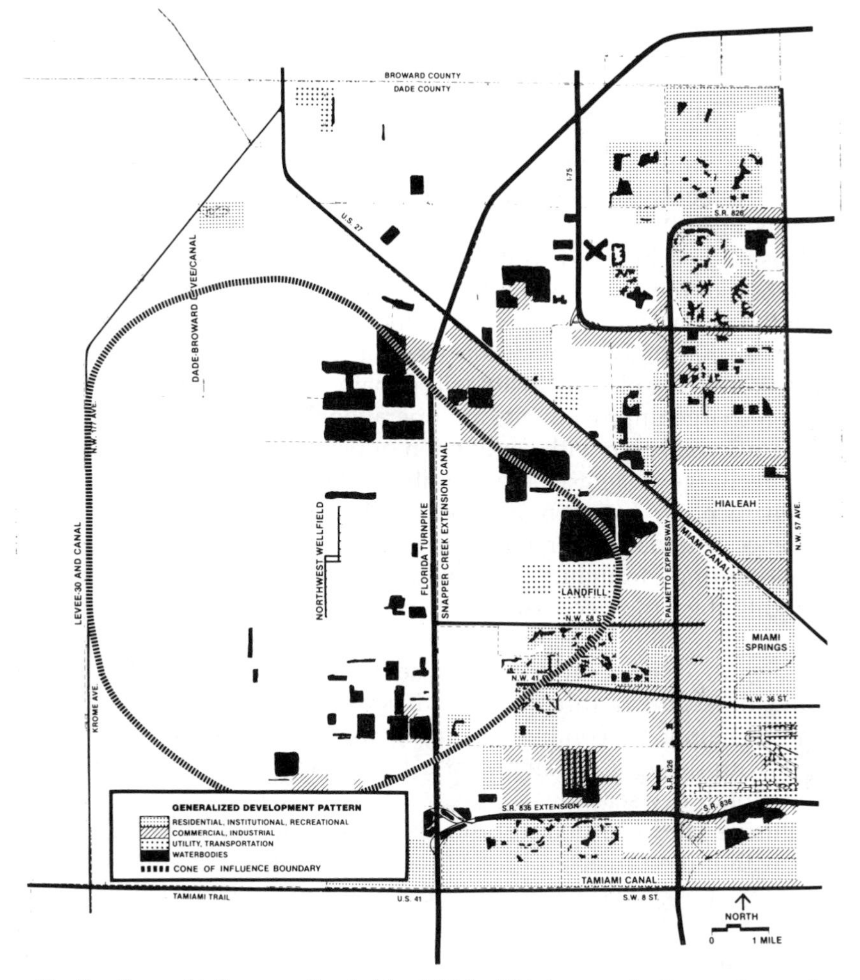

Fig. 9. Cone of influence estimated for 1985 by "third-generation" groundwater flow model assuming 150-mgd pumpage from Northwest Wellfield and 10-mgd pumpage from wellfields in Hialeah and Miami Springs, and average wet season conditions.

wellfield and the landfill area during normal conditions, although not during severe drought. Some 5 to 8 years could be required to obtain regulatory approval and to construct the canals. Use of the older wellfield will, again, be phased down at that time.

Implementation of these two measures will cause the cone of influence boundary to contract in two steps. Accordingly, the management program formulated for the Northwest Wellfield area includes an initial management area boundary plus two additional management area boundaries that are based on the two projected time-stepped contractions in the cone of influence boundary. The time steps are referred to in the plan as time phases.

The first phase began with the adoption of the Wellfield Protection Study's recommendations in Novemer 1985. The new regulatory boundary was established, as a matter of policy, at the west side of the suspect landfill (Fig. 10).

Phase 2 is scheduled to begin by 1989. This corresponds with the operation of the air stripping facilities and the resumption in pumping from the old wellfields. The cone of incluence boundary is predicted to retract at that time to a location west of the landfill site. Phase 2 and operation of the old wellfields will remain in effect until the proposed canal recharge improvements are operational.

Phase 3 will begin with completion of proposed improvements to the canal recharge system in the vicinity of the new wellfield. This is expected to occur between 1991 and 1994. By doing this, the cone of influence boundary is predicted to retract to the location of the canal, except during drought. To provide safety during such periods of drought when the necessary canal recharge would not occur, an area east of the canal has been included in the Phase 3 regulatory area. The area included east of the canal represents the distance that groundwater would flow during three consecutive years characterized by average hydrometeorologic conditions during the 5-month wet season alternated with drought conditions of a 1:20 year return frequency during the 7-month dry season.

Ultimate land use policies and regulations are scheduled to govern land within the projected Phase 3 boundary beginning with the initiation of the program (Phase 1), while the land located between the Phase 1, Phase 2, and Phase 3 boundaries will be subject to interim land use policies and regulations. Provisions of the county's protection program for the Northwest Wellfield and provisions of its overall wellfield protection strategy are described more fully in the following section.

III. WELLFIELD PROTECTION PROGRAM

Since the original Wellfield Protection Ordinance was adopted in 1981, the county's Wellfield Protection Program has been evolving into one that is comprehensive. With the measures that grew out of the Wellfield Protection Study the program consists of five elements, two of which were introduced in the previous paragraphs. Some provisions will change as the program enters its future time phases. Most notably,

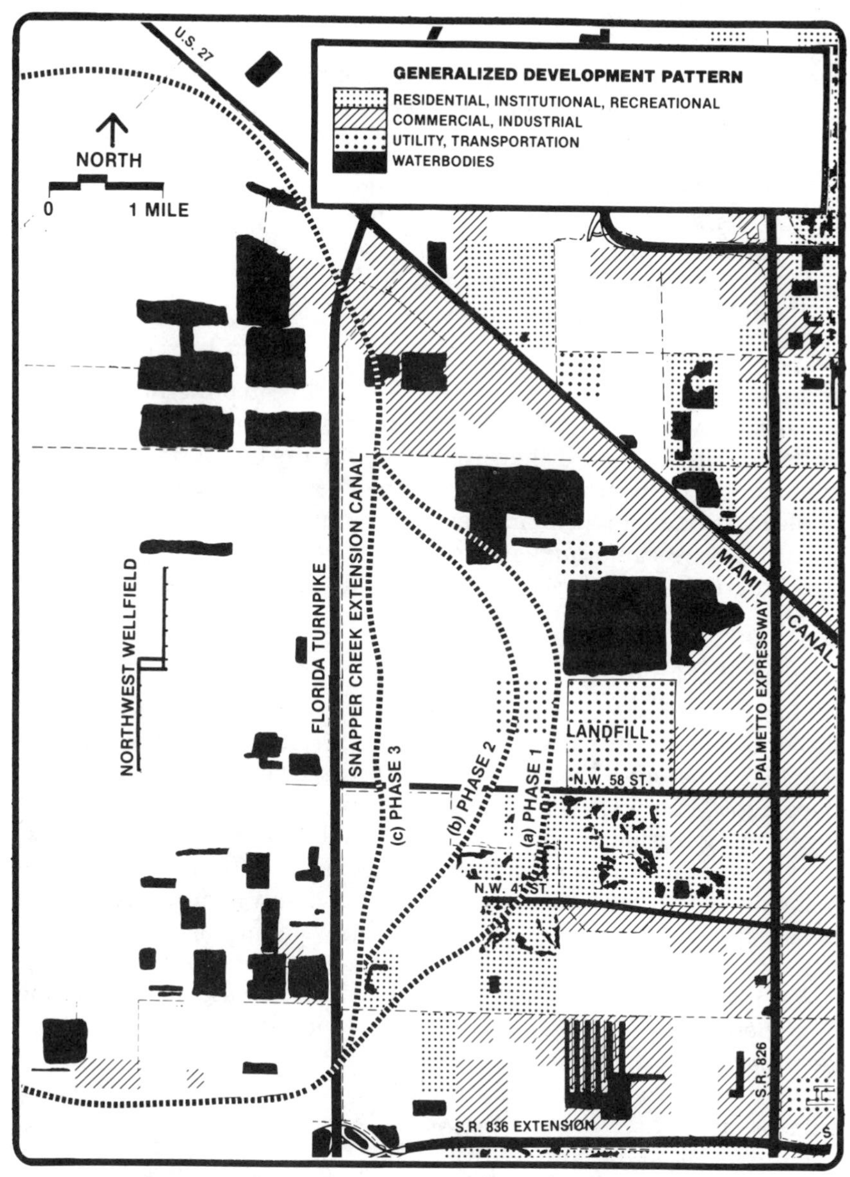

Fig. 10. Three-phased Wellfield Protection Program boundaries. (a) Phase 1 boundary is located at a 5-year groundwater travel distance east of the Phase 2 boundary, computed assuming current pumpages referenced in Fig. 9 and extreme drought conditions. (b) Phase 2 boundary is located at a 5-year groundwater travel distance west of the landfill based on projected pumpage of 120 mgd at Northwest Wellfield and 50 mgd from wellfields in Hialeah and Miami Springs, and average wet season conditions. (c) Phase 3 boundary is based on pumpage of 220 mgd at Northwest Wellfield assuming average wet season conditions with an added safety zone east of the recharge canal based on three consecutive drought dry seasons.

the geographic area governed by the wellfield protection regulation and land use policies will change with each future time phase. The five elements of the current wellfield protection program are:

1. Water Management and Monitoring
2. Water and Waste Treatment Facilities
3. Land Use Policies and Controls
4. Environmental Regulations and Enforcement
5. Public Awareness and Involvement

While many of the provisions apply only within the wellfield areas, some of the measures that will be very supportive in protecting wellfield water quality apply countywide. Together, all of the provisions are expected to protect not only new wellfields, such as the Northwest, which have not yet been surrounded by urban development, but the program is also expected to help improve water quality at older urbanized wellfields. The significant measures of the five-element Wellfield Protection Program are discussed in the following sections.

A. Water Management

Management of the water regime has been used in south Florida to protect water quality since the 1940s and 1950s. It was during that period that the numerous drainage canals that had been constructed to provide flood protection were retrofitted with coastal salinity dams to prevent the inland intrusion of salt water. Based on the permeability and coastal thickness of the Biscayne Aquifer, a 2.5-foot head of fresh water must be maintained on the inland side of the coastal salinity dams to prevent the denser salt water from moving inland. This water management strategy has for years, prevented saltwater contamination of many wellfields.

A similar water management strategy will now be used to prevent the migration of groundwater contaminated with VOCs and other toxic chemicals toward the Northwest Wellfield. The canal system that traverses the area around the Northwest Wellfield will be improved to enhance aquifer recharge on the east side of the cone of influence (Fig. 11). The computerized groundwater flow model indicates that deepening the recharge canal, which is located approximately midway between the wellfield and the previously identified polluting landfill, to intersect highly permeable strata 35 feet below, and maintaining stages in the canal a few tenths of a foot higher than the adjacent water table elevation, will establish a hydrologic boundary and prevent the expansion of the cone of influence into the degraded area east of the canal even when the wellfield is pumping at its designed maximum capacity of 220 mgd.

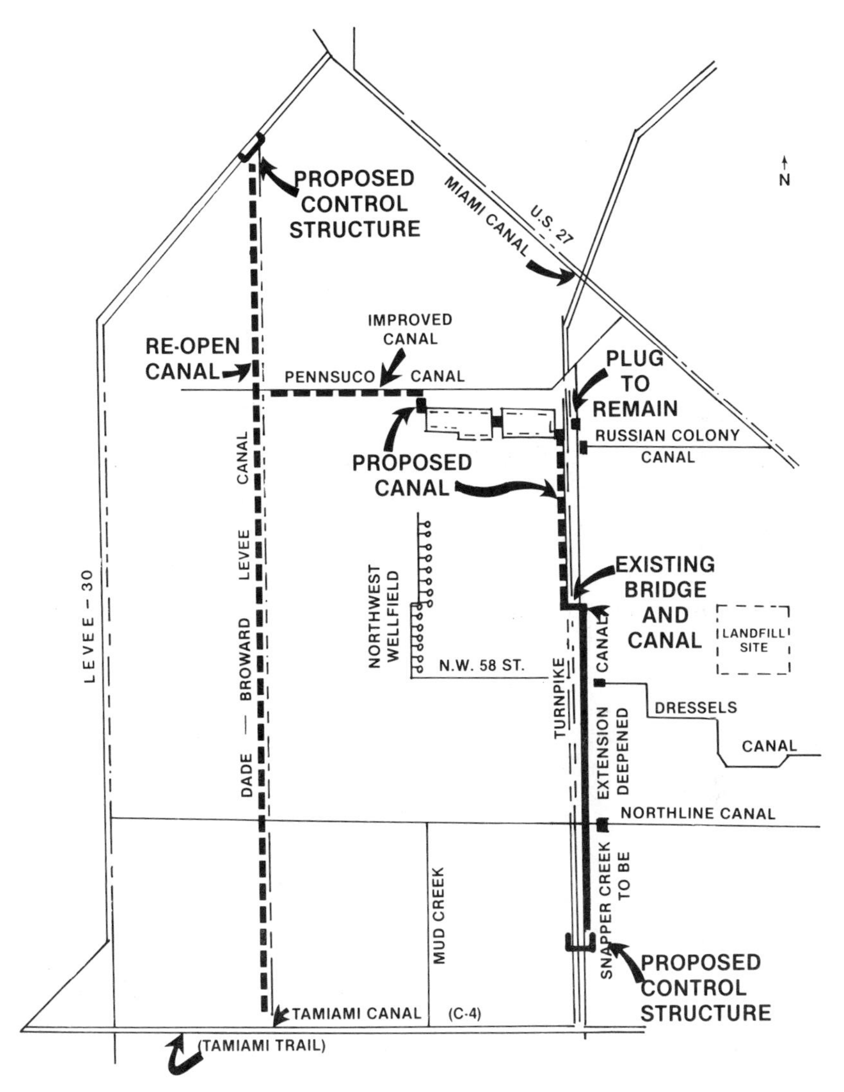

Fig. 11. Proposed canal improvements.

This will, in turn, prevent the flow of degraded groundwater toward the wells under the influence of the hydraulic gradient. The plan includes provision to isolate the canal from adjacent development and its attendant runoff. Also, a safety zone is provided east of the recharge canal to account for reversed groundwater movement during droughts when the desired water surface elevation cannot be maintained in the recharge canal (this is described more fully in Section III,C). The estimated cost of these improvements is $12.6 million in 1985 dollars, and is expected to take 5 to 8 years to design, acquire permits, and complete.

A second water management strategy that will be employed periodically to protect the pristine quality of the Northwest Wellfield is the increased use of the old wellfields in Hialeah and Miami Springs. As described previously, when both the new Northwest and the Hialeah–Miami Springs wellfield complexes are pumping at moderate or high rates, a water divide is established in the region between the two wellfields. Water on the west side of the divide will flow toward the Northwest Wellfield and water on the east side will flow toward the Hialeah–Miami Springs wellfields. Since the Northwest Wellfield began its operation in 1983, groundwater in the vicinity of the polluting landfill is believed to have been migrating in the direction of the new wellfield, albeit at a slow rate. Phase 2 of the Northwest Wellfield Protection Program will begin as soon as air stripping facilities for removal of VOCs are operational. The significant feature of Phase 2 is that pumpage at the Hialeah–Miami Springs Wellfield will be increased to establish a water divide at a location west of the area where groundwater is estimated to have been contaminated by the landfill. Phase 2 and the proportioning of pumpage will continue until the recharge canal improvements are operational.

After Phase 3 begins, use of and blending of water from the Hialeah and Miami Springs Wellfield are expected to occur only during extended drought periods when the recharge canal cannot function as desired. During such drought periods, pumpage from the Hialeah wellfield complex will be increased, while pumpage will be decreased at the Northwest Wellfield to prevent the Northwest Wellfield's cone of influence from expanding into the area where groundwater is estimated to have been contaminated by the landfill.

Reductions in system pumpage and water use restraint comprise the third water management strategy. The Governing Board of the South Florida Water Management District presently has the authority to require mandatory reduction in pumpage when a water shortage exists in the district or a portion of it. Dade's Wellfield Policy Advisory Committee has recommended that the director of the Department of Environmental Resources Management be provided with similar authority to mandate reductions in water use and/or pumpage from the Northwest Wellfield when necessary to prevent its contamination. A comprehensive study is also proposed to determine measures that the county can take to conserve or augment the water supply obtained from the Biscayne Aquifer.

Related closely to the above three water management strategies is an extensive program to monitor the quality and surface elevaton of the canals and groundwater in the Northwest Wellfield area. Areas along the urbanizing eastern margin of the cone of influence will receive particular attention. A network of 21 multidepth well clusters and 6 surface water stations (and 2 of the production wells) will be monitored to detect contamination and trace its migration. Analyses will include indicator parameters, EPA Series 601 and 602, and mass spectrometer analysis, supplemented with an annual areawide geophysical survey using electromagnetics. The network of monitoring stations will also be used to measure the elevation of the canal and groundwater surface to monitor the location of the boundary of the cone of depression and to provide additional data with which to improve the calibration of the county's groundwater flow model. To supplement countywide funding sources, some of the financing for this monitoring program will be generated by fees from permits to be required of nonresidential establishments in the Northwest Wellfield protection area. Other funding will be derived from a 2% surcharge on water/sewer utility service fees that finances countywide water quality monitoring among other pollution control programs.

B. Water and Waste Treatment Facilities

Provisions for the treatment of wastes and water supply constitute the second of the five elements. As referenced in the discussion of the first element, periodic pumping from the Hialeah and Miami Springs wellfields is instrumental in manipulating the eastern boundary of the Northwest Wellfield's cone of influence. Because the water withdrawn at Hialeah and Miami Springs is blended for consumption, improved treatment for removal of VOCs will enable a full range of choice in proportioning pumpage between the two wellfield complexes. The method to be employed is air stripping by countercurrent packed aeration towers. (Editor's note: see Chapter 4 for discussion of aeration towers.) The U.S. Environmental Protection Agency has evaluated the feasibility of removing VOCs by air stripping water from one of the most contaminated production wells in Miami Springs. Vinyl chloride, *cis*-1,2-dichloroethylene, chlorobenzene, 1,4-dichlorobenzene, and 1,2-dichlorobenzene were detected in prior studies in concentrations greater than 1 ppb (highest value was 41.6 ppb for vinyl chloride). The EPA study, according to the Dade County Department of Environmental Resources Management (1984) determined that removal efficiencies at air to water ratios of 100 : 1 ranged from 99.92% for vinyl chloride to 78% for *cis*-1,2-dichloroethylene.

The potential effectiveness in removing trihalomethanes (THMs) was also estimated. The water tested contained only two THMs in concentrations greater than 1 ppb, chloroform (13–17 ppb) and bromodichloromethane (2–3 ppb). Both chemicals were effectively reduced to less than 1 ppb during the test, at moderate air flows. The removal efficiency for THMs at a 100 : 1 air to water ratio was calculated to be approximately 99% by EPA. Table III illustrates the predicted effectiveness of air stripping water treated at the Preston Treatment Plant. The removal rates for individual chemicals as determined by the EPA air stripping study were applied to the contaminants detected in the treated water at the Preston facility in 1983 when it was relying exclusively on the Preston Wellfield. The removal rates assume a 100 : 1 air to water ratio. The results indicated that air stripping would sufficiently reduce volatile contaminants in the Preston Wellfield to a degree that would enable the water after treatment to comply with current drinking water standards. Total reduction of VOCs, as reported by the Dade County Department of Environmental Resources Management (1984), would be from 35.02 to 4.69 ppb (Table III).

Wastewater disposal practices constitute the second area addressed under this element. The original wellfield protection regulation limited the sewage loading of discharges into septic tanks and into sewers. The limitations were based on the distance from the public wells, the type of land use, the type of soil, and the availability of public water service to the site. The loading factors, in turn, translate into limitations on residential density and commercial floor area as illustrated in Table I. The September 1983 revision to the Northwest Wellfield regulations prohibited from the full extent of the cone of influence any new land uses that would use septic tanks. This ban has now been replaced with an allowance for septic tank usage by residences at a density of one per 5 acres provided that the drain field lies above the 1-in-10-years flood elevation. New nonresidential land uses within the cone of influence must utilize public sewers.

As an outgrowth of the September 1983 amendments to the wellfield protection ordinance, the countywide septic tank regulations were also amended to prohibit the approval of any new commercial or industrial zoning or building permits unless the availability of sewers is imminent. A unified countywide sewage system consistent with the comprehensive land use plan is now substantially completed. Industrial pretreatment is also required countywide.

Sewer exfiltration standards have also been strengthened for areas within the cones of influence. The countywide allowance for exfiltration from gravity sewers is 200 gallons per day per inch of pipe diameter per

TABLE III

Effectiveness of Air Stripping Water from Hialeah, Preston, and Miami Springs Wellfields (100 : 1 air to water ratio)

Parameter	Concentration after conventional treatment (ppb)	Standard (ppb)	After air stripping		Concentration after blending with 70% Northwest Wellfield (ppb)
			Percentage removal	Concentration (ppb)	
Vinyl chloride	3.79	1	99.92	0.003[b]	0.001[b]
Vinylidene chloride	0.51	—	98.9[a]	0.006[b]	0.002[b]
trans-1,2-Dichloroethene	8.71	—	78.0[a]	1.92	0.58
1,1-Dichloroethane	3.06	—	98.9[a]	0.03	0.01[b]
cis-1,2-Dichloroethene	12.12	—	78.0	2.67	0.80
1,1,1-Trichloroethane	0.00	200	98.9[a]	—	—
1,2-Dichloroethane	0.00	3	78.0[a]	—	—
Tetrachloromethane	0.00	3	—	—	—
Trichloroethylene	2.97	3	98.9[a]	0.03	—
Tetrachloroethylene	0.00	3	98.9[a]	—	—
Chlorobenzene	0.99	—	98.9	0.01[b]	0.02[b]
O,M,P-Chlorotoluene	1.12	—	98.9	0.01[b]	0.01[b]
M,P-Dichlorobenzene	0.99	—	99.92	0.001[b]	0.01[b]
O-Dichlorobenzene	0.76	—	98.4	0.01[b]	0.003[b]
Benzene	—	1	—	—	
Ethylene dibromide	0.00	0.02	—	—	
THMs	22.5	100	99	0.23	0.23
Totals	57.52			4.92	

Source: Dade County Department of Environmental Resources Management (1984).

[a] Presumed percentage removal determined by comparing Henry's coefficient of each compound to Henry's coefficient of compounds evaluated during EPA air stripping demonstration in February 1984.

[b] Below analytical detection limit for that compound.

mile. In wellfield protection areas, the allowance has been cut in half to 100 gal/day/inch/mile. This suggests that a 10-inch-diameter gravity sewer could leak some 40 gal/hr/mile. As an outgrowth of the Wellfield Protection Study, the standard will be substantially strengthened for the Northwest Wellfield cone of influence to require essentially the same efiltration standard as that required of force mains. This will reduce exfiltration by more than an order of magnitude.

Finally, under this element, the county is seeking to provide facilities for the collection and transfer or disposal of hazardous wastes. Six acceptable sites for collection/transfer facilities have been identified outside of environmentally sensitive areas and incineration is under consideration.

C. Land Use Policies and Controls

The third element in Dade County's program to protect wellfield areas involves the body of land use controls that are not contained within the environment regulations. The county's Comprehensive Development Master Plan (CDMP) provides the foundation of the land use control program. The CDMP Land Use Plan map and accompanying text identify the special use restrictions that prevail in wellfield areas and will be enhanced to specifically depict the regulated wellfield areas.

The long-term land use policy recommended for the Northwest Wellfield protection area will provide for continued limestone quarrying (resulting in numerous deep lakes) between the wellfield and the recharge canals and beyond. Near the up-gradient perimeters of the protection area, areas of residences on 5-acre parcels may occur. In the urbanizing area east of the wellfield, further limitations on commercial and industrial development will be imposed in Subarea 3 (the drought safety zone) immediately east of the improved recharge canal, and temporarily in Subarea 2 (see Fig. 10). When the air stripping facilities enable increased pumpage from the urban wellfields and, subsequently, when the recharge canal improvements become operational, the cone of influence boundary is expected to contract and release Subarea 1 and, subsequently, Subarea 2 from the cone of influence caused by pumping water from the wellfield.

The revised land use policy will allow commercial and office development in Subarea 3 and, additionally, some industrial development in Subarea 2. Permitted uses, however, will be limited to prohibit the handling of hazardous materials, including those uses that might involve hazardous materials as a logical extension of their activity. For example, commercial establishments occasionally exist that install tires and perform wheel balancing but perform no other repair work. This type of business is, however, a rarity. Typically, in Dade County, where a specific land use is permitted, the business evolves over time to expand into related products and services. Tire dealerships, in this instance, will not be permitted because they typically expand into other automotive activities that involve the use of solvents, lubricants, and other hazardous materials. Because of the public imperative in protecting groundwater quality, this policy of prudence seeks to avoid future potential for noncompliance with the prohibition on use of hazardous materials within thousands of acres formerly eligible for warehouse-type developments. This policy will be implemented through the creation of a wellfield protection zoning overlay district that enumerates permitted uses. For a use to be allowed, it must appear in the list of uses permitted in both the wellfield overlay district and the underlying zoning district. Industrial zoning is prohibited in all of the areas except Subareas 1 and 2. Provisions of the underlying zoning will govern all other nonuse parameters such as height, bulk, setbacks, and parking requirements. By including wellfield protection provisions in the zoning code, it is expected that prospective investors/developers will make better informed decisions about use limitations than previously, when all wellfield protection provisions were contained only in the county's environmental regulations.

During the brief period that Subarea 1 remains in the cone of influence, it will be subject to essentially the same land use policy as it has been since 1983, that is, commercial and industrial uses that do not involve the use or handling of hazardous materials will be permitted. The methods of administering this policy will also remain unchanged. Industrial and business zoning may be granted in the area but deed restrictions must be recorded that contain the prohibition on the use or handling of hazardous materials and the requirement that public sanitary sewers must be utilized.

D. Environmental Regulation and Enforcement

Dade County's environmental regulations have historically provided most of the protection for land areas within the boundaries delineated by the wellfield cones of influence. Where the comprehensive plan permits land uses that could have potentially degrading effects of groundwater quality, the environmental regulations minimize these impacts. Many significant measures have recently been expanded to govern development activity countywide, inside and outside of wellfield protection areas. For example, no new industrial or heavy commercial zoning or building permits may be granted anywhere in the county unless public sanitary sewerage is to be used for collection and disposal of wastewater. Pretreatment of industrial or commercial effluent before it reaches the sewer may also be required.

Activities that would generate as little as 50 gallons of hazardous waste per year or that involve a potentially polluting industrial process require an industrial waste (IW) or annual operating (AO) permit from the Dade County DERM. These permits finance periodic inspections and provide the necessary incentive to comply wth prevailing county regulations. In its review of site or building plans, DERM may require design and operational safeguards to minimize the accidental release of contaminants to air or water.

Notable among other countywide requirements are recently enacted regulations governing new and existing underground storage tanks (Dade County Board of County Commission, 1983). The regulations established requirements and criteria for design, installation, modification, repair, replacement, and operation of underground tanks and pipes. Requirements now include corrosion-resistant materials, liners, automatic leak detection devices, and inventory control. Dade County has also enacted a program to permit and regulate liquid waste transporters. The county requires monthly reports from haulers that enable DERM to cross-check origins, destinations, and amounts of wastes hauled. The federal manifest forms are used.

As mentioned previously, the wellfield protection ordinance is a significant component of the environmental regulation. The wellfield protection ordinance prohibits any new use involving hazardous materials; it limits discharge into sewers and into septic tanks, thereby limiting residential density and commercial intensity in close proximity to wellfields (Fig. 12); it requires sewers to be constructed to standards exceeding countywide standards; and it limits the discharge of storm water into the ground.

Protection area boundaries are determined by use of a computerized groundwater flow model. The protection areas initially extended to the 210-day groundwater travel distance from the wells computed under assumed drought conditions. Special wellfield protection area programs have subsequently been established for newer wellfields and tailor boundaries and rules to fit the specific conditions and opportunities in each area. The Northwest Wellfield area has received the greatest scrutiny to date because it remains largely undeveloped. Another wellfield complex in the southwestern part of the county has also had a special program formulated to protect it. Much of the land above its cone of influence has already been urbanized albeit in a suburban residential character. Under prospects of pending industrialization, an expanded

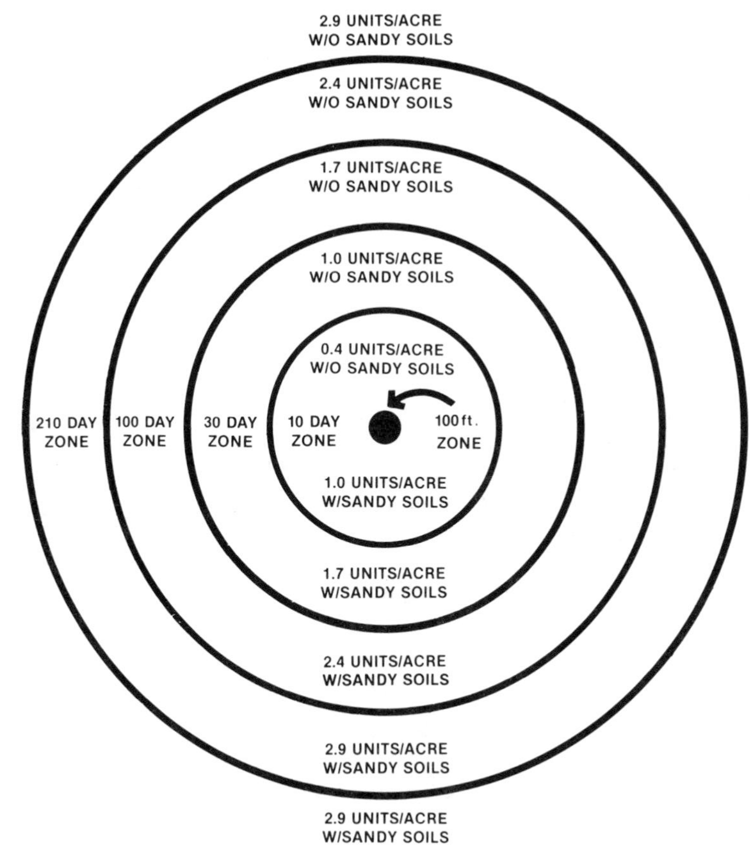

Fig. 12. Density of residential development served by septic tanks permitted in Wellfield Protection areas.

protection program was established similar to that established for the Northwest Wellfield, and limitations on activities that would handle hazardous materials or generate hazardous wastes were imposed on the expanded protection area. Protection programs for other urbanized wellfields are being upgraded using the Northwest and Southwest wellfield protection programs as models.

E. Public Awareness and Involvement

The fifth element of Dade's program seeks to increase involvement of the public. Since many traditional household and business practices are no longer acceptable near wellfields if future degradation is to be averted, a concerted effort at public education and involvement is considered to be more important than ever. Groundwater pollution is essentially invisible to the general public; minute concentrations of many chemicals can render a water supply unfit for potable use; and many activities that can degrade the resource may have previously been accepted as normal business or household practice.

In a concerted effort to improve public awareness of groundwater quality/water supply issues, the Dade County DERM maintains a full-time public information officer to work with business and industry groups, the media, and the school system in informing the public about the need and means for protecting environmental resources—groundwater in particular. The county is also posting information signs along major roadways entering wellfield protection areas to notify the public that the area is a recharge area for their drinking water supply and who to call if polluting activity is witnessed. DERM maintains a 24-hour "pollution hotline" to which anyone who witnesses an apparent pollution violation can call to report the information and receive follow-up instruction and to have an inspector dispatched to the site.

The county, in cooperation with the private sector, has begun a waste oil recycling program, in which a number of publicized facilities such as gas stations and auto parts stores serve as collection stations to receive waste oil from "do-it-yourselfers." The program has been publicized by notices sent out in water/sewer utility bills.

During 1985 the State of Florida conducted a program of "Amnesty Days," in which mobile collection points were established where hazardous waste materials could be brought free of charge. The state then sorted the material and transported it to approved disposal sites. The program was so successful at increasing public awareness as well as collecting hazardous materials that Dade County is considering establishing a similar program of its own. The program would address household materials as an offshoot of the proposed hazardous waste management program directed at commercial waste generators.

IV. CONCLUSION

Dade County, like many other rapidly growing areas, is faced with difficult choices. Ways must be found to accommodate inevitable urban growth. But in doing so, encroachment on areas amenable to development of future public water supplies, areas in the Everglades that are ecologically sensitive, or areas that are devoted to productive agricultural activities may be necessary. In making long-range land use decisions, Dade County encounters these competing objectives whichever way it turns. So a decision to prevent or discourage expansion into one area to achieve one objective causes trade-offs in other areas.

It is, however, becoming evident how much more difficult the problems that will have to be solved will become in the future. Dade County is fortunate that it started early in its effort to protect its new wellfield. Dade is often subject to criticism and might be guilty of overkill, but on the other hand, we would be equally subject to criticism if we do too little to ensure long-term public health. Only time will tell.

REFERENCES

Camp Dresser and McMcKee, Inc. (1982). "Wellfield Travel Time Model for Selected Wellfields in Dade, Broward and Palm Beach Counties, Florida" (unpublished report). Prepared for Dade County Department of Environmental and Resources Management, Broward County Planning Council, and Palm Beach County Area Planning Board, Fort Lauderdale, Florida.

Camp Dresser and McKee, Inc. (1985). "Groundwater Flow Model for the Northwest Wellfield, Dade County, Florida" (unpublished report). Prepared for Metropolitan Dade County Department of Environmental and Resources Management, Fort Lauderdale, Florida.

Dade County Board of County Commissioners (1981). Ordinance 81–23, codified as Sect. 24–12.1. Code of Metropolitan Dade County, Florida.

Dade County Board of County Commissioners (1983a). Ordinance 83–82 amending Sect. 24–12.1. Code of Metropolitan Dade County, Florida.

Dade County Board of County Commissioners (1983b) Ordinance 83–108 creating Sect. 24–12.2. Code of Metropolitan Dade County, Florida.

Dade County Department of Environmental Resources Management (1980). "Protection of Potable Water Supply Wells Program." Dade County DERM, Miami, Florida.

Dade County Department of Environmental Resources Management (1983). "A Survey of Major Public Utilities in Dade County for Twenty-One Volatile Synthetic Organic Compounds." Dade County DERM, Miami, Florida.

Dade County Department of Environmental Resources Management (1984). "Water Quality Considerations for Drinking Water Protection in Dade County, Florida," Wellfield Protection Study. Dade County DERM, Miami, Florida.

Klein, H., and Hull, J. E. (1978). Biscayne aquifer, southeast Florida. *Water-Resour. Invest.* (U.S. *Geolo. Surv.*) **78–107.**

Symons, J. M., Bellar, T. A., Carswell, J. K. *et al.* (1975). "National Organics Reconnaissance Survey for Halogenated Organics in Drinking Water." U.S. Environmental Protection Agency, Cincinnati, Ohio.

8

Wausau, Wisconsin, Case Study

G. WILLIAM PAGE
Department of Urban Planning
and Center for Great Lakes Studies
University of Wisconsin–Milwaukee
Milwaukee, Wisconsin 53201

I. BACKGROUND

The city of Wausau is located in north-central Wisconsin (Fig. 1). The city had a 1980 population of 32,426 and occupies a total area of about 9 square miles divided between both banks of the Wisconsin River. Gently rolling plains comprise the topography with some areas having a marked increase in relief as the result of downcutting by the Wisconsin River and its tributaries. Wausau is the primary retail, financial, medical, and general business center of Marathon County and of north-central Wisconsin. The Wausau metropolitan area includes ten suburban communities. The metropolitan area's population grew 16% between 1970 and 1980 and projections show a growth rate of 1.7% per year to the year 2000 (Wausau Chamber of Commerce, 1984).

Paper manufacturing, the largest industry in Wausau, employs 2500 people in seven firms. These seven paper manufacturers have a payroll of about $64.7 million, 25% of the metropolitan area's payroll (Wausau Chamber of Commerce, 1984). The other major industries in Wausau are metal fabrication, lumber, cheese, and electrical.

The city founded the Wausau Water Works in 1885 with a single well 40 feet in diameter and 32 feet deep (Wausau Water and Sewerage Utilities, 1985). Wausau's municipal water utility operates six high-yield wells positioned in the alluvial deposits of the Wisconsin River flood plain, characterized as the glacial sand and gravel aquifer (Lawrence *et al.*, 1984). These Pleistocene deposits are the remnants of outwash sand and gravel carried into the Wisconsin River valley by the Rib and Eau Claire rivers. All Wausau municipal wells are screened in the outwash deposits and do not extend into the lower bedrock aquifer (Westin,

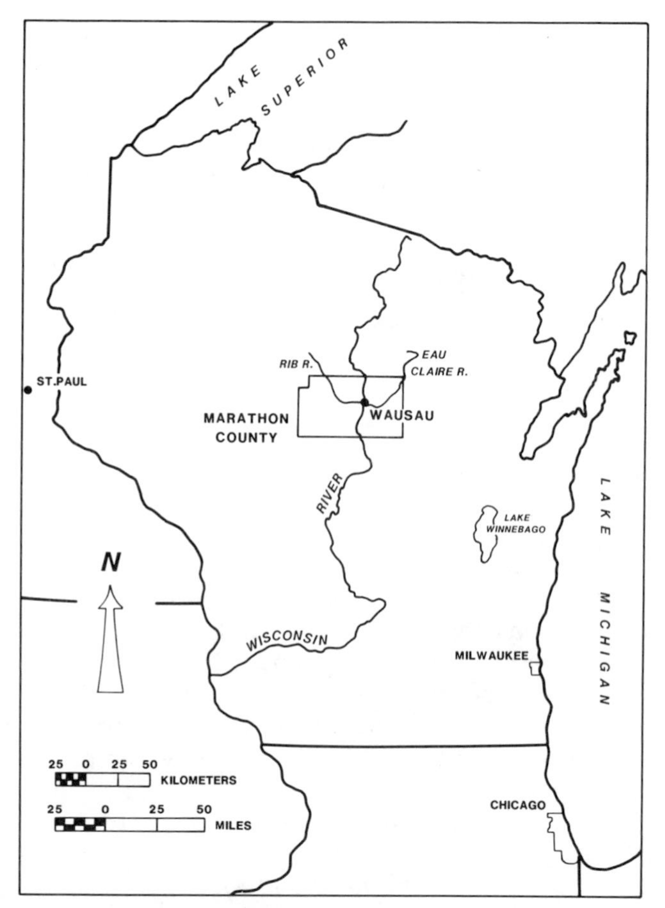

Fig. 1. Location of Wausau, Marathon County, Wisconsin.

1985). The groundwater in the glacial deposits receives recharge from direct infiltration from the land surface and discharges as base flow to the Wisconsin River. Induced recharge from the Wisconsin River has occurred in isolated areas as a direct result of pumping by municipal and private wells (Westin, 1985). Three of the municipal wells are located on the east bank of the Wisconsin River and three wells are located on the west bank. The six wells range in capacity from 500 to 1900 gallons per minute (gpm) and provide the utility with a maximum capacity of 12.9 million gallons per day (mgd) (Westin, 1985; Obey, 1984). Average daily consumption is about 4.72 million gallons with peak demands of 8 million gallons per day (Obey, 1984).

II. SEQUENCE OF EVENTS

Toxic organic contaminants were first discovered in the Wausau water supply accidentally. In early 1982, Zimpro Inc., a local analytical laboratory, installed some new laboratory instruments. One of the employees tested a water sample from his residence and discovered the occurrence of volatile organic contaminants (VOCs). The City of Wausau was notified in March 1982 (Westin, 1985).

In March 1982 the city of Wausau tested for volatile organic compounds in all six of its municipal wells. Initial testing revealed wells 3 and 6 to be contaminated. Later that year, after using well 4 extensively that summer to replace well 3, the city discovered that well 4 was seriously contaminated. These three public supply wells were contaminated with one or more of the following volatile organic compounds: tetrachloroethylene (maximum concentration of 100 parts per billion (ppb)), trichloroethylene (maximum 260 ppb), *trans*-1,2-dichloroethylene (maximum 110 ppb), and toluene (maximum 120 ppb) (Westin, 1985). There is evidence of carcinogenicity for each of these chemicals. The results of sampling in wells 3, 6, and 4 are presented in Tables I, II, and III.

Initial attempts at determining the extent of the groundwater contamination and removing the contaminants through water treatment were unsuccessful. In 1982 and early 1983 Wausau contracted for the installation of seven monitoring wells and for a survey of commercial firms in an attempt to establish the location and possible sources of the contamination. The results suggested a complex problem that would require a much more extensive study. The city also modified one of its water treatment plant aerators, on a trial basis, as a means to volatilize the contaminants. This attempt using aeration proved ineffective at removing volatile organic compounds and was discontinued (Westin, 1985).

Well 4 had the most serious contamination problem as well as being located in the immediate vicinity of known contamination sites. Because of high levels of iron in the water, well 4 is normally used only during periods of high water demand. Well 4 was constructed in July 1966 at a site 100 feet east of the Wisconsin River. The 132-foot-deep well tested at 2000 gallons per minute for 12 hours with 36 feet of drawdown. A significant portion of its recharge is from the Wisconsin River (Westin, 1985). In addition to the river, other potential sources of contamination are the Wausau Chemical Company, the Marathon Box Company, and property owned by the Chicago, Milwaukee, St. Paul, and Pacific Railroad. Well 4 is located 300 feet south of storage tanks on the property of the Wausau Chemical Company, which sells bulk solvents to the dry cleaning industry. In May 1975 during excavation of a foundation on a site immediately adjacent to the Wausau Chemical Company, construc-

TABLE I
Summary of Selected Volatile Organic Contaminant Concentrations in Wausau Well 3, March 1982–November 1984

Collection date	Concentration (ppb[a])		
	PCE[b]	TCE[b]	DCE[b]
3-16-82	100	100	50
4-07-82	90	140	110
4-15-92	80	130	90
6-14-82	40	80	—[c]
6-14-82	50	20	80
6-14-82	60	120	40
6-16-82	50	70	60
7-13-82	60	110	70
8-09-82	60	90	50
9-09-82	20	190	20
9-24-82	30	130	20
9-27-82	20	140	30
9-28-82	30	100	30
10-28-82	20	90	10
2-07-83	10	80	10
5-03-83	10	100	10
6-30-83	10	80	10
6-30-83	10	70	10
7-06-83	10	70	10
11-30-83	10	120	10
1-10-84	10	90	10
1-17-84	10	100	10
2-06-84	10	130	10
2-20-84	10	150	10
5-07-84	10	140	10
9-17-84	10	150	20
10-01-84	0	150	10
10-03-84	10	160	10
10-17-84	0	110	0
11-08-84	0	210	10

Source: Westin (1985).

[a] Reported values have been rounded to two significant figures.

[b] PCE, Tetrachloroethylene; TCE, trichloroethylene; DCE, dichloroethylene.

[c] Nondetectable concentration.

tion workers complained of strong odors. The Wisconsin Department of Natural Resources (DNR) tested the groundwater in the excavation and found traces of toluene, xylene, tetrachloroethylene, and trichloroethylene. In December 1982, the City of Wausau complied with the Depart-

TABLE II

Summary of Selected Volatile Organic Contaminant Concentrations in Wausau Well 6, March 1982–October 1984

Collection date	Concentration (ppb[a])		
	PCE[b]	TCE[b]	DCE[b]
3-16-82	–[c]	80	–
4-07-82	–	110	–
4-15-82	–	110	+[d]
6-14-82	–	120	+
6-14-82	–	70	+
6-14-82	–	140	+
6-16-82	–	110	+
8-09-82	–	120	+
11-15-82	–	190	–
2-07-83	–	120	–
5-03-83	+	150	+
11-30-83	+	180	+
2-20-84	+	210	+
5-07-84	+	140	+
6-30-84	–	210	+
7-03-84	–	220	+
7-04-84	–	220	+
7-05-84	–	220	+
7-06-84	–	220	+
7-07-84	–	220	+
7-08-84	–	230	+
7-11-84	–	190	+
7-18-84	–	80	+
7-26-84	–	160	+
8-02-84	–	180	+
8-08-84	–	170	–
8-15-84	–	190	+
8-22-84	–	220	+
8-30-84	–	260	+
9-05-84	–	220	+
9-12-84	–	200	–
9-17-84	–	140	–
9-19-84	–	170	–
9-25-84	–	170	–
9-26-84	–	190	–
10-01-84	–	110	–
10-17-84	–	190	–
10-24-84	–	150	–

Source: Westin (1985).

[a] Reported values have been rounded to two significant figures.

[b] PCE, Tetrachloroethylene; TCE, trichloroethylene; DCE, dichloroethylene.

[c] Nondetectable concentration.

[d] Less than 10 ppb.

TABLE III
Summary of Selected Volatile Organic Contaminant Concentrations in Wausau Well 4, March 1982–January 1985

Collection date	Concentration (ppb[a])				
	PCE[b]	TCE[b]	DCE[b]	Toluene	Xylene
3-16-82	–[c]	–	–	–	#[d]
4-15-82	–	–	–	–	#
6-14-82	–	+[e]	–	–	#
6-14-82	–	–	–	–	#
6-16-82	–	–	–	–	#
10-28-82	50	60	340	–	#
11-02-82	30	10	300	–	#
6-30-83	20	+	160	–	#
6-30-83	20	+	200	–	#
7-06-83	10	+	200	–	#
10-30-83	20	+	80	10	#
11-30-83	130	80	500	120	#
1-10-84	110	180	410	–	#
1-17-84	90	170	340	90	#
2-20-84	80	190	210	–	#
2-27-84	150	320	380	100	#
5-07-84	80	90	80	30	#
9-17-84	50	70	80	30	#
10-01-84	40	60	90	30	#
10-17-84	40	60	70	–	#
10-29-84	60	70	70	30	#
11-08-84	40	80	70	–	#
12-11-84	40	60	80	20	#
12-19-84	30	70	110	20	#
1-02-85	30	40	80	10	#

Source: Westin (1985).
[a] Reported values have been rounded to two significant figures.
[b] PCE, Tetrachloroethylene; TCE, trichloroethylene; DCE, dichloroethylene.
[c] Nondetectable concentration.
[d] Less than 10 ppb.
[e] Less than 20 ppb.

ment of Natural Resources request that Wausau take well 4 off-line (Kannenberg, 1984). The utility pumped water from well 4 to waste (pumped into the Wisconsin River) in an attempt to remove the contamination and to limit the migration of the plume of contaminated groundwater (Wu, 1984).

After taking well 4 off-line, the remaining wells in the system provided an adequate supply of water through blending water from contaminated wells 3 and 6 with water from contaminated wells 7 and 9. Well 8 was available only as a backup source because of its location, distance from the water plant, and historically high iron content.

Because of high chemical concentrations, Wausau shut down well 3 on April 22, 1982, in response to a request from the Department of Natural Resources regarding the exceeding of the EPA Suggested No Adverse Response Levels (SNARLS) (Westin, 1985). The utility pumped well 3 and a nearby private well to waste 24 hours a day in an attempt to control the plume of contaminated groundwater. By December 1982, the concentrations of tetrachloroethylene and *trans*-1,2-dichloroethylene had declined to trace levels and the Department of Natural Resources gave Wausau permission to put well 3 back on-line (Gehin, 1985). Pumping to waste during 1982 cost an estimated $15,000–$20,000 in electricity (Gehin, 1985).

In early 1984 three things happened to precipitate a crisis in the Wausau water utility. The utility had continued blending water from 1982 into 1984. Volatile organic compound concentrations in the distribution system, after blending, occasionally exceeded the existing health advisory level for trichloroethylene of 45 ppb (Westin, 1985). In early 1984 the National Academy of Sciences Carcinogen Assessment Group established a new health advisory concentration of 28 ppb (10^{-5} excess cancer risk) for trichloroethylene in potable water. Both the U.S. EPA and the Wisconsin Department of Natural Resources recognized this new health advisory level (HAL). Most samples collected throughout the Wausau water distribution system had concentrations of trichloroethylene greater than this new health advisory level (Westin, 1985). The second factor in the crisis was the increasing concentration of volatile organic compounds in wells 3 and 6. At Department of Natural Resources (1984) request, Wausau took well 3 off-line for 2 months and well 6 for 1 month in 1984 (Gehin, 1985). The third factor in the crisis was the deteriorated condition of the pumping system in well 7, which along with well 9 were the only two usable wells without serious volatile organic compound contamination. The pump in well 7 appeared to be in a deteriorated state and would likely require repair. Thus at the beginning of the high water demand summer months, Wausau faced the prospect of being short of potable water. The Wausau Water and Sewerage Commission requested citizens and industry to conserve water. The public received daily information on water consumption rates and the effectiveness of the conservation efforts through the news media (Kannenberg, 1984; *Daily Herald*, 1984a,b). Even with reduced water demand because of the water conservation efforts, the water contamination problems required outside help.

III. SOURCES OF OUTSIDE ASSISTANCE

The state of Wisconsin through its Department of Natural Resources actively attempts to protect groundwater from toxic contamination and to respond to groundwater contamination events. Wisconsin has enacted one of the most thorough and comprehensive groundwater management acts (1983 Wisconsin Act 410) in the nation (Wisconsin Legislative Council, 1984). Unfortunately, Wausau's groundwater contamination problems predated the Wisconsin legislation. When Wausau's contamination problems started, Wisconsin's Department of Natural Resources could provide only limited assistance.

In response to the 1975 discovery of volatile organic compounds in the groundwater of an excavation site adjacent to the Wausau Chemical Company, the Department of Natural Resources ordered the Wausau Chemical Company to stop all activities that would release contaminants to the soil. The Department of Natural Resources ordered the Wausau Chemical Company to pressure test all their solvent storage tanks to identify any leaks and required the company to file a report on the extent of the contamination and the parameters of the groundwater flow patterns under the Wausau Chemical Company and surrounding properties. Detailed plans and engineering reports of steps to eliminate leakage of contaminants into the soil waters on and around its facilities were also required.

The Wausau Chemical Company was a prime focus of the investigations because of the contamination of groundwater in the excavation site in 1975 and because of two reported spills of tetrachloroethylene. The first spill occurred in February 1983 when a valve broke on a 285-gallon delivery truck during off-loading, in which approximately 135 gallons spilled (Westin, 1985). In December 1983, a second spill released 500 gallons of tetrachloroethylene. Because of these spills, the Wisconsin Department of Natural Resources requested that Wausau Chemical Company contract to have monitoring wells installed and samples collected and analyzed.

Wausau contacted the U.S. Environmental Protection Agency for help early in 1982, soon after discovery of the first volatile organic compound contamination. Wausau applied for and was awarded a U.S. EPA Cooperative Agreement for developing a long-term treatment system for the removal of volatile organic compounds from Wausau's water. The Cooperative Agreement specified that U.S. EPA would provide 90% and Wausau 10% of the costs (Westin, 1985). The Agreement called for the design, installation, and testing of both a granular activated carbon filtration unit and a packed-column air stripping tower. EPA Cooperative

Agreements are contracts for research activities conducted by the Office of Research and Development (ORD) Laboratories of EPA. Though Wausau encountered a major crisis because of toxic contamination of its water supplies, the timing of the crisis was propitious. The Drinking Water Research Division of the Municipal Environmental Research Laboratory of EPA was interested in finding a research project to study the effectiveness and cost of different technologies for removing volatile organic compounds from municipal water supplies containing a complex mixture of organic contaminants. From the EPA research laboratory perspective, Wausau offered the right problem at the right time. Wausau, Wisconsin became the site of EPA's rigorously controlled scientific analysis of volatile organic compound removal from municipal water supplies. EPA completed the evaluation of air stripping (Hand *et al.*, 1986), of a small 1 to 2 gallons per hour granular activated carbon unit, and a field GAC unit with about 100 gallons per minute capacity started in the fall of 1985 (Gehin, 1985).

Wausau again became a study for an unresolved problem concerning volatile organic compound removal from municipal water supplies in 1985 when the American Water Works Association Research Foundation agreed to fund a study of toxic air contaminants released from air stripping towers. The American Water Works Association agreed to provide 90%, $150,000, and the Wausau Water Utility will provide 10%, $10,000, of the funding for this study. This research project will last into 1986 and will help determine the effectiveness and costs of using a granular activated carbon filtration unit to remove volatile organic compounds from the exhaust air flow from an air stripping tower (American Water Works Association Research Foundation, 1985).

When the water supply crisis of the demand exceeding the available noncontaminated supply threatened in the spring of 1984, the Wisconsin Department of Natural Resources, on Wausau's behalf, again turned to the U.S. EPA for help. Wausau notified the Region V office of EPA of the impending crisis in May of 1984. Assessment by the Technical Assistance Team contractor confirmed that the water utility might not provide an adequate supply of safe water during the summer of 1984 (Westin, 1985). With this result, EPA started an Emergency Response action under the Comprehensive Environmental Response, Compensation, and Liability Act (CERCLA), popularly known as the Superfund act (Pub. L. 96–510). There are two categories of government response under CERCLA: (1) Emergency Response actions, which are life-threatening situations that need immediate EPA involvement but normally will last less than 6 months and cost less than 1 million dollars, and (2) Remedial Response actions, which are usually extremely serious long-term cleanups, but not an immediate threat to life. In Wausau, the Emergency Response status enabled EPA to work simultaneously for two objectives: (1) to assure an uninterrupted supply of uncontaminated water to the residents of Wausau until completion of a permanent air stripping facility or granular activated carbon filtration facility as a result of the EPA Cooperative Agreement research project and (2) to characterize the extent of the groundwater contamination, identify potential contamination sources, and evaluate groundwater rehabilitation alternatives.

To achieve the first objective, EPA brought in a mobile granular activated carbon (GAC) filtration system for use on municipal well 6. EPA selected well 6 because it combined high yield with accessibility for retrofitting the facility with granular activated carbon. The mobile granular activated carbon system consisted of four contact vessels, each containing 20,000 pounds of granular activated carbon (Westin, 1985). This system was capable of treating 1.8 million gallons of water per day. The mobile granular activated carbon unit operated from July 2, 1984, to October 29, 1984, when the air stripping unit began operation (Westin, 1985). Original estimates were that this unit could successfully remove the concentrations of volatile organic compounds in Wausau's groundwater for 1 year before the carbon would need regeneration. After 5 months, the unit reached breakthrough with measurable concentrations of volatile organic compounds reported after treatment by the mobile granular activated carbon unit (Gehin, 1985).

In June 1984, in pursuit of the second Superfund objective, EPA assigned the Technical Assistance Team contractor, Roy F. Westin, Inc., the task of investigating the hydrogeology of the Wausau contamination problems. This investigation included (1) a seismic refraction survey to map the bedrock configuration and the aquifer configuration, (2) an industrial survey to identify potential users and/or source areas for volatile organic compounds, (3) a survey of past and present landfills, (4) the use of monitoring wells to identify possible source areas, (5) the sampling of river sediments for volatile organic compounds, and (6) an effort to monitor contaminant concentrations in groundwater.

The investigation of groundwater contamination near Wausau well 4 revealed complex groundwater flow patterns and multiple sources of toxic contaminants. The size of the well 4 cone of influence is a function of river stage and pumping regime. The interaction of the fluctuating elevation of the Wisconsin River recharge boundary and the pumping rates is a major factor in the concentration and movement of toxic contaminants. Induced recharge caused by high river levels dilutes contaminants in the cone of influence of well 4 (Westin, 1985), resulting in an inverse relationship between river stage and concentration of toxic contaminants in groundwater. Sampling data from monitoring wells show that a plume of contaminated groundwater emanates from the Wausau Chemical Company property and reaches Wausau well 4 (Westin, 1985). Contaminated groundwater emanating from the Marathon Box Company and possibly also from the Chicago, Milwaukee, St. Paul, and Pacific Railroad property also reaches Wausau well 4 (Westin, 1985).

Wausau municipal well 3, also on the east bank of the Wisconsin River, had been closed as a result of toxic contamination problems. A short-term release of volatile organic compounds is presumed to have occurred at an unknown source up-gradient from well 3 in an east-southeast direction. The existence of a large industrial well located between the presumed source and well 3 acts as a partial barrier and minimizes the impact of the contamination on Wausau municipal well 3 by intercepting most of the plume of contaminated water (Westin, 1985).

Wausau municipal well 6 on the west bank of the Wisconsin River has had contamination problems with trichloroethylene. The EPA Superfund-sponsored investigation could not positively identify the source (Westin, 1985). The source may be a slow leak from a bulk storage tank or landfilled material, or it may be the result of a one-time spill.

The volatile organic compound contamination problems in Wausau, Wisconsin, are also being considered for a Remedial Response action under CERCLA. In 1985 the Wausau site is proposed for the National Priority List awaiting action to designate Wausau for long-term cleanup under Superfund. With designation as a Remedial Response site, EPA will undertake cleanup, further investigation of the sources of contamination, and legal proceedings to recover costs.

In 1983 the Wausau Water Utility built a second air stripping tower to remove volatile organic compounds from its groundwater sources of potable water. The first air stripping tower had been constructed as part of the EPA Cooperative Agreement. The 23-foot tower is 8.5 feet in diameter with a 15-horsepower blower. The blower forces air up through the tower as the water falls down through the tower. The tower contains 6-inch plastic saddle-shaped objects that produce turbulence when the water cascades over them. The first air stripping tower has a rating of 1500 gallons per minute (0.09464 m/sec), an estimated total capital cost of $120,107, and a total daily operational cost of $25 based on a power cost of $0.55/kw-hour (Hand *et al.*, 1986). It should be noted that a large portion (35%) of the capital cost was for a support structure designed so that the treated water from the air stripping tower would be gravity fed into the treatment plant. The estimated total annual cost of treatment is $0.0586 per 1000 gallons of treated water (Hand *et al.*, 1986). Independent water lines from the wells into the water treatment plant dictated a second air stripping tower. The second air stripping tower stands 23 feet high, but with a 9.5-foot diameter and a 25-horsepower blower. The second air stripping tower has a rating of 2000 gpm (0.1261 m/sec) (Hand *et al.*, 1986). These two units treat about one-third of Wausau's water supply and remove about 99% of the volatile organic compounds (Boers, 1985). Figure 2 is a photograph of the two air stripping towers.

The air stripping towers have performed well, but some problems encountered have required modifications to the water treatment system. Winter operation of an outdoors air stripping tower in northern Wisconsin has not been a problem. The tower has a special valve to drain the influence line so that water does not stand in the pipe when not in operation. Even when not in operation overnight in subzero temperatures, the system started without problems (Boers, 1985). Naturally high concentrations of iron in the groundwater have caused some problems. The plastic packing saddles that produce turbulence in the air stripping towers receive a buildup of primarily iron oxide (Hand *et al.*, 1986).

Fig. 2. The two air stripping towers in Wausau, Wisconsin.

Washing an acid solution through the towers every 2 or 3 years should keep this inorganic buildup from interfering with the operation of the towers (Boers, 1985). The increased oxidation of the naturally occurring iron in Wausau water also caused a problem in the flocculation stage since the iron floc did not readily settle. The utility solved the problem by increasing the pH of the water before alum addition (Boers, 1985). The remaining steps of the treatment sequence consist of lime and silica addition to achieve flocculation, recarbonation to lower pH, chlorination, and finally filtration.

The Wausau Water Utility wants to develop a new high-volume well to ensure an adequate supply of safe drinking water. Wausau floated a $2.3 million bond issue to cover the cost of well development. Together with an Urban Development Action Grant (UDAG) of $650,000, the money will fund improvements in the distribution system (Gehin, 1985). The Wisconsin Public Service Commission granted a small rate increase in 1984, and in 1985 another rate increase application was prepared by the Wausau Water Utility (Gehin, 1985).

IV. PROTECTION PLAN ELEMENTS

Several elements of a groundwater protection plan for the Wausau area exist. There is no single groundwater protection plan and the existing groundwater protection elements have been implemented by different government bodies. A description of the different protection plan elements follows.

The Wisconsin Department of Natural Resources in 1985 began developing a Statewide Groundwater Management Plan. Marathon County, with Wausau as the county seat, and Rock County will be the first areas of the state with detailed, area-specific groundwater management plans. (Editor's note: see Chapter 5 for a description of the Rock County plan.) Because of limited financial resources, the Department of Natural Resources will prepare these detailed groundwater management plans only for those counties scoring high on three criteria: (1) having the highest susceptibility/potential for groundwater contamination, (2) having numerous cases of groundwater contamination, and (3) having high population potentially at risk as a result of groundwater contamination. The Department of Natural Resources selected Marathon County as the first county in Wisconsin to receive a Groundwater Management Plan because it has serious groundwater problems meeting all three of the state's criteria. Much of the county is within the Central Sands physiographic region of north-central Wisconsin with highly permeable superficial aquifers. Nitrate and the pesticide aldicarb contaminate groundwater in agricultural areas. Serious volatile organic compound contamination of groundwater exists in several municipalities in addition to the city of Wausau. Parts of the detailed, area-specific groundwater management plan for Marathon County are in place. Completion of the plan is expected in early 1986.

The groundwater management plan for Marathon County has five major sections. The first is a Physical Resource Definition. The team collected data for this section of the plan from a variety of sources, including Department of Natural Resources, U.S. Geological Survey, Wisconsin Geological and Natural History Survey, the U.S. Soil Conservation Service, and Marathon County. The data include groundwater countour maps, aquifer potential maps, irrigable lands projections, soils descriptions, bedrock geology, areas with greater than 40 feet of unconsolidated materials over bedrock, wetland maps, floodplain maps, locations of landfills, locations of all gravel pits, surface water drainage, and high-capacity and drilled well logs. The team will develop groundwater contamination potential maps for Marathon County from these data.

The second element of the plan is an Inventory of Contamination Sources. This element will include a listing of all known sources of toxic substances that potentially could contaminate groundwater in the county. The team will develop this listing from a variety of sources, including Wisconsin Pollution Discharge Elimination System (WPDES) permit files, hazardous waste manifest system records, spill files, agricultural chemical use data, chemical storage information, municipal and industrial sludge spreading files, nonpoint source discharge information, fuel storage site data, abandoned landfill inventory, and other sources. The team will use this data in developing a ranking system that will take into account the potential for groundwater contamination severity based on source, pollutant toxicity, effect of the physical environment on contamination potential, and estimated environmental fate of the pollutant.

The third element of the groundwater management plan for Marathon County is Monitoring. This element will list all current and planned groundwater monitoring and set priorities for future groundwater monitoring in the county. An analysis of physical resource data, the inventory of potential contamination sources, and the estimated severity of potential impacts on humans and the natural ecosystem will determine future monitoring priorities.

The fourth element of the plan is Management Recommendations. The team will analyze each problem area identified and develop a set of recommendations specific to that problem area. The recommendations will include regulatory options applicable at the state, county, and municipal levels and site-specific "best management practices."

The fifth and final element of the Marathon County Critical Area Groundwater Plan is Public Involvement. The importance of groundwater protection in Wisconsin makes public support for the plan essential for effective implementation. The county will encourage public involvement through frequent media coverage and through a series of public information meetings.

Marathon County and the city of Wausau are also cooperating on developing techniques to protect groundwater from contamination with toxic chemicals. This cooperation is institutionalized at some levels. For instance, the city of Wausau does not have its own planning department but has an arrangement with the Marathon County Planning Department, which is located in Wausau. As part of the bylaws that created the Marathon County Planning Commission, the city receives priority attention equivalent to one full-time planning staff position. There have been numerous discoveries of serious organic chemical contamination of groundwater supplies in Marathon County in the past few years, and developing groundwater protection plans has become a major concern of the Marathon County Planning Department. While the major effort of the Planning Department in this area has been in participating in the development of a county groundwater management plan in cooperation with the statewide efforts of the Wisconsin Department of Natural Resources, the Marathon County Planning Department also has been the initiator of important groundwater protection actions.

The Marathon County Planning Department has developed the first model zoning ordinance for protecting groundwater in Wisconsin. This ordinance was developed initially for the town of Rib Mountain and adopted by the town board in September 1985 (Marathon County Planning Department, 1984). The ordinance protects groundwater resources by imposing appropriate restrictions on land use activities located within the groundwater recharge area of the town's municipal wells. These restrictions are a special form of zoning imposed on top of all existing residential, commercial, or industrial zoning districts. The ordinance establishes two zoning districts within the 1400-acre area designated as the recharge zone for municipal wells. Zone A is considered critical to the municipal well fields and Zone B is still important to the well field, but poses a reduced potential for influencing the wells. In Zone A, 34 specific activities have been prohibited (see Table IV). In Zone B, only underground petroleum storage tanks for industrial, commercial, residential, or other uses are prohibited. In both zones, any nonprohibited commercial or industrial use must obtain a special permit before locating in the area. A permit will be granted or denied by the town board based on the likelihood that the proposed use could cause groundwater contamination problems. The County Planning Depart-

TABLE IV

Restricted Activities in Zones A and B[a]

Activities restricted in both Zone A and Zone B

1. Areas for dumping or disposal of garbage, refuse, trash or demolition material
2. Asphalt products manufacturing plants
3. Automobile laundries
4. Automobile service stations
5. Building materials and products sales
6. Cartage and express facilities
7. Cemeteries
8. Chemical storage, sale, processing or manufacturing plants
9. Dry cleaning establishments
10. Electronic circuit assembly plants
11. Electroplating plants
12. Exterminating shops
13. Fertilizer manufacturing or storage plants
14. Foundries and forge plants
15. Garages—for repair and servicing of motor vehicles, including body repair, painting or engine rebuilding
16. Highway salt storage areas
17. Industrial liquid waste storage areas
18. Junk yards and auto graveyards
19. Metal reduction and refinement plants
20. Mining operations
21. Motor and machinery service and assembly shops
22. Motor freight terminals
23. Paint products manufacturing
24. Petroleum products storage or processing
25. Photography studios, including the developing of film and pictures
26. Plastics manufacturing
27. Printing and publishing establishments
28. Pulp and paper manufacturing
29. Residential dwelling units on lots less than 15,000 square feet in area. However, in any Residence District, on a lot of record on the effective date of this ordinance, a single family dwelling may be established regardless of the size of the lot, provided all other requirements of the Rib Mountain Zoning Ordinance are complied with
30. Septage disposal sites
31. Sludge disposal sites
32. Storage, manufacturing or disposal of toxic or hazardous materials
33. Underground petroleum products storage tanks for industrial, commercial, residential or other uses
34. Woodworking and wood products manufacturing

Conditional uses in both Zone A and Zone B

The following conditional uses may be allowed in the Municipal Well Recharge Area Overlay District, subject to the provisions of Article VI, Section 6.10

1. Any other business or industrial use not listed as a prohibited use
2. Animal waste storage areas and facilities
3. Center - pivot or other large scale irrigated agriculture operations

TABLE IV (*Continued*)

Prohibited uses in Zone B only
- Underground petroleum products storage tanks for industrial, commercial, residential or other uses

Conditional uses in Zone B only
- The following conditional uses may be allowed in the Municipal Well Recharge Area Overlay District, subject to the provisions of Article VI, Section 6.10
  - Any business or industrial use

[a] Prepared by Joe Pribanich, Marathon County Planning Department, November 14, 1984, and revised April 25, 1985.

ment actively encourages other municipalities in the county to adopt their own modified versions of this model ordinance.

The Marathon County Health Department has also been active in planning for groundwater protection as part of its objective of encouraging "preventive health" programs. This activity has been concentrated in three areas: (1) a water quality education program, (2) a groundwater testing program, and (3) a high-risk area identification and analysis program (Wittkopf, 1985). The County Health Department hired a groundwater specialist to accomplish these projects. The perspective of the Health Department was broad. It has been looking at private wells in addition to municipal wells, and has been interested in nontoxic water quality contaminants such as nitrates and bacteria as well as volatile organic compounds, pesticides, and heavy metals.

The educational program aspects of the Marathon County Health Department activities are especially noteworthy. The program teaches county residents about groundwater. The program operates as a two-step process in selected towns. In the first step volunteer participants have water samples from their private wells tested. The second step involves an educational meeting with discussion of the analytic results and explanation of groundwater systems. The goal is providing the participants with an understanding of why improper use and disposal of potential contaminants can affect their own drinking water. Participation has been free and both phases of the program have had good numbers of volunteers. The hazardous health risk identification and assessment program, which includes a contaminants trend mapping component, has not progressed as far as the educational program, but represents an advance planning approach to groundwater contamination problems that attempts to protect the public's health from the risk of exposure to toxic chemicals in drinking water.

The Marathon County Office of Emergency Preparedness, in cooperation with the County Health Department and the City of Wausau Fire Department, has developed a hazardous materials inventory for the county. This inventory resulted from a 5-month survey of all hazardous materials stored, tank capacities, locations, and procedures needed to respond to fires, spills, or leaks of these materials. The county updates inventories every 2 years.

The last element in a groundwater protection plan for the Wausau, Wisconsin, area has to do with the development of new well fields and future expansion of the water systems. The City of Wausau and other municipalities in the area have experience toxic contamination problems in their groundwater supplies. Developing new wells is a more difficult problem in Wausau because the city is almost completely developed and has few locations for a new well field. The City of Wausau is testing potential sites for a new high-capacity well, and if an adequate site is not found, the city may investigate developing a well field outside its city boundaries in a separate municipality (Gehin, 1985). At least three adjacent municipalities also have experienced toxic contamination problems and face increased operating costs to contend with the contamination. All these municipalities face the prospect of developing new well fields in the near future. These towns have discussed the possibility of a regional water system. Advantages of a regional system include economies of scale in the treatment of water to remove toxic chemicals, developing new well fields, and protecting the recharge areas of the new well fields from activities that may release potential contaminants.

REFERENCES

American Water Works Association Research Foundation (1985). "Recovery of Volatile Organic Air Stripping Towers Using Gas Phase Adsorption" (Res. Proj. in City of Wausau, Wisconsin), Contract No. 83–84. AWWARF, Wausau, Wisconsin.

Boers, Richard. (1985). Personal interview with Wassau Water Utility Water Plant Supervisor. October 11, Wassau, Wisconsin.

Daily Herald (1984a). June 9, Wausau, Wisconsin.

Daily Herald (1984b). June 16, Wausau, Wisconsin.

Department of Natural Resources (1984). News release, June 15. DNR, Madison, Wisconsin.

Gehin, J. (1985). Personal interview with Director of Wassau Water and Sewerage Utilities. October 11, Wassau, Wisconsin.

Hand, D., Crittendon, J., and Gehin, J. (1986). Design and economic evaluation of a full scale air stripping tower for treatment of VOCs from a contaminated groundwater. *J. Amer. Water Works Assoc.* (in press).

Kannenberg, J. L. (1984). Letter to Wisconsin Department of Natural Resources, June 25. John Kannenberg, Mayor, City of Wausau, Wisconsin.

Lawrence, C. L., and Ellerson, B. (1982). Water use in Wisconsin, 1979. (*U. S. Geol. Surv.*) *Water Resour. Invest.* **82–444,** 1–100.
Lawrence, C. L., and Ellerson, B., and Cotter, J. (1984). Public supply pumpage in Wisconsin, by aquifer. *Geol. Surv. Open-File Rep.* (*U.S.*) **83–931.** 1–40.
Marathon County Planning Department (1984). Rib Mountain Municipal Well Recharge Area Overlay District Zoning Ordinance. Wausau, Wisconsin.
Obey, D. R. (1984). Letter to the Appropriations and Joint Economic Committees of the U.S. Congress, David Obey, Congressman.
Wausau Chamber of Commerce (1984). Marathon County economic development kit. WCC Wausau, Wisconsin.
Wausau Water and Sewerage Utilities (1985). "Annual Report 1984." WWSU, Wausau, Wisconsin.
Westin, R. F. (1985). "Hydrogeological Investigation of Volatile Organic Contamination in Wausau, Wisconsin, Municipal Wells" (Report prepared for U.S. EPA Region V), Contract No. 68-95-0017. Wausau, Wisconsin.
Wisconsin Legislative Council (1984). "The New Law Relating to Groundwater Management," Info. Memo. 84–11. WLC, Madison, Wisconsin.
Wittkopf, T. (1985). Personal interview with Director of Environmental Health, Marathon County Health Department, October 11.
Wu, B. (1984). Immediate Removal Request for the City of Wausau, Wisconsin (Letter to the Wisconsin Department of Natural Resources).

9

Urban Growth Management and Groundwater Protection: Austin, Texas

KENT S. BUTLER
Department of Natural Resource Policy and Programs
Lower Colorado River Authority
Austin, Texas 78767

I. INTRODUCTION

Austin, Texas, has recognized that the greatest opportunity to prevent declining groundwater quality is through proper location, design, construction, and maintenance of new urban development and its associated drainage systems. Increasing development in the environmentally sensitive watersheds south and west of Austin resulted in the promulgation of a series of innovative watershed development ordinances in the early 1980s (Butler, 1983). These regulations are designed to protect the water quality of the Edwards Aquifer, a unique karstic limestone system. This process of planning for the protection of groundwater is the subject of this chapter.

Specifically, this chapter concerns the effects that the ordinances for several watersheds and the associated aquifer will have on suburban land development and water quality protection for the Edwards Aquifer. These ordinances are salient examples of a new, largely untested venture into urban runoff pollution control using a combination of engineering and land use management techniques. The immediate motivation for adopting these techniques is protecting groundwater and spring discharges from degradation and contamination with nontoxic chemicals. Austin also recognized that toxic contaminants may become a threat to groundwater. The combination of engineering and land use management techniques as adopted in Austin's ordinances should protect the groundwater from organic and inorganic microcontaminants as well as the more common pollutants associated with urban storm runoff.

This chapter addresses three key questions: Why is the Edwards Aquifer area so deserving of special protection in the face of urban expansion? How do these development standards affect the planning of new subdivisions and site developments? And how do groundwater quality protection standards operate in the broader context of growth management in this rapidly urbanizing region of central Texas?

II. THE EDWARDS AQUIFER AND ASSOCIATED WATERSHEDS

A. Geohydrological Setting

The Edwards Aquifer is a complex geologic and hydrologic unit that extends across parts of south-central and west Texas. Today, the aquifer is a karst system—a series of water-bearing layers of cavernous, fractured, and honeycombed layers of limestone (see Figs. 1 and 2) that hold and bear vast quantities of remarkably clean water. Originally deposited about 100 million years ago as sedimentary layers in a shallow sea, the Cretaceous-aged limestone has been reworked by earthquakes, the dis-

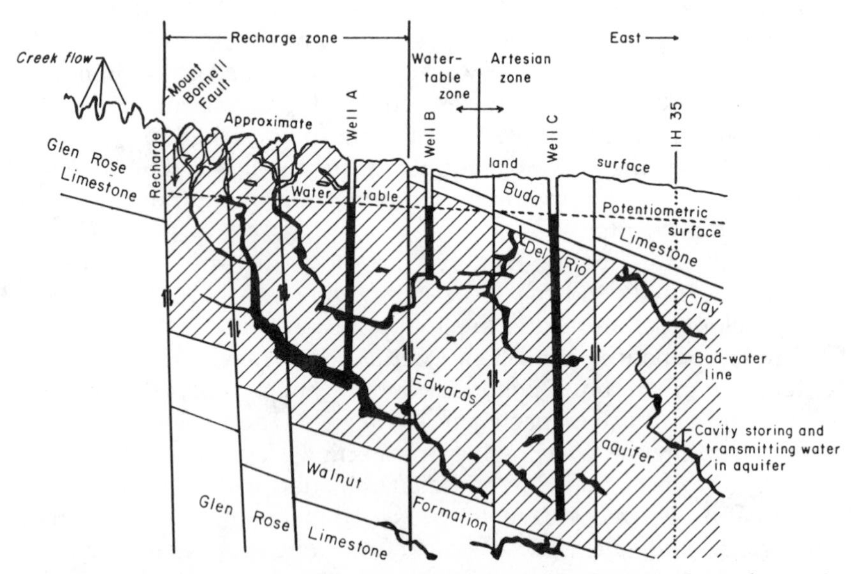

Fig. 1. Conceptual cross section of the Edwards Aquifer depicting the geology, solution cavities, and water table conditions (after Woodruff and Slade, 1985).

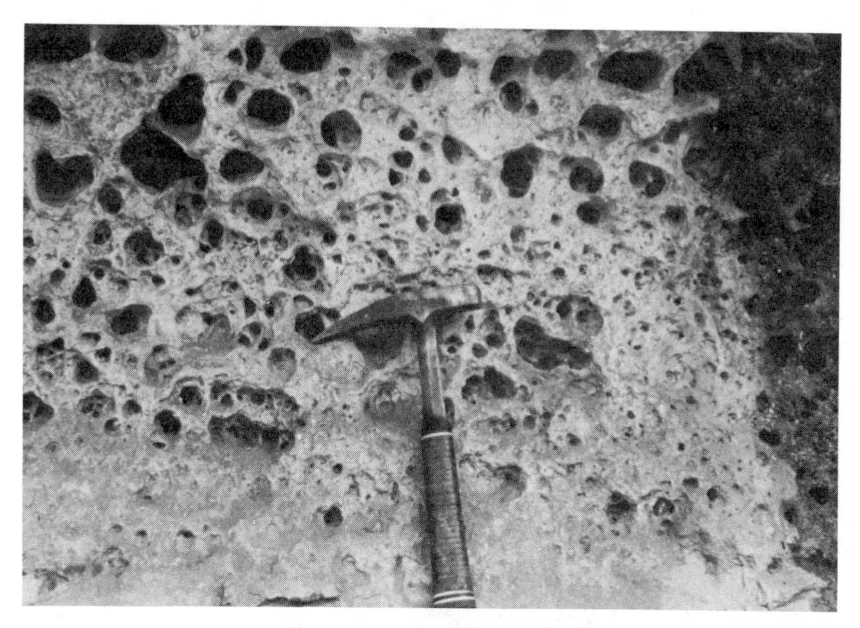

Fig. 2. Exposed surface of the Edwards Limestone showing honeycombed porosity.

solving forces of carbonic acid in rainwater, and downward cutting by streams. Extensive seismic activity during the Miocene epoch created an area of extensive faulting, which today is known as the Balcones Fault Zone, a series of stair-stepped fault blocks with vertical displacements of as much as 200 feet. The hydrologically discrete Austin unit of the aquifer encompasses approximately 155 square miles, extending from northeastern Hays County into south-central Travis County, generally following the Balcones Fault Zone.

The Edwards Limestone crops out in the western half of the aquifer, creating a recharge zone with free water table conditions. There are five surface watersheds that are in hydrologic communication with the aquifer by virtue of the fact that the creeks traverse the aquifer recharge zone from west to east (see Fig. 3). When creeks are flowing in the watersheds that traverse the Austin unit of the aquifer, much of the surface flow is lost through faults and fractures into the creek beds, thereby recharging the aquifer (see Figs. 4 and 5). Approximately 85% of the recharge occurs in the beds of main channels of creeks, while the remainder occurs on tributaries to the main channels or through diffuse upland infiltration (Slade *et al.*, 1985).

The formation extends and dips to the east of the fault zone. It is confined by relatively impermeable clay and limestone layers, creating

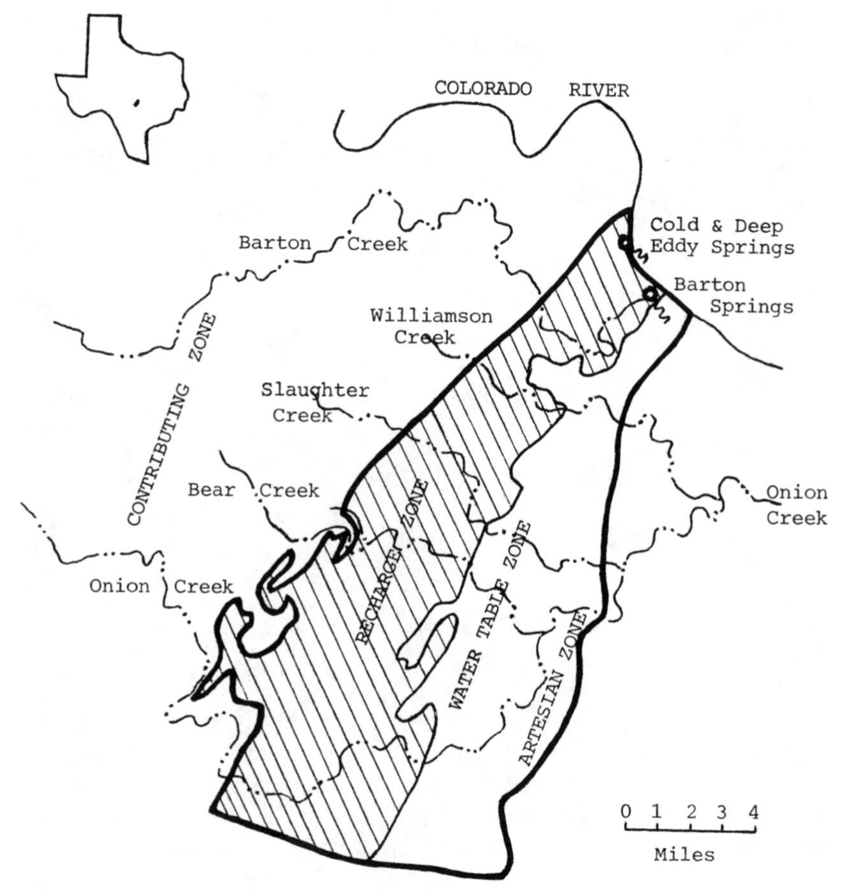

Fig. 3. Map of creeks and watersheds in the Barton Springs-associated Edwards Aquifer area.

an artesian zone (see Figs. 1 and 3). Wells drilled into faulted areas or solution cavities may produce sustained flows of several thousand gallons per minute with minimal drawdown. The Austin unit of the aquifer drains to the northeast, releasing approximately 85% of the total recharge volume through Barton Springs. The average discharge through Barton Springs, which is located in a pool in central Austin, is about 30 million gallons per day. The remaining discharge is accounted for by well pumpage and discharges through other natural springs in the area. For the most part, the quality of water in the aquifer is excellent because the upstream drainage and recharge areas are relatively undeveloped.

Fig. 4. Photograph of Barton Creek, upstream of the recharge zone, under normal flow conditions.

B. Urban Development and Its Possible Consequences

Urban storm water is a significant source of nonpoint pollution. Its adverse impact on both surface and underground receiving waters is widely appreciated by professionals and public officials involved in land use and water resource programs (American Public Works Association, 1981). The control of storm water runoff and pollution necessarily involves the regulation of land uses that degrade water quality or increase the volume of runoff.

Water quality in the Edwards Aquifer and its contributing watersheds is particularly sensitive to urban expansion and nonpoint pollution. Runoff from urban streets, construction sites, and other areas of intensive use often contain high bacterial concentrations, nutrients, suspended solids, petroleum residues, metals, pesticides, and other deleterious substances. High concentrations of pollutants are generally associated with densely populated and/or industrial areas and drainage areas with construction activity. Recent studies conducted in the Austin area, under the Nationwide Urban Runoff Program, have shown that the amount of pollutants in storm runoff from medium-density residen-

Fig. 5. Photograph of Barton Creek, below the recharge zone, under normal no-flow conditions.

tial areas can be five to ten times greater than that from low-density residential areas (City of Austin and Engineering Science, Inc., 1982). Most of the urban runoff reaches surface and ground waters through diffuse flows and seepage and not through a pipe or other point source of pollution.

The Edwards Aquifer is probably as sensitive and vulnerable to contamination as any in the south-central United States. The recharge area is typified by exposed rock surfaces and faulted dry stream beds with little or no soil mantle to filter potential contaminants in surface waters. Contamination of aquifers such as the Edwards is quite simple in such cases—recharge waters flow directly through crevices and cracks into the aquifer without being filtered. Rates of infiltration and diffusion of liquids through the unsaturated zone and the aquifer are possibly faster and more extensive than any other in the state of Texas.

Fortunately, there have been few reported events of groundwater contamination in the aquifer. One set of incidents of unsafe bacterial concentrations in Barton Springs Pool was caused by a leaking sewer line in the aquifer area. In a study of 38 wells that were tested for water quality constituents, the U.S. Geological Survey identified 12 wells with abnormally high counts of fecal streptococci. In most of those cases, the specific source was not identified (Andrews *et al.*, 1984).

Even in less densely developed areas, disease-carrying organisms may be carried to the aquifer from animal wastes or septic tank drain fields where the soil is shallow and the water can move rapidly through the fractured limestone. As a general rule, though, increased densities of human habitation over a recharge zone will lead to a heightened potential for aquifer contamination (Edwards Underground Water District and Edwards Aquifer Research and Data Center, 1981).

Because states delegate the authority to promulgate land use regulations to local governments, the responsibility for storm water management rests most heavily with municipalities and counties. This notion of reliance on local government has prevailed at least since 1972, when areawide water quality management and planning was incorporated into Public Law 92-500, the Federal Water Pollution Control Act (U.S. Code, 1979).

III. CONTRIBUTING FACTORS TO ENACTMENT OF THE ORDINANCES

A large segment of the Austin community and city government has been engaged in discussion and debate about nonpoint pollution and aquifer water quality for the past several years. Among the many issues raised by the participants, the following were key to legislative action taken by the Austin City Council.

A. Governmental Authorities

The State of Texas is among the few remaining states that recognize the English "Absolute Ownership" doctrine of groundwater rights. Under this doctrine, a landowner has the right to pump and use water as an inalienable property right much like a farmer's right to use his soil (Johnson, 1982). (Editor's note: see Chapter 2 for a more thorough discussion of water rights).

The state government has not undertaken any comprehensive program to manage or protect groundwater quality. The state statutes do provide, however, for local initiatives to protect groundwater resources. One such statute, which is part of the enabling authority for Austin's groundwater protection ordinances, authorizes municipalities to adopt areawide water pollution abatement programs to protect their water supplies from urban runoff pollution (Texas Water Code, 1977). This authority can be extended beyond the corporate limits of the municipality and into the area of extraterritorial jurisdiction (ETJ) that extends radially for five additional miles around cities with more than 100,000 population, including Austin. The City of Austin's authorization to adopt and enforce urban runoff control policies in the ETJ was sustained by the courts in the case of *City of Austin* vs. *Jamail*, 662 SW 2d 641 (1983).

B. Citizens' Opinions on Water Quality

A recent opinion survey in the Austin area confirmed a sustaining citizen concern for environmental quality, particularly water quality protection. The survey of 1000 randomly selected registered voters was performed by Opinion Analysts, Inc., as part of the Nationwide Urban Runoff Program–Austin Area Study (City of Austin, Department of Public Works, 1982).

When the survey asked respondents to characterize surface water and groundwater quality problems in their area, 59% regarded groundwater quality as a "serious" or "somewhat serious" problem, and 69% responded accordingly about surface water quality. When asked who should bear the responsibility (pay the costs) for controlling pollution from development sites, 72% felt that "the polluter" should pay the entire cost, 7% felt that "all utility users" should pay, and 13% felt that costs "should be divided" between polluters and utility users. When asked what would be the best method of controlling water pollution, the most common response (47%) was the use of "legal restrictions."

These survey responses are indicative of the widespread citizen support for aggressive programs to manage water quality problems associated with urban development (Butler, 1983).

C. Barton Springs Pool

The citizens of Austin reap substantial benefits from the aquifer, both as a recreational resource and as a municipal water supply. The unique and natural qualities of Barton Springs Pool are nationally acclaimed as an urban recreational resource (see Fig. 6). The seminatural spring-fed pool has always been the most popular swimming spot in the city, attracting more than 300,000 bathers each year. Except for infrequent episodes of degraded water quality (characterized by high levels of coliform bacteria and suspended sediment) that are associated with storm runoff events, the water quality is very good.

The natural discharge from Barton Springs contributes to the city's water supply at a treatment plant on the Colorado River. During certain months of the year, water is not released from reservoirs on the river above Austin, thereby making the springs an important source of fresh water for customers of Austin's water utility. The average historical rate of discharge from the springs is 32 million gallons per day and has

Fig. 6. Photograph of Barton Springs Pool in central Austin.

ranged from 6 million to 107 million gallons per day. The average rate of water treatment and production at the plant that periodically receives water from Barton Springs is 28 million gallons per day.

Public concern and support for protection of the springs and the pool has been persistent and widespread. Citizens formed organizations such as the Save Barton Creek Association and Zilker Park Posse out of concern about the possible degradation of Austin's water resources, particularly Barton Springs. These organizations and other interested citizens were instrumental in persuading the city council in the late 1970s and again in the mid-1980s to address urban expansion over the aquifer. City council hearings provided a forum for considerable debate about water quality protection strategies, some of which became confrontational. Ultimately, the council instituted strong measures to manage and limit the intensity and drainage patterns of future development (Butler, 1983).

D. Austin's Master Plan

The Austin Tomorrow Comprehensive Plan (City of Austin, Department of Planning, 1980), adopted in 1979 and 1980, represented an extensive citizen input into urban policy planning. More than 3500 citizens provided direct input to the plan-making process through public hearings and workshops. Many of the policies and objectives in the Comprehensive Plan address urban water quality and are strikingly consistent with the 1982 survey results described in the preceding section. The following excerpts from the Comprehensive Plan are representative of the many stated objectives and policies that lend support to the aquifer-related watershed ordinances:

> Policy 322.3—Provide guidelines for drainage and runoff control that reduce erosion, peak flows, and poor water quality.
>
> Objective 331—Improve the quality of water runoff and lessen peak discharges.
>
> Policy 331.1—Assure that development in the more environmentally sensitive watersheds meets water quality and drainage standards.

E. A "Land Rush"

Development in the aquifer-related Williamson Creek Watershed burgeoned during the late 1970s and early 1980s. As is often the case, roadways and sewer line extensions led the way. Highways that already bisected the watershed and impending southwestern expansions of other expressways provided ready access to the region from the city.

Austin extended sewer and water lines into the upper reaches of the watershed in the late 1970s, making higher-density subdivision developments feasible.

The ensuing "land rush" is documented in an accounting of subdivision plan and plat approvals for the two years prior to enactment of the aquifer-related ordinances. By the time of ordinance enactment in 1980, almost 40% of the 7000-acre aquifer-related Williamson Creek Watershed area had been committed to urban development (see Fig. 3). As of January 1981, there were 5320 undeveloped lots in pending and recently approved subdivisions on approximately 1775 acres of land (City of Austin, Department of Planning, 1982). By 1985, approximately 75% of the entire Williamson Creek Watershed had been subdivided or developed (Williams, 1986).

The issue of expanding utility service to allow for complete buildout in the watershed has been earnestly debated since the late 1970s. The means by which Austin would protect groundwater quality in the face of urban-scale development has also been a thoroughly discussed topic.

F. Process Leading to Enactment

Each set of ordinances was conceived, drafted, and reviewed following the same general sequence of events. As outspoken citizens and environmental organizations stirred controversies surrounding land development over the aquifer in the late 1970s, public discussion and debate ensued in the meetings of the City Planning Commission and the Environmental Board. Within a few weeks, the forum shifted to city hall. Citizens pressured the city council to freeze pending developments in the fall of 1980 with subdivision platting moratoria while the ordinances were drafted.

At different times, the city council appointed two special task forces, one for the Barton Creek Watershed and the other for Williamson Creek and the Lower Edwards (Slaughter, Bear, Little Bear, and Onion Creek) watersheds. The city council selected members from the City Planning Commission and the Environmental Board, the Save Barton Creek Association, and representatives of the housing development industry. The city council directed the task forces to gather and review information, consider such related issues as utility extension policy, and recommend appropriate development standards for the various watersheds. City staff, a consulting firm under contract to the city, university faculty, and other experts assisted the task forces.

Considerable debate on alternative engineering and legal methods of protecting aquifer water quality accompanied the drafting of ordinances for each watershed. Reliability and enforceability were key considerations, as were the social, economic, and environmental trade-offs of different levels of development control. As one unnamed landowner commented, it was "democracy in its raw form."

The Barton Creek Watershed Task Force logged more than 250 hours and the Williamson Creek and Lower Edwards Watersheds Task Force held over 400 hours of scheduled meetings. Following these deliberations, public discussion and debate on appropriate development controls continued at public hearings before the city council. The council finally acted on the task force proposals, making compromises among interest groups such as the Austin Association of Builders and the Save Barton Creek Association that were involved in the process (Butler, 1983). The council adopted the Barton Creek Watershed Ordinance in April 1980. The Williamson Creek Watershed Ordinance was first adopted in December 1980 and the Lower Edwards Watersheds Ordinance was adopted in May 1981. In 1981, 1982, and 1986 the council adopted various revisions to the ordinances.

IV. WATERSHED DEVELOPMENT STANDARDS

Nationwide, municipalities have incorporated storm water runoff management principles into subdivision ordinances, site development codes, zoning ordinances, building codes, water pollution regulations, plumbing and sewer ordinances, and general plans and policies (American Public Works Association, 1981; Thurow *et al.*, 1975). In a similar manner, Austin adopted aquifer-related watershed development standards as an element of the subdivision and site development regulations of the City of Austin Code (1980). They contain requirements for planning and platting, delineation of special watershed zones, controlled development intensity, storm water runoff management and water quality protection, and compliance.

A. Plan and Plat Review Process

The first stage of subdivision and site development review for projects in the aquifer-related watersheds is usually the preapplication conference, in which the applicant and city officials review conceptual plans for drainage, land use, and utility and street layouts. The next stage is the submittal and review of a preliminary subdivision plan. The ordinances call for the submittal of various maps, data surveys, and special plans as part of the preliminary plan, many of which are directly relevant to aquifer water quality protection. Such plans include the delineation of watershed zones, projected land development intensities, location of special terrain features, erosion–sedimentation control plans, drainage plans, water quality control plans, and utility layouts. The planning commission and city council are authorized to grant variances from specific ordinance requirements if the literal enforcement would cause a deprivation of privileges or safety enjoyed on similarly situated property with similarly timed development.

The last stage of administrative processing is the submittal, review, and recordation of the final plat. Most of the platting requirements apply to all subdivisions within the corporate limits as well as the ETJ of the city. Some special final plat requirements in the aquifer-related ordinances are the completion of erosion–sedimentation and drainage plans, designation of watershed zones and building setbacks, delineation of critical environmental features such as sinkholes and exposed faults, certification of plans by a registered professional engineer, special assessments for water quality monitoring and maintenance programs, and the attachment of plat notes (covenants) that incorporate the special ordinance requirements.

Parallel with the subdivision ordinances are the site development ordinances for proposed construction on existing, legally platted lots. Essentially the same planning and permitting requirements specified for subdivision plats are required for site developments. The ordinances exempt single- and two-family residential construction from site development ordinances.

Through this process of administrative review, city staff evaluate the impacts of development plans and work with developers and their planners and engineers to ensure compliance with the appropriate ordinance provisions. The review of engineering and planning requirements under the ordinances is a major challenge for all parties. Austin is still "feeling its way" into the issues of water quality protection as they relate to urban growth, and the administrative rules are likely to be revised repeatedly in the coming years.

B. Watershed Zones

A primary strategy for groundwater protection in the ordinances is a three-tiered system of watershed zones established along waterways. The zones represent an attempt to incorporate known hydrologic features of the aquifer and the contributing watersheds into a scheme to mitigate water quality and quantity problems associated with urbanization. Consistent with observations that aquifer recharge takes place primarily within the creek beds overlying the recharge zone (Slade *et al.*,

1985), the ordinances establish progressively tighter development controls in zones closer to the creeks. The three tiers of zones are summarized in Table I and illustrated in Fig. 7.

The Critical Water Quality Zone overlays the creek beds and extends for considerable distances on either side. In most cases, the width of the Critical Zone is defined as the width of the 100-year floodplain, with maximum and minimum widths set on the basis of the size class of waterway. For example, along "intermediate waterways" the Critical Water Quality Zone encompasses the 100-year floodplain, provided that it shall not extend less than 100 feet nor more than 200 feet on either side of the centerline of the waterway.

Critical Water Quality Zones must remain free of almost all development and land alteration, with the exception of certain utilities and other facilities such as infrequent street crossings that cannot be reasonably placed elsewhere. The purpose in designating these development-free zones is to establish a "last line of defense" of unaltered open space to mitigate pollutant loads in stormwater before they enter the creek and the aquifer. It is presumed that vegetation in these zones acts as a natural filter, removing sediments and other pollutants through overland sheet flow. Several studies indicate that such zones are extremely effective in preserving the quality of receiving waters (Koenig, 1979; Wong and McCuen, 1981).

TABLE I
Waterways and Watershed Zones

Waterway designation	Size of drainage area	Width of Critical Water Quality Zone[a] (measured from centerline of waterway)	Width of Water Quality Buffer Zone[a] (measured from outer edges of Critical Water Quality Zone)
Minor	64–320 acres	100-yr floodplain BUT 100-ft max.	100 ft
Intermediate	320–640 acres	100-yr floodplain BUT 100-ft min. AND 200-ft max.	200 ft
Major	640 acres	100-yr floodplain BUT 200-ft min. AND 400-ft max.	300 ft

[a] The Upland Zone comprises all lands not included in the Critical or Buffer zones.

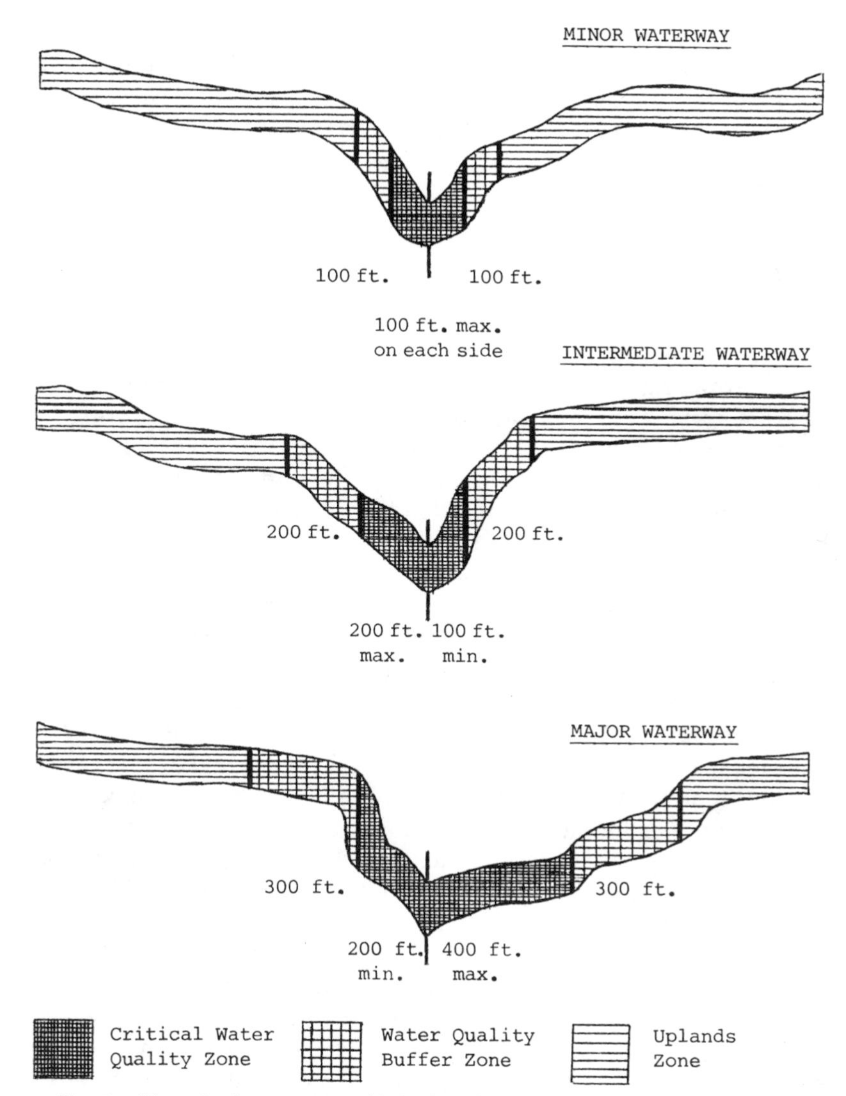

Fig. 7. Watershed zones as established in the aquifer protection ordinances.

The second tier in the ordinances is the Water Quality Buffer Zone, in which only low-density development is allowed. Buffer Zones range in width from 100 to 300 feet, increasing in width as the size of the drainage basin upstream increases. The bases for designating low-density Buffer Zones are twofold: activities close to receiving waters or aquifer recharge areas have a greater potential for transmitting pollutants into the water or aquifer than activities farther removed, and runoff water quality is significantly improved by maximizing the amount of overland flow. The use of buffer zones is one of the more common methods of surface water quality protection in the United States (Thurow *et al.*, 1975).

The Uplands Zone is the third tier and constitutes all remaining lands in the aquifer-related watersheds.

C. Development Intensity Controls—Density versus Impervious Cover

The intensity of land development is one of the most significant factors determining the amount of storm water runoff and pollutant loading from a site. Commonly accepted principles of civil engineering indicate that the amount of impervious cover—rooftops, streets, drives, and parking areas—is directly proportional to the rate and volume of runoff (Marsh, 1983). For many urban land uses, impervious cover is also highly correlated with the loading of such pollutants as total nitrogen, total phosphorous, suspended solids, volatile organic compounds, pesticides, oils, and chemical oxygen demand (Lager, 1977). Recent studies conducted in the Austin area indicate that the runoff pollutant loading per acre of surface area from medium-density residential areas with 39% impervious cover is several times greater than that from low-density residential areas with 21% impervious cover (City of Austin and Engineering Science, Inc., 1982).

In the typical case of a single- or two-family residential subdivision, the density (usually measured in living units per acre) and percentage impervious cover of the development are closely interrelated. For example, a medium-density residential development with three living units per acre corresponds to 25–30% impervious cover; land development controls addressing one such measure will effectively determine the intensity of the other. The correlation for more intensive land uses, such as multifamily or commerical developments, however, is not as clear.

The ordinances establish maximum allowable development intensities in each watershed. There are progressively more restrictive standards in zones closer to the creeks, consistent with the tiered system of watershed zones (see Table II).

A comparison of the ordinances for the various watersheds reveals significant differences in the degree and method of controlling development intensity. In the Williamson Creek Watershed, prior commitments to urban densities and sewer and water service in the late 1970s made the application of more stringent controls economically and politically untenable. Consequently, the maximum impervious cover standards are

TABLE II

Watershed Zones and Land Development Intensity Controls

Zone	Development controls
Critical Water Quality Zone	Development-free zone except for certain utilities, road crossings, and open space uses
Water Quality Buffer Zone	Residential: ranges from 3 acres/unit ave. and 2 acres/unit min.[a] to 30% impervious cover[b]
	Commercial and multifamily: ranges from 15% imp. cover[a] to 40% imp. cover[b]
Upland Zone	Residential: ranges from 2 acres/unit ave. and 3/4 acre/unit min.[a] (or 1 acre/unit ave. and 1/2 acre/unit min. with transfers of development intensity) to 40% imp. cover (or 55% imp. cover with transfers of development intensity)[b]
	Commercial and multifamily: ranges from 20% imp. cover (or 25% with transfers of development intensity)[a] to 60% imp. cover for 0–10 slopes (or 70% imp. cover with transfers of development intensity)[b]

[a] Barton, Slaughter, Bear, and Onion Creek watershed standards.
[b] Williamson Creek and Lower Slaughter Creek watershed standards.

of moderate intensity. On commercial sites the maximum allowable impervious cover is 70% and in single-family residential subdivisions the maximum is 40%.

The other watershed ordinances establish restrictive density controls as the primary means of attenuating pollutant loads. The standards in the low-density Buffer Zone range from a prohibition on housing or commercial development to one living unit per three acres of average density. The standards in the Uplands Zone range from one unit per two acres to two units per acre average density.

D. Transfers of Development Intensity

An important feature of all the ordinances is the progressively more lenient development controls in zones further from the creeks and, consequently, from the primary recharge areas. Reinforcing this tiered system is the "transfer of development intensity"—an incentive mechanism that allows development rights to be transferred from the Buffer Zone to the Uplands Zone. By dedicating land in the Buffer Zone as permanently undisturbed open space, developers are granted proportionately higher development allowances (including bonus allowances) in the Uplands Zone. The ordinances also allow bonus transfers for the dedication of land in the Critical Zone as public parkland. The tiered density standards and transfer provisions indirectly serve to protect ecological and aesthetic qualities of the creek corridor, as well as mitigate pollution of recharge waters.

Some of the ordinances provide an incentive for clustered development in the Upland Zone by reducing the minimum lot size to 5750 square feet per unit in consideration for the preservation of corresponding amounts of open space. Another provision found in all the ordinances is a minimum density standard of one living unit per acre for developments not served by collective sewerage systems, or the required installation of an alternative type of on-site waste treatment and disposal system.

E. Storm Water Runoff Controls

A key strategy for water quality protection is the development of guidelines for conveying, detaining, and filtering storm water runoff before it enters the creek and the aquifer. Unlike traditional urban storm drainage system designs that route storm water through sewer pipes and lined channels rapidly and efficiently to the next water body, a multipurpose storm water management approach may employ both nonstructural and structural measures to control the quantity and quality of runoff leaving a site.

Collectively, the aquifer-related watershed ordinances represent an important departure from traditional urban drainage criteria. A common objective of the ordinances is to ensure that postdevelopment storm water runoff quality and quantity are no worse than predevelopment conditions. There are differences among the ordinances, however, in the degree of reliance on engineering approaches to reduce pollutant loads.

The Williamson Creek ordinance relies on engineered storm water quality control measures as the primary means of aquifer protection. If planned development intensities exceed 20% impervious cover, the ordinance requires the specific construction of storm water detention–sedimentation basins and the release of water from the basins through filter media in accordance with specific performance criteria (see Fig. 8). The basins are designed to isolate and detain the "first flush" (0.5 inch) of runoff following a storm, which carries most of the pollutants.

The ordinances for the less developed watersheds also require the use of water quality basins when developments exceed specified intensities, but they also employ more generalized storm water management guidelines, calling for overland drainage and the preservation of undisturbed

Fig. 8. Storm water detention–filtration basin receiving runoff from a retail shopping mall.

surfaces for drainage purposes. Clustered housing and commercial developments must leave large portions of the sites in a natural state and position the open space to receive overland runoff from the development site.

F. "Source" Controls for Groundwater Quality Protection

In addition to the requirements for treating storm water runoff as it leaves the development site, the ordinances also establish standards to control water pollution at the point of origin (City of Austin Code, 1980, 1981, 1982). A key "source" control is the requirement for on-site control of erosion and sedimentation, especially during construction stages. Studies have documented that the amount of sediment yielded from bare ground during the construction of buildings and highways may be hundreds of times greater than the amount eroded from farms and woodlands in a comparable period of time (Wolman, 1964).

All subdivision and commercial development activity in the aquifer-related watersheds must submit detailed erosion and sedimentation control plans. The plans must include a program for sequencing the installation of control measures in accordance with a construction schedule. Some of the construction-stage erosion control measures routinely used in the Austin area include brush berms, siltation basins, reinforced tubes covered with filter fabrics, and staked hay bales (City of Austin, Department of Public Works, 1983). The city employed several environmental inspectors in 1985 to provide on-site construction inspection and exercise authority to issue stop-work orders if needed in the case of noncompliance.

The ordinances also address on-site sewage treatment and disposal systems at the points of discharge. On-site systems overlying the aquifer recharge zone must utilize "alternative" methods of treatment or disposal, such as lined evapotranspiration beds, to minimize the risk of aquifer contamination.

There are several other required engineering-oriented source controls. All commercial parking areas with more than 5000 square feet of surface must employ vacuum street-sweeping equipment at least three times each week. Sewer systems constructed in the aquifer recharge zone must meet special standards to minimize the risk of releasing raw sewage through exfiltration from sewer lines and manholes. The ordinances prohibit the construction of permanent waste disposal sites for spoils or organic or toxic materials. Facilities used for storage of hydrocarbon products and other hazardous chemicals must be of double-wall construction or be designed to capture accidental spillage in impervious basins.

There are many other provisions that are not identified or discussed herein but that also serve to control water quality and protect the aquifer (City of Austin Code, 1980, 1981, 1982).

G. Compliance Measures

The ordinances address the issue of compliance in three ways: enforcement, routine inspection and maintenance, and long-term water quality monitoring. The enforcement provisions state that the city has the right to enjoin violations and bring suit in a court of competent jurisdiction. The ordinances also require the issuance of a revocable contractor's permit before a contractor may work in the watershed. In the case of noncompliance, city inspectors may issue stop-work orders.

The need for inspection and maintenance is largely a function of the specific water quality control strategies employed by each ordinance. Strict density controls, by their very nature, require minimal inspection or maintenance following subdivision plan review. Once a subdivision project is constructed within prescribed density limits, compliance is all but guaranteed. On the other hand, engineering-oriented storm water runoff controls require routine inspection and maintenance to ensure that they perform according to design standards. The ordinances make extensive reference to maintenance and inspection—who pays, who is responsible, and how often it is done. Developers must pay a special one-time fee for maintenance of water quality basins ($80 per lot) at the time of final plat approval. The ordinance also contains several stipulations indicating when and under what conditions individual owners, homeowners associations, or the city must assume the maintenance responsibilities.

The city has initiated an extensive water quality monitoring program to evaluate the various control strategies in each ordinance. The next section of this chapter contains a description of the monitoring program.

H. Alternative Approaches Illustrated by the Ordinances

The ordinances can be characterized as representative of one of two generic approaches to water quality protection: the "land use regulatory" approach or the "engineering management" approach.

1. *The Land Use Regulatory Approach*

An underlying assumption of the land use regulatory approach is that the greatest opportunity for preventing further water quality degradation is through proper management of the location, design, and density of new urban development. By significantly limiting housing density and impervious cover in several of the ordinances, conceivably the effect will be to limit total population, traffic generation, waste generation, and related factors associated with environmental degradation.

2. *The Engineering Management Approach—Williamson Creek Ordinance*

The performance effectiveness of such management practices as storm water detention basins, sweeping of streets and parking lots, sewer line construction criteria, and other standards is largely untested in the Austin area, but studies elsewhere indicate that highly significant water quality treatment is achievable (Biggers and Hartigan, 1980; Lager, 1977). A particular challenge for the city in relying on this approach in the Williamson Creek Watershed is the need for routine inspection, maintenance, and monitoring, at considerable cost, to ensure that each water quality control device performs according to design criteria (Butler, 1983).

3. *The Integrated Best Management Approach—A Comprehensive Code*

The city council of the City of Austin appointed a special task force in January 1986 to develop a comprehensive watersheds ordinance that would combine the various aquifer-related standards into a single code as well as establish water quality protection standards for other watersheds in the city's 1240-square-mile corporate and ETJ area. The task force is combining nonstructural land use controls found in the Lower Edwards Watersheds Ordinance with the same engineering–best management practices specified in the Williamson Creek ordinance. By this process, it is likely that a combinational, "integrated best management" approach will ensue that better balances the benefits of groundwater protection with flexibility and economy in land development.

The motivations for employing such an integrated approach are twofold. First, there is considerable uncertainty in relying on either the engineering management or the land use regulatory approach when local data on their effectiveness do not exist. Second, each approach has obvious advantages and disadvantages; conceivably, a workable combination of the two would be the most appropriate in the long run.

V. CURRENT EVENTS AND LONG-RANGE ISSUES

At this time, it is not clear whether the ordinances' goals of aquifer water quality and areawide environmental protection will be realized. The City of Austin is now struggling with decisions about how to manage and cope with rapid urban expansion in the aquifer-related watersheds and in other presently less developed areas to the west of the city. The effectiveness of the ordinances will be determined in part by the nature of development overlying the recharge areas. In the face of these uncertain events, many questions remain unanswered about the ordinances and the aquifer itself.

A. Unknowns Concerning the Aquifer System

In recent years the U.S. Geological Survey has conducted an extensive program of monitoring groundwater hydrology and water quality in the Edwards Aquifer (Slade *et al.*, 1980, 1982). But the intricate surface water and groundwater flow characteristics may never be understood or modeled in sufficient detail to resolve questions as to which protective measures are more appropriate at a site-specific level. No amount of research can disclose the three-dimensional, labyrinthine flow patterns in the cavernous and fractured limestone aquifer. On an areawide basis, it is exceedingly difficult to predict the long-term, cumulative impacts of urban-density expansion, or low-density development with a proliferation of septic tank systems, or the likelihood that such scenarios will materialize.

It is difficult to assess the risk of spills or gradual releases of toxic and hazardous chemicals from industrial facilities, commercial establishments, or even private residences in the recharge zone or upstream contributing areas. Some prominent examples include volatile organic carbon compounds, pesticides, petroleum-based products, and heavy metals. Ordinances have not been drafted to cover every eventuality. For all the data, research, and expert testimony reviewed during the preparation of the ordinances, the standards enacted into law are only a "best effort" in the face of considerable technical uncertainty.

The ordinances address the need for a water quality monitoring program to evaluate the specific storm water quality management techniques mandated by the ordinances. The city council authorized funds to implement the program in 1982 and the Department of Public Works–Watershed Management Section initiated the monitoring program in 1984. The program complements the extensive background data on water quality from the Nationwide Urban Runoff Program (City of Austin and Engineering Science, Inc., 1982) and the ongoing U.S. Geological Survey–City of Austin Cooperative Monitoring Study (Slade *et al.*, 1980; 1982).

The city monitors the pollutant removal efficiencies of control structures such as sedimentation and filtration basins. There are currently sixteen stations equipped with automated water quality samplers and flow meters. Each monitoring site typically includes two stations, one located upstream and the other located downstream from the water quality control structure (e.g., filtration basin). The monitoring sites include residential and commercial developments and undeveloped areas, representing a range of building densities and amounts of impervious cover. City personnel can activate and receive data from flow meters and samplers remotely by a minicomputer via telecommunication lines. Automated water quality samplers can obtain data at frequent intervals during the course of a storm runoff event.

The city's water quality monitoring program should establish a clearer and more accurate predictive relationship between development intensity and pollutant loading in creeks that recharge the aquifer. To objectively determine whether the ordinances are effective in protecting water quality, it will be necessary to collect and analyze at least two or three years of data, covering numerous runoff events.

B. Challenges for Managing Urban Growth

The aquifer area exhibits all the symptoms of what will ultimately be a major urban expansion—new industrial locations, regional shopping malls, multiphased subdivision developments, major expressway extensions, and the like. As a consequence, the key questions about managing growth will center on municipal annexation and the provision of utilities and other services—by what entity, over how large an area, at what prescribed development density, how quickly, and at whose cost.

The City of Austin has provided centralized water and wastewater services in the northern section of the aquifer. Development to the south of Austin's service area has relied on the aquifer as a water supply. The reliability and quality of Austin's utility services in the aquifer area have gradually declined during the past few years due to decisions not to fund major improvements. As a result, urban expansion in the south Austin service area has advanced slowly in recent years while the rate of rural subdivision development further south has proceeded very rapidly.

The city recently decided, however, to extend utility services to lands in the upper Williamson Creek and central and lower Slaughter and Bear Creek watersheds and significantly expand the capacity of its water and wastewater treatment plants. This expansion will be very costly on a per capita basis because of the long distances westward, higher elevations, and significant deficiencies in utility line sizes throughout the southwest Austin service area. But the rate and density of development over the aquifer are likely to increase substantially.

While the city extends water and wastewater services in the Williamson, Slaughter, and Bear Creek watersheds, sewer lines may never be extended into the upper reaches of Barton Creek, and probably are several years away from the other watersheds. The city and county are currently evaluating their sanitary codes relating to on-site wastewater systems, hoping to anticipate impending water quality problems if low- to medium-density development proliferates throughout the hilly limestone area. The ultimate consequences of the buildout of residential and commercial development on the quality and quantity of the aquifer are not known.

VI. CONCLUDING OBSERVATIONS

The elected officials of many communities would not have taken such aggressive measures as adopting Austin's watershed ordinances to protect groundwater quality and the natural environment. But the citizens of Austin are unusually earnest about retaining a high quality of life in the face of an expanding metropolitan area. For many Austinites, there is no better benchmark of environmental quality than the purity of Barton Springs Pool and its aquifer.

The watershed ordinances are an amalgam of land and water management techniques, many of which have been applied individually in many other communities. But Austin's ordinances are probably unique in terms of relating urban stormwater runoff and aquifer resource protection. Given the sensitive nature and karst characteristics of the Edwards Aquifer and the groundwater recharge process, it is essential to integrate storm water management policies with land development controls and preservation of creekside open space to mitigate the adverse impacts of urbanization. The distinctive approaches taken in each ordinance, particularly with respect to density versus engineered runoff controls, make them a noteworthy case study for other units of government. Results from the city's water quality monitoring programs during the next year or two may corroborate the benefits of one method over the other.

To be workable and achieve the goal of groundwater quality protection, storm water runoff controls should be integrated into a comprehensive planning, management, and regulatory program. The drainage system that results from urbanization should be designed to modify the volume, rate, and quality of runoff in such a way that it can be released into the natural drainage system with minimum adverse impacts. Austin's aquifer-related watershed ordinances are a major step in this direction.

The ordinances are only part of a medley of growth management instruments that currently are utilized in Austin. Others include a municipal annexation plan, zoning ordinance, utility service policies, capital improvements programs, and other ordinances. The watershed ordinances must operate compatibly with these other growth management programs and must address the economic realities of housing affordability and commercial markets. The city cannot expand its jurisdictional limits to encompass and manage development throughout the entire watershed areas to the west because of the prior existence of other municipal jurisdictions. Consequently, Austin needs to share its experiences with adjoining communities located in the aquifer area and work cooperatively toward the extablishment of regional policies and rules. Significant progress in interlocal cooperation is just now being realized.

The quantity of water in the Austin unit of the Edwards Aquifer is also becoming a primary concern. On a statewide basis, groundwater use has increased 18-fold in 50 years and today represents almost 70% of all the water used in Texas (Texas Department of Water Resources, 1984). Many communities in the San Antonio region rely solely on the southern portion of the Edwards Aquifer for their water supply. They are struggling with dire projections of insufficient groundwater flows within the next 20 years (Edwards Underground Water District, 1981). The City of Austin is working with adjoining communities which overlie the Edwards Aquifer to create an underground water conservation district, that will be able to address water quantity as well as quality concerns throughout the growing metropolitan area (Butler, 1986).

REFERENCES

American Public Works Association (1981). "Urban Stormwater Management," Spec. Rep. No. 49, p. 255. APWA, Chicago, Illinois.

Andrews, F. L., Schertz, T. L., Slade, R. M., Jr. and Rawson, J. (1984). Effects of stormwater runoff on water quality of the Edwards aquifer near Austin, Texas. *Water Resour. Invest. U.S. Geol. Surv.* **84-4124.**

Biggers, D., and Hartigan, J., Jr. (1980). "Urban Best Management Practices (BMP's): Transition from Single-Purpose to Multi-purpose Stormwater Management" (International Symposium on Urban Stormwater Runoff). University of Kentucky, Lexington.

Butler, K. S. (1983). Managing growth and groundwater quality in the Edwards aquifer area, Austin, Texas. *Public Affairs Comment* **29** (2), 1–10.

Butler, K. S. (1986). "Background Report on the Necessity and Benefit of the Proposed Barton Springs–Edwards Aquifer Conservation District," Unpublished report prepared for the Texas Water Commission, Austin.

City of Austin Code (1980, 1981, 1982). Each of the following ordinances is codified as an amendment to Chapter 13, Section 3 and Chapter 9, Section 10 of the Austin City Code:

Barton Creek Watershed Subdivision Ordinance, Ord. No. 810430 (April 30, 1980).

Williamson Creek Watershed Subdivision Ordinance, Ord. Nos. 801218-W and 810212-K (December 18, 1980, and February 12, 1981).

Slaughter, Bear, Little Bear, and Onion Creek Watershed Subdivision Ordinance, Ord. No. 810514-S (May 14, 1981).

Slaughter, Bear, Little Bear, and Onion Creek Watershed Site Development Ordinance, Ord. No. 810514-T (May 14, 1981).

Williamson Creek Watershed Development Ordinance, Ord. Nos. 810319-M and 810507-K (March 19, 1981, and May 7, 1981).

Barton Creek Watershed Site Development Ordinance, Draft Ordinance (June 22, 1982).

City of Austin and Engineering Science, Inc. (1982). "Final Report of the Nationwide Urban Runoff Program in Austin, Texas," p. 183. City of Austin, Austin, Texas.

City of Austin, Department of Planning (1980). "Austin Tomorrow Comprehensive Plan."

City of Austin, Department of Planning (1982). Unpublished files and memoranda relating to the status of subdivision plan and plat applications.

City of Austin, Department of Public Works (1982). Unpublished public opinion survey data collected and compiled by Opinion Analysts, Inc., as part of the Nationwide Urban Runoff Program–Austin Area Study.

City of Austin, Department of Public Works (1983). "Erosion and Sedimentation Control Manual."

Edwards Underground Water District and Edwards Aquifer Research and Data Center (1981). "Water, Water Conservation and the Edwards Aquifer" pp. 21–22. Edwards Underground Water District, San Antonio, Texas.

Johnson C. (1982). Texas groundwater law: a survey and some proposals. *Nat. Resour. J.* **22,** 1017.

Koenig, L. (1979). "Costs of Development: A Portion of the Preventative and Remedial Measures Phase of the Edwards Aquifer Study." Metcalf and Eddy, Inc., San Antonio, Texas.

Lager, J. A. (1977). "Urban Stormwater Management and Technology: Update and User's Guide," Rep. No. EPA-600/8-77-014, pp. 90–92. U.S. Environmental Protection Agency, Cincinnati, Ohio.

Marsh, W. M. (1983). "Landscape Planning," pp. 130–146. Addison-Wesley, Reading, Massachusetts.

Slade, R. M., Jr., Dorsey, M. E., Gordon, J. D., and Mitchell, R. N. (1980). "Hydrologic data for urban studies in the Austin, Texas metropolitan area, 1978. *Geol. Surv. Open-File Rep. (U.S.)* **80–728.**

Slade, R. M., Jr., Gaylord, J. L., Dorsey, M. E., Mitchell, R. N., and Gordon, J. D. (1982). Hydrologic data for urban studies in the Austin, Texas metropolitan area, 1980. U.S. *Geol. Surv. Open-File Rep. (U.S.)* **82–506.**

Slade, R. M., Jr., Ruiz, L., and Slagle, D. (1985). Simulation of the flow system of Barton Springs and associated Edwards aquifer in the Austin area, Texas. *Water Resour. Invest. U.S. Geol. Surv.* **85–4299.**

Texas Department of Water Resources (1984). "Water for Texas," Vol. 2, Technical Appendix. Dwp. Austin, Texas.

Texas Water Code (1977). Sec. 26.177.

Thurow, C., Toner, W., and Erley, D. (1975). "Performance Controls for Sensitive Lands: A Practical Guide for Local Administrators." American Society of Planning Officials, Chicago, Illinois.

U.S. Code (1979). Vol. 33, sec. 1288 (Supp. II, 1978); see also C.F.R., secs. 35.2513-5, 1521-4.

Williams C. S. (1986). Environmental Planner III. Office of Land Development Services, City of Austin, (personal communication).

Wolman, M. G. (1964). "Problems Posed by Sediment Derived from Construction Activities in Maryland." Maryland Water Pollution Control Commission, Annapolis.

Wong, S. L., and McCuen, R. H. (1981). "The Design of Vegetative Buffer Strips for Runoff and Sediment Control." University of Maryland, Department of Civil Engineering, College Park.

Woodruff, C. M., Jr., and Slade, R. M., Jr. (1985). "Hydrology of the Edwards Aquifer–Barton Springs Sement," Guidebook No. 6. Austin Geol. Soc., Austin, Texas.

10

Perth Amboy, New Jersey, Case Studies

G. WILLIAM PAGE
Department of Urban Planning
and Center for Great Lakes Studies
University of Wisconsin–Milwaukee
Milwaukee, Wisconsin 53201

I. BACKGROUND

Perth Amboy is an old industrial city in the heart of the urban-industrial corridor in northeastern New Jersey. Located in the northeastern corner of Middlesex County, Perth Amboy is across the Arthur Kill from New York City (see Fig. 1). Perth Amboy occupies 3532.6 acres with 26% devoted to industrial activity (Middlesex County Planning Board, 1978). Perth Amboy had a 1980 population of 38,951 (U.S. Bureau of the Census, 1981). The population has remained stable since the 1920s.

Perth Amboy first experienced water supply problems before the end of the nineteenth century. At that time Perth Amboy was already an industrial city of over 10,000 population. The many private, domestic, and industrial wells lowered the water table and drew salt water from the Arthur Kill into the aquifer beneath Perth Amboy. In 1882 the city formed the Perth Amboy Water Company to provide water to those citizens with private wells contaminated by salt water intrusion (City of Perth Amboy, 1973).

The Perth Amboy Water Company faced the responsibility of supplying water to a rapidly growing city while its supply of potable water decreased because of growing salt water intrusion. In planning for an adequate water supply, the supply of endogenous water available to Perth Amboy was clearly insufficient for their needs. The present and future water supply needs of Perth Amboy required an exogenous source of water.

The Perth Amboy Water Company acquired a source of water by purchasing land in an undeveloped watershed in a rural portion of

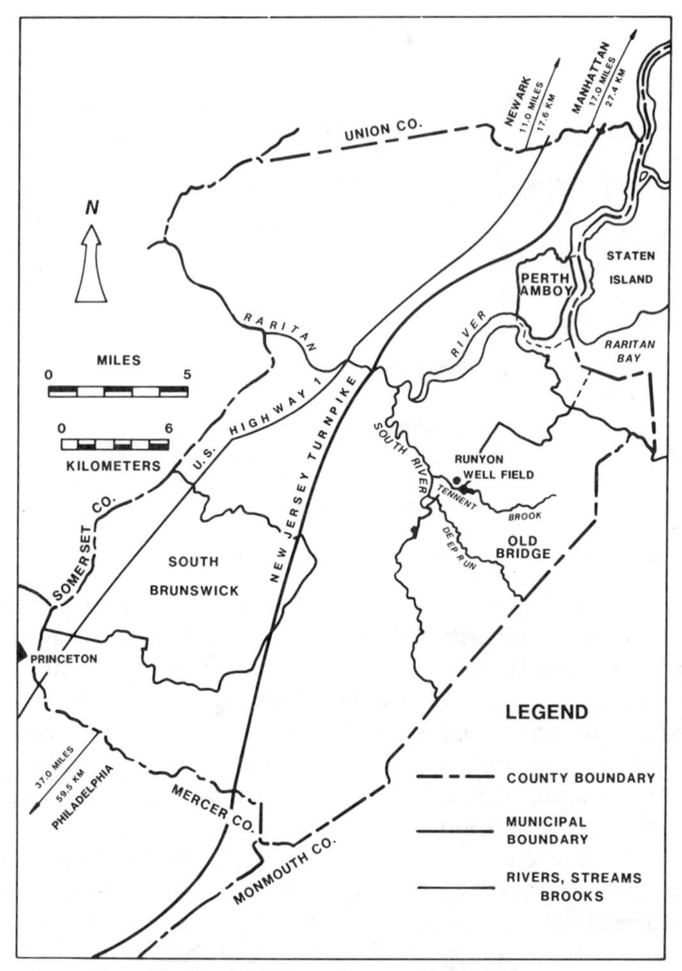

Fig. 1. Middlesex County, New Jersey.

Middlesex County about 45 miles away. In 1893, Perth Amboy purchased 1300 acres of land along Tennents Brook in what is now Old Bridge Township (City of Perth Amboy, 1973). In the period around the turn of the century, other neighboring cities were purchasing undeveloped watersheds, which were often considerably distant and provided a sufficient, high-quality source of water supply. Some of these neighboring cities include New York City, Newark, and Jersey City.

Soon after its purchase, the Runyon property became the sole source of water supply for the City of Perth Amboy. Originally, Perth Amboy pumped some surface water directly from Tennents Brook, but soon came to rely on a system of wells and induced recharge from Tennents Brook and its impoundments. The Runyon Well Field averages about 44 inches of precipitation per year (Middlesex County Planning Board, 1979a). Located within the inner coastal plain physiographic province of New Jersey, the well field draws from several extensive high-yield and excellent water quality aquifers. The two most important aquifers for the Runyon Well Field, Middlesex County, and extensive areas to the east and south of Middlesex County are the Old Bridge Sand and the Farrington Sand members of the Potomac–Raritan–Magothy formation. Over the years wells have been added so that today the Runyon Well Field has 108 wells tapping the two aquifers (Rakowski, 1985).

By the turn of the century, Perth Amboy had established a well-protected source of water supply with apparently unlimited supplies of high-quality water. The Runyon Well Field consisted of a contiguous parcel of 1300 acres of land in its natural state in a small watershed in an undeveloped municipality.

Old Bridge Township consists of 24,769 acres of land and at the turn of the century had a population of less than 1700 (Middlesex County Planning Board, 1978). The development of the Runyon Well Field in the 1890s was a fine example of foresighted planning by the City of Perth Amboy. Even today developing a water supply source comparable to the Runyon Well Field at that time would be outstanding environmental planning.

II. NONTOXICS PROBLEMS AT THE RUNYON WELL FIELD

No matter how inexhaustible the supply of water or well protected the source of supply appeared in 1900, the years have brought changes. While Old Bridge Township occupies a large area and remains about 80% open space, the population has grown rapidly since World War II. The population of this rapidly growing suburban community exceeded that of Perth Amboy in the mid-1960s. Competition for water has increased with the growth of Old Bridge Township and neighboring municipalities. The Sayreville Water Department and the Duhernal Water Company established large-scale water supply sources using groundwater near the Runyon Well Field (Middlesex County Planning Board, 1979b). Considerable industrial and residential development continues in many areas of Old Bridge Township.

The quantity of groundwater beneath the Runyon Well Field has declined dramatically. When users pump more water from an aquifer than precipitation recharges, then an overdraft condition results and the quantity of available groundwater declines. When Perth Amboy first established the Runyon Well Field, strong artesian conditions existed in the Farrington Sand Aquifer. A potentiometric surface or "head" 30 to 40 feet above the land surface existed in 1897. Because of overdraft conditions over the years, the head is now more than 70 feet below sea level (Geraghty and Miller, Inc., 1976).

When the head falls below sea level in a coastal plain, intrusion of salt water often becomes a serious problem. In Middlesex County salt water intrusion in the Farrington Sand Aquifer is a serious problem. The State of New Jersey deepened the Washington Canal in 1929 and exposed the Farrington Sand Aquifer to brackish water from the tidal waters of the South River via the canal (Barksdale, 1943). As increasing quantities of water have been pumped from the Farrington Sand since 1940, the front of the salt water intrusion has moved south and first reached the Runyon Well Field in 1972 (Geraghty and Miller, Inc., 1976).

The Perth Amboy Water Company has relied increasingly on the Old Bridge Sand Aquifer for its water supply. The serious overdraft conditions and salt water intrusion in the Farrington Sand Aquifer have caused a decrease in the quantity of water available. The Old Bridge Sand Aquifer is the surface formation in the Runyon Well Field and large areas of Middlesex County. Because of a decreased quantity of fresh water in the deeper aquifer, Perth Amboy pumped increasing quantities of water from the Old Bridge Sand.

Increased dependence on the Old Bridge Sand Aquifer has produced many problems including overdraft conditions. The Perth Amboy Water Department uses surface water to recharge the Old Bridge Sand that outcrops on the Runyon Well Field property. Recharge increases the quantity of groundwater available. The practice of artificial recharge is commonly used in Europe and other parts of the world and uses the filtering capacity of the soil to cleanse the surface water of sediments and bacteria. As early as 1941 this system recharged about 5 million gallons per day (mgd) of water to the Old Bridge Sand (Geraghty and Miller, Inc., 1976). Virtually the entire low flow of Tennents Brook as well as a system of canals and ponds recharged the aquifer. Plans exist for a 100-acre recharge pond on Deep Run within the runyon Well Field that will recharge an additional 7 million gallons of water per day (Middlesex County Planning Board, 1979a). In 1985, a 70-acre pond with two large-capacity wells (4 mgd each) is nearing operational status (Rakowski, 1985).

Dependence on a surficial aquifer has exposed the Perth Amboy water supply more directly to the effects of human-caused pollution. The rapid growth of housing developments in the headwaters of the Deep Run watershed has markedly increased surface runoff with high concentrations of bacteriological and chemical pollutants directly into Deep Run (City of Perth Amboy, 1973). The surface water in Deep Run recharges into the Old Bridge Sand Aquifer. Perth Amboy's Special Water Counsel and the head of the mayor's Water Advisory Board have protested applications for housing developments in Old Bridge Township. Obviously neither public officials nor the private sector in Old Bridge Township are pleased by attempts by the City of Perth Amboy to interfere in the internal affairs of Old Bridge Township. The mayor of Perth Amboy has announced that he will oppose all future developments proposed near Perth Amboy's Runyon watershed (*News Tribune,* 1977b). While the efforts of Perth Amboy may not have been effective at slowing development and pollution in the vicinity of its Runyon Well Field, regional water supply problems have delayed development in Old Bridge Township in the vicinity of Perth Amboy's Runyon Well Field.

The availability of an adequate water supply is at present a severe obstacle to future growth in much of Middlesex County, New Jersey. Since 1947 New Jersey has regulated the amount of groundwater that may be withdrawn. Domestic wells are not covered, but permits are required for pumping more than 100,000 gallons per day of groundwater (Geraghty and Miller, Inc., 1976). The state Water Policy and Supply Council uses the permit power to allocate rights to groundwater. Because municipalities pump more than 18 million gallons per day of water from the Farrington Sand while the estimated optimum yield of the aquifer is only 16.2 million gallons per day (Middlesex County Planning Board, 1979a), developing municipalities such as Old Bridge Township are rapidly approaching their maximum diversion right without authorization for additional supplies. (Editor's note: see Chapter 1 for discussion of optimum yield.) One of the anticipated recommendations of the South River Water Supply Feasibility Study is for the diversion rights of many municipalities using groundwater in Middlesex County to be reduced by 30–50% from present levels to contend with overdraft conditions and salt water intrusion (Cesnik, 1985). Old Bridge Township had an actual average daily use of 5.346 million gallons of water per day in 1983 with projected average daily uses of 8.414 million gallons per day in 2000 and 11.820 million gallons per day in 2020 (Middlesex County Planning Board, 1984). At present Old Bridge Township has applications for a total of about 20,000 housing units under consideration. These applications are delayed pending the provision of an adequate water supply (Cesnik, 1985). This scale of development in Old Bridge Township potentially could have severe detrimental impacts on Perth Amboy's Runyon Well Field.

III. TOXICS CONTAMINATION PROBLEMS

Perth Amboy's water supply has experienced several toxic substances contamination problems. Some of these problems have been short term and the result of a single pollution episode. Other toxic contamination problems have been chronic.

A chronic toxic contamination problem seems to be the result of the location of several industries in Old Bridge Township near the Runyon watershed. Several industries are located along Pricketts Run, which flows into Tennents Pond, a major source of artificial recharge to the Old Bridge Sand Aquifer within the Runyon Well Field. Runoff into Pricketts Run has been found to cause the stream to be high in biological oxygen demand (BOD), chemical oxygen demand (COD), zinc, lead, aluminum, cadmium, and iron (Geraghty and Miller, Inc., 1976). The toxic heavy metals among these pollutants, especially zinc, lead, and cadmium, have been found in some of Perth Amboy's water supply wells by the city and the New Jersey Department of Environmental Protection (DEP) (City of Perth Amboy, 1973). This toxic heavy metals pollution caused the Department of Environmental Protection to close 6 wells in the Runyon Well Field in 1971 and an additional 24 wells in 1973 (Langenohl, 1979). The 30 wells that have been closed because of toxic contamination pumped about 2.5 million gallons per day of water (Middlesex County Planning Board, 1979a). Since 1973 the Perth Amboy Water Department has purchased water from the Middlesex Water Company, a private purveyor, to replace the water lost because of these sources of toxic substances contamination (Langenohl, 1979). The legal proceedings to recover some of the expenses connected with this chronic toxic substances contamination against the industries accused of being the sources of the contamination are complex (*Home News*, 1979a). In 1980, the courts awarded the City of Perth Amboy compensation, but the technical details and cost of a slurry wall to encase the plume of contamination and to pump the contaminated groundwater to waste have not been decided (Rakowski, 1985).

Details regarding conditions at one of the industrial facilities implicated in the pollution of heavy metals suggest some of the ways the Old Bridge Sand Aquifer may have become contaminated. When Perth Amboy closed the first six wells in the Runyon Well Field on March 17, 1971, it identified Food Additives, Inc., as one of several potential sources of pollution. The firm, located about one-half mile from the well field, produces zince flouride and other chemical compounds. Inspections of the plant found extremely poor practices of storage and housekeeping and a lack of protection against spillage in the processing and storage areas (Geraghty and Miller, Inc., 1976). The firm claimed to discharge all its liquid waste into the Old Bridge Township sewer line via a pipe connection after pretreatment. Investigation of the township sewer line led to the discovery of a second lateral pipe leading from Food Additives, Inc., that bypassed the firm's pretreatment facility and was either ruptured or not properly connected to the township's sewer line (Geraghty and Miller, Inc., 1976). This lateral pipe discharged at a greater rate than the primary lateral pipe (Geraghty and Miller, Inc., 1976).

The City of Perth Amboy's Runyon watershed has also experienced isolated episodes of toxics pollution that have produced severe short-term effects on their water supply system. On the evening of June 22, 1977, the police apprehended the driver of a tank truck discharging chemicals from the truck as he drove along the infrequently traveled access road to the Runyon Well Field property (*News Tribune,* 1977a). The truck carried transformer oil containing large quantities of polychlorinated biphenyls (PCBs). Both the city and the New Jersey Department of Environmental Protection undertook intensive monitoring of the Perth Amboy water supply. On July 1, 1977, the Department of Environmental Protection ordered all 105 wells in the Runyon Well Field closed because of contamination by PCBs (*Home News,* 1977). Although at the time no established national or state health standard for PCBs existed, the New Jersey Department of Environmental Protection closed the wells because of concentrations in excess of one part per billion (ppb) and the extreme toxicity and suspected carcinogenicity of PCBs.

Perth Amboy faced a serious crisis because of this contamination. Except for the 3 million gallons of water per day the city was purchasing, because of the chronic heavy metals contaminations problems described earlier, the city depended on the Runyon Well Field for the rest of the 123 million gallons per day of potable water Perth Amboy used each day. The city instituted strict conservation measures and purchased an additional 5 million gallons per day of water from the Middlesex Water Company at a cost of about $5000 per day (*News Tribune,* 1977a).

On July 11, 1977, some 10 days after the Department of Environmental Protection ordered the wells in the Runyon Well Field closed, the DEP gave Perth Amboy permission to resume the use of the 105 wells in the Runyon Well Field. Perth Amboy officials estimated that this short episode of PCB contamination had cost the city about $250,000 (*News Tribune,* 1977a). In 1979 the state convicted the truck driver of spraying chemicals along the road to the Runyon watershed and sentenced him to 18 months in the Middlesex County Adult Correction Center. The court fined Gold Leaf Transportation, Inc., $50,000 (*Home News,* 1979b).

IV. GROUNDWATER PROTECTION PLAN COMPONENTS

The City of Perth Amboy has a long history of planning for the protection of its water supply. More than 80 years ago the city established a model water supply system. The purchase of an undeveloped watershed that retains its natural vegetation in a rural area far from an urban center remains to this day an ideal of sound environmental planning (Burby, *et al.*, 1983). Despite following the best precepts of environmental planning, the city of Perth Amboy has experienced severe problems of groundwater contamination with toxic substances. In purchasing a pristine water supply source in a distant area, the City of Perth Amboy implemented the central component of its water supply protection program. In evaluating Perth Amboy's water supply protection program, it seems that the city put all its eggs in one basket. While the city can control its Runyon Well Field, it cannot control activities in the municipality in which its Runyon Well Field is located.

Controlling access to the Runyon Well Field is one of the components of the groundwater protection plan developed by the City of Perth Amboy in response to toxic contamination problems. A right-of-way dirt road owned by Old Bridge Township runs through the Runyon Well Field. The road receives considerable local traffic and Old Bridge Township has refused requests to control access with a gate (Rakowski, 1985).

Fencing the portions of the Runyon Well Field not intersected by the public road is a component of the groundwater protection program that has been implemented. The Deep Run portion of the Runyon Well Field, about one-third of the total area, has been enclosed by a fence. The fenced-in area includes the recently developed 70-acre pond/recharge system, two high-capacity wells, and water transmission lines. This system and the fence were paid for with a grant of $2.5 million from the U.S. Economic Development Agency. The grants are part of a jobs creation program to be achieved by strengthening basic infrastructure and stimulating the economic prospects of urban areas. The Middlesex County Planning Board played a major role in assisting the City of Perth Amboy in its application for this Economic Development Agency grant (Rakowski, 1985). It is of interest that some of the effectiveness of the fence decreased when Old Bridge Township refused permission for Perth Amboy to use barbed wire along the top of this fence (Cesnik, 1985).

Most components of groundwater protection plans used by other municipalities are not available to the City of Perth Amboy because its Runyon Well Field is located in a different municipality. Perth Amboy has no control over activities in Old Bridge Township. The city's attempts at controlling land use or new development near the Runyon Well Field have not succeeded. A Superfund site, Burnt Fly Bog, is located 3 to 4 miles up-gradient of the Runyon Well Field (Cesnik, 1985). Perth Amboy is not even able to control the disposal, storage, or handling of hazardous materials in the areas surrounding its Runyon Well Field. Nor can Perth Amboy implement and operate a surveillance program for hazardous wastes in Old Bridge Township.

The options available to the City of Perth Amboy in planning for a dependable and safe supply of drinking water in the future are limited. Many of the most effective elements of groundwater protection programs are not available to the city because its source of water is located within a different municipality. At present the City of Perth Amboy is meeting its requirements for potable water through skillful operation of its artificial recharge system within the Runyon Well Field and the purchase of a small quantity of surface water from a private purveyor. Experience has proven that the Runyon Well Field is susceptible to serious groundwater contamination with microcontaminants. Strong pressure for growth and development in Old Bridge Township will increase the potential for future groundwater contamination problems.

One option available to the City of Perth Amboy would be to build sophisticated treatment facilities capable of removing microcontaminants. These facilities are not presently needed. Perth Amboy could wait for a contamination event to stimulate a crisis before constructing treatment facilities, or the city could construct a treatment facility in anticipation of another crisis caused by microcontaminants. Under the latter scenario, Perth Amboy would be able to avoid a crisis caused by the necessity of closing the Well Field in response to a serious toxic contamination problem. Such a treatment system would be expensive because the toxic contaminants threatening the Runyon Well Field include heavy metals and heavily chlorinated hydrocarbons as well as volatile organic chemicals. The threat presented by this array of potential contaminants suggests that sophisticated and new technological approaches to water treatment would be required and that several new treatment technologies might be needed to remove the toxic contaminants that Perth Amboy has experienced in the past and may experience again.

The City of Perth Amboy could develop or sell its valuable Runyon Well Field property and purchase treated surface water from one of the private purveyors in the area. A sufficient supply of surface water is probably available because of the efforts of the New Jersey Department of Environmental Protection in developing its Water Supply Master Plan. Developing or selling the Runyon Well Field would generate financial resources far greater than needed for purchasing water. While this is obviously a drastic step, it may be an option other municipalities will also be forced to evaluate as groundwater contamination problems with toxic contaminants become widespread.

REFERENCES

Barksdale, H. C. (1943). "The Ground-Water Supplies of Middlesex County, New Jersey," Spec. Rep. No.8. New Jersey State Water Policy Commission, Trenton, New Jersey.

Burby, Kaiser, E., Miller, T., and Moreau, D. (1983). "Drinking Water Supplies: Protection through Watershed Management. Ann Arbor Sci. Publ., Ann Arbor, Michigan.

Cesnik, W. (1985). Interview with Middlesex County Planning Board, Senior Environmental Planner, June 25.

City of Perth Amboy (1973). "A Study of Deep Run Dam, Reservoir, and Watershed" (unpublished report). Department of Municipal Utilities.

Geraghty and Miller, Inc. (1976). "Task 8 Ground Water Analysis, Lower Raritan/Middlesex County 208 Water Quality Management Planning Program" (unpublished).

Home News (1977). July 14, New Brunswick, New Jersey.

Home News (1979a). July 27, New Brunswick, New Jersey.

Home News (1979b). October 25, New Brunswick, New Jersey.

Langenohl, M. (1979). General Engineer and Superintendent of Public Utilities and City Engineer, City of Perth Amboy (telephone interview, October 6).

Middlesex County Planning Board (1978). "General Statistics for Middlesex County." p. 28. MCPB, New Brunswick, New Jersey.

Middlesex County Planning Board (1979a). "Policies and Practices for Managing Middlesex County's Groundwater Resources," P. 58. MCPB, New Brunswick, New Jersey.

Middlesex County Planning Board (1979b). "Water Supply Planning Alternatives," p. 135. MCPB, New Brunswick, New Jersey.

Middlesex County Planning Board (1984). "Preliminary Population and Housing Projections, Implied Household Sizes, 1990 and 2000" MCPB, New Brunswick, New Jersey.

News Tribune (1977a). July 12, Woodbridge, New Jersey.

News Tribune (1977b). July 28, Woodbridge, New Jersey.

Rakowski, J. S. (1985). Interview with City of Perth Amboy Assistant Engineer, June 24.

U.S. Bureau of the Census (1981). "1980 Census of Population and Housing." U.S. Govt. Printing Office, Washington, D.C.

11

Santa Clara Valley (Silicon Valley), California, Case Study

THOMAS LEWCOCK
City of Sunnyvale
Sunnyvale, California 94086

I. OVERVIEW, PAST AND PRESENT, OF THE SANTA CLARA VALLEY

Located some 50 miles south of the City of San Francisco and nestled against the southern portion of San Francisco Bay, the Santa Clara Valley is today wrestling with groundwater problems and other problems created by rapid urbanization (see Fig. 1). In general, these problems pose a dilemma for those who live and work in the valley who wish to preserve a high quality of life while at the same time profiting from the economic dynamics that have brought wealth to many and a high standard of living to most.

The Santa Clara Valley, formed by the Diablo mountain range to the east and the Santa Cruz range to the west, has through its history benefited from its benign Mediterranean climate. The valley floor, created over eons by rich alluvial materials washing from its adjacent mountain ranges, attracted native Americans, then Spanish rancheros, and later a thriving agricultural economy.

The more recent history of the valley is one of rapid social and economic transition. In the not too distant past, the area was known as the "Valley of the Heart's Delight," a reflection of mile after mile of fruit orchards. Valley residents of 30 years ago could walk uninterrupted through cherry, apricot, and plum orchards. A major portion of the population was employed in the agriculturally based economy of the valley. As post-World War II development began, one could find inexpensive land and affordable housing. With its remarkable climate, rich soils, and abundant water supplies, the valley attracted immigrants from Japan, Yugoslavia, China, Ireland, Germany, etc., which, when

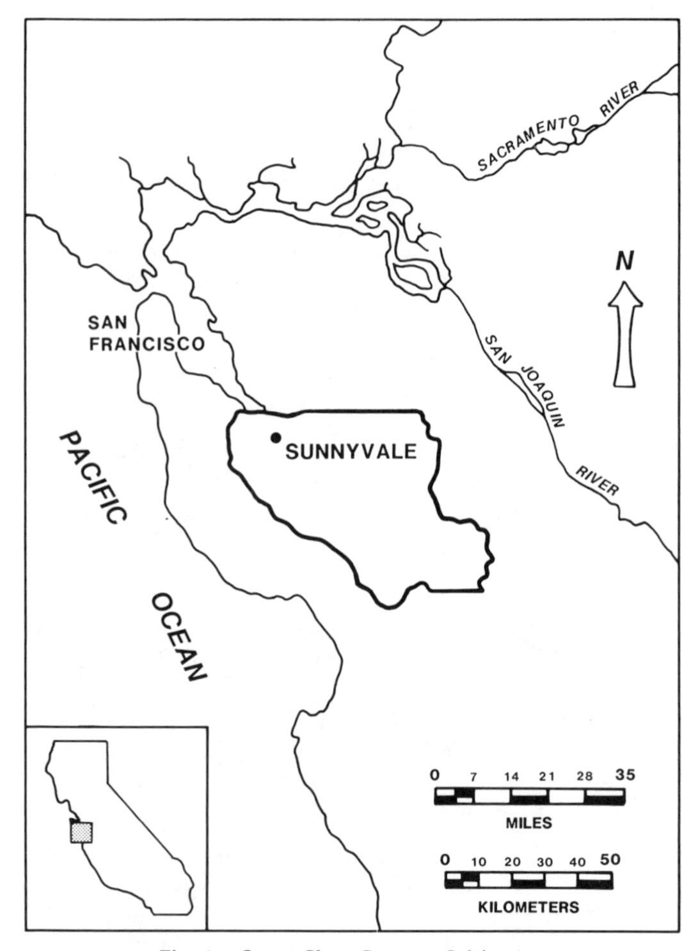

Fig. 1. Santa Clara County, California.

coupled with its Mexican-American population, made for an incredibly diverse mix of cultures, languages, and peoples.

With the agricultural period of the Santa Clara Valley came the first concern regarding water management. In the mid-1850s farming of the valley started in earnest. Irrigation was required because of the semiarid conditions of this part of California. With the invention of the deep-well turbine and a dry period from 1898 to 1908, the water table began to drop. By 1933, the valley floor had begun to subside because of excessive water extraction. By the 1940s, water management in the valley had to respond to a new challenge, rapid industrialization and urbanization (Santa Clara Valley Water District, 1978).

Today's scene is very different from that of 100 years ago. Where cherry orchards once stood, mile after mile of industrial parks now stand. Where apricot orchards once flourished, housing tracts now provide the shelter for the workers who have come to the valley. Santa Clara County's population stands at 1.4 million, now the largest county in the nine-county San Francisco Bay Area Region. This transition from rural to urban parallels the rapid post-World War II transition of not only California but the entire Sunbelt. In once central way, however, it reflects a uniqueness of national and international importance. Namely, during this transition period Santa Clara Valley has become the Silicon Valley, a mecca of high technology industrial development, and with this distinction have come groundwater contamination problems.

In several respects, knowledge of this transformation is necessary to gain an understanding of the valley's groundwater issues, as well as the approaches that have been taken to resolving them. During midcentury, the Lockheed Corporation, located in Burbank, California, created a new company called The Lockheed Missiles and Space Company that was dedicated to a new mission beyond the production of aircraft, namely, missiles, space, and other new and high technologies. Shortly after their formation, corporate officials decided that it would be wise to seek a new location that afforded the opportunity of a well-trained labor supply, cheap land, affordable housing, and room to grow. They selected the city of Sunnyvale, adjacent to Moffett Naval Air Station, a military air installation being a prerequisite in their decision-making process. With reasonable rapidity, the new company grew to over 20,000 employees and remains the largest employer in the Santa Clara Valley today. As high technology began to blossom in the 1960s and 1970s, the valley served as a magnet in attracting start-up companies that later evolved into the well-known high technology companies of today. These include the major integrated circuit companies, such as Intel, Signetics, and Advanced Micro Devices, and the mainframe computer companies, such as IBM, Amdahl, and Hewlett–Packard. There are also those companies that launched the successful revolution in office equipment, personal computers, and home computers, such as Apple and Atari. There are the plethora of companies primarily dedicated to high technology defense interests, such as Lockheed, ESL, and Ford Aerospace. NASA/Ames Research Laboratory and the Stanford Research Institute are also located in the valley. The list goes on and on, with the significance being that today the Santa Clara Valley has built its economic foundation on high technology.

The political landscape has changed as well. Today the valley is composed of fifteen cities, each of which maintains considerable independence from one another with considerable latitude over controlling its own destiny. The original high technology cities of Sunnyvale, Santa Clara, Palo Alto, and Mountain View have now come close to their maximum potential of growth. The city of San Jose, representing approximately one-half of the valley's population and now one of the fastest-growing cities in the United States, is blessed with substantial additional vacant land to allow further industrial growth and is the primary source of additional land for housing.

II. THE WATER SYSTEM OF THE SANTA CLARA VALLEY

Water use and water management in the Santa Clara Valley reflect the delicate balance required to support an industry and population whose demand for water far exceeds the natural supply available within the county. While at one time the groundwater basin was capable of naturally replenishing itself, this has not been the case for at least 80 years. In 1913, farm interests attempted unsuccessfully to secure a federal water project for the valley. After two unsuccessful votes, the voters approved formation of the Santa Clara Valley Water District in 1929. By 1933 the valley floor began to subside, prompting the water district to construct the first six reservoirs to capture runoff from the adjacent mountains and then recharge the water table through a series of percolation ponds (Santa Clara Valley Water District, 1978). Between 1940 and 1950 the valley's population doubled and industry began to develop. Subsidence again began to occur in the 1940s. Local reservoir capacity was doubled with the addition of two more reservoirs in the early 1950s. Even with these additions, demand exceeded locally available sources. In 1965 deliveries of water through the South Bay Aqueduct from the Sacramento/San Joaquin River Delta commenced. In 1977 the district executed an agreement with the United States Bureau of Reclamation to import water from their San Felipe reservoir. The aqueduct to carry this water is expected to be completed in the late 1980s. The San Francisco Water Department also supplies wholesale water to certain portions of the valley. The net result is that today approximately 50% of the valley's water needs are met through withdrawal from the underground water basin (California Department of Health Services *et al.*, 1984).

In summary, drinking water in the Santa Clara Valley comes from three sources: (1) surface water imported through the South Bay and Hetch Hetchy aqueducts; (2) water drawn from groundwater wells; and (3) impounded surface water (California Department of Health Services *et al.*, 1984). Large volumes of local surface water and imported water are used to recharge the groundwater basin artificially. This is necessitated because the draw on groundwater supplies greatly exceeds the capability of the groundwater basin to replenish itself. Surface waters are impounded by a series of reservoirs rining the valley, capturing water runoff from the adjacent mountains. While the valley floor itself receives approximately 14 inches of rain annually, predominantly between November and April, the adjacent mountains receive upward of 30 to 40 inches of rainfall annually.

The City and County of San Francisco operate the Hetch Hetchy aqueduct that brings water from the Sierra Nevada to parts of the Bay Area. Hetch Hetchy water supplies about one-fifth of the valley's potable water. It is treated at four facilities between Yosemite National Park and the Santa Clara Valley, where it is sold to water retailers (California Department of Health Services *et al.*, 1984).

The Santa Clara Valley Water District operates two treatment plants for water imported from the Sacramento River Delta through the South Bay Aqueduct and sells the treated water to water retailers. Approximately one-fourth of the valley's water supply comes directly from the South Bay Aqueduct. The Santa Clara Valley Water District is also responsible for the overall management of the valley's groundwater resources, including recharge of the groundwater basin (California Department of Health Services *et al.*, 1984).

Residents and businesses in different parts of the valley are served by different water sources, or combination of water sources. The majority of the valley utilizes at least some groundwater for its water needs. Some areas are served only by groundwater and others only by surface water. Water is distributed by nineteen different water retailers. Eleven of the water retailers are city water systems. Eight are large, private water companies. The nineteen major water retailers operate approximately 300 groundwater wells. In addition, there are more than 200 small water systems that serve more than three and less than 199 users. It is estimated that in the northern part of the county there are approximately 1200 wells serving three or less connections, these wells being shallow in contrast to the deep wells of the large retailing agencies. In the southern portion of the county, it is estimated that there are 4000 private wells. The Santa Clara Valley Water District estimates that approximately 14% of the groundwater used in the valley for nonagricultural purposes is derived from 5200 private wells (Santa Clara Valley Water District, 1978).

All imported surface water is thoroughly treated. The majority of groundwater is supplied to the end user without treatment, although a few of the public water supply wells are chlorinated prior to use.

Three primary subbasins exist within the Santa Clara Valley. Two of these subbasins are relatively minor, serving the furthest southern portion of the valley, which remains relatively unurbanized and primarily agricultural. By far the largest subbasin is the Santa Clara Subbasin, which serves the majority of public water supply wells.

Although the Santa Clara Subbasin is geologically divided into several distinct parts, it can generally be described as being divided between a confined zone and a recharge zone. The recharge zone, generally found in the southern portion of the Santa Clara Basin, is composed of materials that are highly permeable and consists of one major aquifer. This area is highly susceptible to contaminant spread from tank leaks, agricultural contaminants, and septic systems because of its high permeability. This is the portion of the subbasin where percolation ponds are utilized to continually replenish the basin. The northern portion of the valley and the groundwater basin is generally the confined zone, which is characterized by a silty clay layer approximatley 100 to 200 feet deep that separates a shallow upper aquifer serving most of the private wells from the lower aquifer that is used for most of the public drinking water systems (see Fig. 2). While this silty clay layer provides a formidable barrier to the spread of contaminants from the surface, at present it is unknown whether this barrier will preclude the spread of contaminants already existing in the shallow aquifers to the deeper levels (California Department of Health Services *et al.*, 1984). One of the critical problems

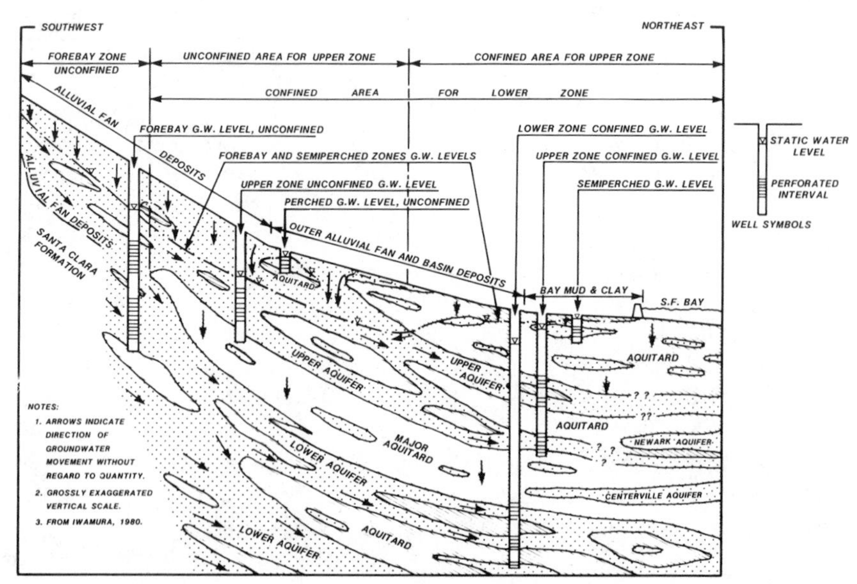

Fig. 2. Diagrammatic geologic profile of groundwater in Sunnyvale, California, area.

is the existence of a large number of abandoned wells once installed for agricultural purposes, whose locations are partially unknown and that could serve as a major conduit to the deeper aquifer.

Drinking water standards are established and regulated by the California Department of Health Services. All water supplies with over 199 connections are directly monitored by the State Department of Health Services for compliance with their regulations. The Santa Clara County Health Department utilizes the same regulations in monitoring water supplies with between five and 199 connections. At present, no governmental agency has responsibility to monitor drinking water quality in private wells with less than three connections (California Department of Health Services *et al.*, 1984).

Historically, drinking water quality in the valley has been high. In the past and even in the present, primary quality problems have derived from water imported by the Santa Clara Valley Water District and not from public drinking water supplies from deep aquifers. Because the Santa Clara Valley Water District derives its water from the Sacramento River, significant treatment is demanded. As a result of the chlorination process, trihalomethanes (THMs) are created and at times approach or exceed federal drinking water standards of 100 parts per billion. There is reason to believe that pesticides may contaminate private drinking wells, particularly in the southern portion of the county as they leach into the shallow aquifers. Further, high nitrate levels exist in the southern portion of the valley. Recently the state legislature approved an extensive private well testing program to determine the nature and extent of contamination.

III. CONTAMINATION ALARM—A NEW AWARENESS

Although the valley's aquifers are susceptible to contamination from a variety of sources such as past and current agricultural contamination, septic systems, etc., the primary focus of concern of local and now state and federal officials resulted from the discovery of leaking underground tanks. Literally thousands of these tanks exist in the valley, as is the case throughout the United States, for the storage and distribution of gasoline. Thousands more in the Santa Clara Valley, however, contain a variety of solvents and other hazardous chemicals used in the manufacturing process of the high technology industries.

At one time, the valley luxuriated in its unique capability to capture the "new" industry of the United States. Although very broad in its work, high technology industries were thought not only to be the most vital and economically progressive of American industries, but also ones that brought, along with jobs, a clean and safe work environment and manufacturing processes. Early in the development of the Santa Clara Valley, cities developed rigorous development standards to protect the visual attractiveness of the burgeoning new industry but also to assure a more fire-safe environment. For both fire safety and esthetic purposes, the majority of hazardous material storage tanks were required to be placed underground. The rate of economic development and technological innovation was so rapid that state building codes (mandatorily enforced by local government) could not keep up with this new and different industry. As a result, industry utilized traditional tank and piping methods in their construction. These methods fully complied with building codes existing at the time and neither industry nor regulatory authorities had any reason to believe that the type of material stored could not be safely contained using these methods. Safety and housekeeping practices were often marginal, resulting in certain chemicals coming into contact and reacting in unknown ways. Substantial quantities of chemicals and reaction products were inadvertently spilled, for example, purity of the chemicals used was essential, so it was not unusual for deliverers of material to initially purge their lines onto the ground prior to putting the hazardous chemicals into underground storage tanks. Additionally, the contamination capability of petroleum products was woefully underestimated. Many believed, and the petroleum industry argued, that their storage methodologies were safe, even though leaking petroleum tanks were known to exist. Further, arguments were presented that even if leaks occurred, they did not present major contamination issues because released gasoline would float on top of the shallow aquifers and therefore not present problems for drinking water supplies. Only later was it learned that the additives in gasoline will separate out and migrate throughout a groundwater aquifer. Some of these additives are highly carcinogenic and do present a major health risk.

The petroleum industry, which has had perhaps the longest history of storing a known product underground, presents a revealing example of why traditional underground storage methods cannot assure that products will be contained. Tank storage technology has dramatically improved and yet time will increase the probability that single-contained tanks and piping systems will breach eventually. It is important to understand that the experience in the Santa Clara Valley with both the petroleum and high technology industries has not solely been one of leaking tanks. The probability of leaks, in fact, is much greater in piping systems than in tanks themselves. Leaks in piping systems are more serious than tank leaks, in that pipe leaks are much more difficult to identify and containment strategies are more difficult to implement.

A. Roles and Responsibilities of Federal, State, Regional, County, and City Governments

Before describing the sequence of groundwater contamination events that initially raised alarm and then compelled action, it is important to gain an understanding of the complexities of intergovernmental relationships regarding groundwater protection in the Santa Clara Valley. In some ways this complexity provided the opportunity of concerted action among certain governmental agencies while other agencies were unable to take effective action. In other words, the complex intergovernmental net may in some ways have facilitated the development and implementation of groundwater protection policies. At the same time, during the period that groundwater protection strategies have been in development, the complex intergovernmental situation has made a unified and rapid approach to solutions difficult.

The Federal Environmental Protection Agency (EPA) has broad authority over groundwater resources. National drinking water standards are established by the EPA, which also has authority for cleanup through the federal Superfund Program. Although EPA's authority is very broad, its tendency in California is to work directly with the State Department of Health Services and the State Water Board. Thus, while its authority is very broad, it tends not to impose its authority over that of state and local agencies. The State Department of Health Services has delegated authority from EPA under federal law and is charged with the responsibility of enforcing the state Hazardous Substance Act that regulates hazardous wastes. It also has authority under both state and federal laws for both the state and federal Superfund cleanup programs. One branch of the Department of Health Services has direct regulatory authority over public water systems with 200 or more connections. In this role, the Department of Health Services establishes drinking water standards and regulates water purveyors to assure that national and state standards are met.

The State Water Resources Control Board has broad authority over surface and groundwaters. Enforcement of their policies is delegated to a series of regional water quality control boards throughout the state. The San Francisco Regional Water Quality Control Board has the Santa Clara Valley within its jurisdiction. Concerning groundwater, the regional board is the responsible agency to abate known underground water contamination and establish requisite cleanup standards. In certain instances, the responsibilities of the Regional Water Quality Control Board and State Department of Health Services overlap.

As noted earlier, the Santa Clara Valley Water District is responsible for flood control and management of groundwater resources in Santa

Clara County. Even though this role is a broad one, it does not include the establishment of standards for either groundwater or drinking water, nor does it have any power to require the abatement of contaminated sources.

The Santa Clara County Health Department monitors small water systems with five to 199 connections in accordance with state drinking water standards. In addition, as a health department covering the entire county, it has reasonably broad discretionary authority in public health areas.

Cities have a substantial latitude of action regarding certain aspects of the groundwater issue. First, cities may adopt requirements that are more restrictive than state law in those circumstances where state law does not preempt city involvement. Generally, cities' authority to act for the abatement of contaminated groundwater has been limited by the state government, although in certain circumstances state agencies permit prosecution by cities under state law through delegation. In such cases, a city attorney's office, when authorized by the State Department of Health, would prosecute a party for illegal discharges (tank leaks or surface spills) to gain criminal penalties as well as a court-ordered abatement program. Cities are charged with the responsibility of enforcing state building codes. Construction and monitoring requirements of the storage of hazardous materials compose an area where cities have considerable latitude to act. Cities are also directly involved as most are purveyors of potable water to the businesses and residents of their communities. In such a capacity, they are required to monitor drinking water quality and report to the Department of Health Services the results of periodic water quality tests. The drinking water quality monitoring requirements applicable to all public water systems are based on EPA's Interim Primary Drinking Water Standards, adopted in the California Drinking Water Standards, California Administrative Code, Title 22.

B. The Problem Is Identified

The first major industrial solvent leak was found at IBM's south San Jose facility in the fall of 1979. In the fall of 1978, an International Business Machines facility on the East coast developed leaks in an underground tank storing organic solvents (see Chapter 12). IBM subsequently began a corporationwide tank monitoring program. Through this program, contaminated soil was discovered around the underground tanks at the San Jose facility. Monitoring wells were installed to determine the extent of contamination and the San Francisco Regional

Water Quality Control Board was notified. In 1981, the storage tanks were removed, the contaminated soils were removed, and additional monitoring wells were installed. Through the intricate set of monitoring wells it was found that the contamination had migrated a great distance. Soon thereafter, a solvent waste tank leak at San Jose's Fairchild Camera and Instrument facility was discovered. Fairchild also removed the contaminated soil and installed a groundwater monitoring system. Again, the monitoring system indicated that contamination had spread several miles from the site. Both of these facilities are located in the recharge area of the Santa Clara Basin.

Investigation of the Fairchild Camera and Instrument Company leak by the San Francisco Regional Water Quality Control Board and State Department of Health Services provided an example of the potential order of magnitude of tank leaks. That investigation revealed that about 60,000 gallons of waste solvent had been released over at least 18 months. When the leak was discovered, the Great Oaks Water Company Well No. 13, serving about 700 people, was found to be contaminated with 1,1,1-trichloroethane (TCA) at approximately 5800 parts per billion (ppb). The well was, of course, immediately taken out of service (California Department of Health Services *et al.*, 1984).

In 1981, the Santa Clara Valley Fire Chiefs' Association formed a task force to study new regulations for the storage of hazardous materials, both above and below ground. The establishment of this task group was a matter of coincidence in relationship to the newly found leaks at IBM and Fairchild. In fact, their original concern was the safety-related aspects of storage of hazardous materials aboveground. As their effort continued, however, it became clear that the storage and management of hazardous materials was a major issue both above and below ground. Quickly, their task force expanded to include industrial, general community, and environmental interests. Their labors extended over a year and a half in the development of new storage regulations (California Department of Health Services *et al.*, 1984).

In March of 1982, the San Francisco Bay Area Regional Water Quality Control Board initiated a leak detection program to define the overall magnitude of the problem (California Department of Health Services *et al.*, 1984). They surveyed over 2500 companies and, based on company responses, initiated subsurface investigations through monitoring wells to determine whether tanks were leaking. Over 70 leaking tanks were found throughout the Bay Area as a result of this program, with most being located in the Santa Clara Valley. Firms selected were those thought likely to house underground storage tanks. Companies were required to bear the cost of the monitoring wells and testing analysis.

This program was voluntary; however, most companies complied with the request.

This program as well as subsequent monitoring efforts (to be described later) revealed that a bewildering number of chemicals had found their way into the soils and water, including TCE, TCB, freon, toluene, methylene chloride, chloroform, acetone, xylenes, PCE, benzene, dichloromethan, DCE, DCA, chromium, hexane, and a host of others (California Department of Health Services *et al.*, 1984). As a result of the activities of the regional board and the monitoring requirements of the Model Storage Ordinance (to be described later), over 700 industrial and petroleum leaks have now been identified in the Santa Clara Valley. The number of leaks is expected to increase substantially over time as the local ordinance takes full effect.

C. Action Begins, Detection and Prevention

With the identification of the IBM and Fairchild leaks and contaminated wells, the public outcry and local government's concern intensified. The quality of drinking water was transformed overnight from a little thought about issue to the major public policy issue in the Santa Clara Valley. The public demanded action and yet the intracacies of intergovernmental relations provided a confounding process of who was responsible, who should act, and who could not. Through the efforts of the Santa Clara Valley Fire Chiefs' Task Force, awareness was built as to the essentialness of much stronger storage and monitoring requirements than had existed before. Over an 18-month period, this task force, representing a broad diversity of interests, hammered away at new local regulations. The Fire Chief's Association included all groups and individuals who wished to join the task force. The Santa Clara County Manufacturing Group, representing the 60 or so largest employers in the county, and the American Electronics Association assigned representatives. The Santa Clara County Toxics Coalition, representing various environmental groups such as the Sierra Club, also actively participated.

The process was not an easy one from a technical perspective or from a political one. Environmental interests not only demanded more rapid action, but the most rigorous of standards. High technology industries provided an active partnership role in the technical development of standards, but did provide some resistance to the full scope of the regulations as they were eventually developed.

It became clear during the development of the new regulations that the method by which the regulations developed by the task force would be applied would be as important as technical requirements. Industries in the valley had plant locations in various cities and rightly saw the need for uniformity in the ordinances adopted from one city to the other. The process of attaining uniformity in the ordinance proposed by the Fire Chiefs' Association Task Force was known to be very difficult given the different political compositions and philosophies of various city councils. As a result, the proposed ordinance that the task force developed was first given to the Santa Clara County Intergovernmental Council. This council has no legislative authority and was established as a part of the county charter to bring together elected officials from each of the county's cities, the county itself, and the various special districts within the county. Its purpose is to review matters of countywide importance. The Intergovernmental Council was believed to be the appropriate body for first review of the proposed Model Ordinance. First, it provided a countywide forum for hearings and debate on the regulations. Second, because its membership included councilmembers from each of the cities, it was believed that these individuals could become highly informed about the nature of the problem and in turn bring this knowledge back to each of their individual city councils. Three hearings were conducted over a several-month period with large turnouts of concerned citizens, environmental groups, local industrial interests, and national petroleum product interests. After conclusion of the hearings, the Intergovernmental Council voted to endorse passage of the Model Ordinance by the county's 15 cities. Each city in turn held their own set of hearings with the same diversity of interest groups present. The petroleum industry argued strongly for exclusion from the Ordinance. They argued that their industry had solved the storage problem and that petroleum products did not pose a threat of groundwater contamination. Their arguments were not persuasive with the various city councils and ultimately they were included as a regulated industry. (It is noted that by 1985 service stations represented over 70% of the known leak sites.)

The strategy of establishing a countywide forum before referral to city councils for aciton was effective. During the spring and summer of 1983, the Ordinance was adopted by all major cities in Santa Clara County. Three small communities chose not to adopt the ordinance. Two of those communities allow only residential development and the third receives protection services from a county fire district and therefore relies on county enforcement. The provisions of the Ordinance have now become state law through the initiative of State Assemblyman

Byron Sher from Santa Clara County. The "Sher" Bill was enacted in *Health and Safety Code* Sections 25280–25289 (by Stats. 1983, c.1046) and as of January 1984 must be enforced by each city or county in California.

Although the Ordinance covers both above and below ground storage, only the underground storage portions of the Ordinance will be focused on for purposes of this case study. There are two central objectives of the Ordinance. The first is detection and the second is prevention.

Materials regulated by the Ordinance encompass the full spectrum of hazardous materials. Included are those materials already regulated under federal and state law, building codes, and fire codes. For example, materials regulated include those defined as hazardous by the California Administrative Code, EPA pollutants lists, flammable liquids, Class II combustible liquids and Class III combustible liquids as classified by the National Fire Protection Association, and materials regulated by the State Department of Industrial Wastes. Specific exclusions are also provided, such as materials appropriately packaged and sold for retail sales and certain metals when stored in a solid state (Sunnyvale Municipal Code, 1983).

Under the Ordinance all business and industrial interests that have any amount of the regulated materials must apply for a permit to store their materials. The initial step of the permitting program is for applicants to provide their respective cities with a full inventory of the materials on their site, the amount, and how they are presently being stored. To assure that all locations would respond, most cities, through their fire services, conducted detailed field surveys of literally all business and industrial locations within their communities (Sunnyvale Municipal Code, 1983). (These inventories are used not only for permit processing purposes, but are also essential to emergency response personnel.)

Following the filing of inventory statements, prospective permit holders are required to file plans explaining how they propose to manage the use of the materials. Volume limits were provided so that small companies using a small amount of material could provide a resonably short action plan. Larger users (more than 55 gallons of liquid product of the same material) were required to file a Comprehensive Hazardous Material Management Plan (HMMP). The HMMP includes a variety of information necessary to process a hazardous material permit as well as information to prepare the business to better handle stored materials. General information includes the name and address of business, names, titles, and phone numbers of responsible company emergency personnel, hours of operation, number of personnel, etc. A general facility description must be provided showing the location of all buildings, chemical loading areas, parking lots, internal roads, storm and sewer drains, and specifications of adjacent property uses. At the direction of the permitting agency, information regarding the location of wells, flood plains, earthquake faults, and surface waterbodies can also be required (Sunnyvale Municipal Code, 1983). Further a "facility storage map" is required. Such maps show the location of all stored hazardous materials, the location of emergency equipment related to each storage facility, and the name, quantity, and location of storage of all regulated materials. In regard to storage tanks, the capacity of each must be provided and, of course, the inventory of materials stored. Each HMMP must contain a description of the methods to be utilized to ensure separation and protection of materials from factors that may cause fire or explosion, production of hazardous vapors, or deterioration of the containment device (Sunnyvale Municipal Code, 1983).

For all existing industries and businesses that store underground materials, monitoring devices are required. Most cities allow only vapor monitoring devices or groundwater monitoring wells. These wells are required to be located in such a fashion that they are capable of detecting leaks that presently exist from either tanks or piping systems. The number of monitoring wells and their location is determined by the Santa Clara Valley Water District in cooperation with the various cities. The district's hydrologists ensure than an appropriate number of wells are required to assure detection of contaminants but also to ensure that appropriate drilling methods are used so that monitoring wells will not become a conduit for the spread of contamination. The number of wells varies substantially based on the size of the tank field and the hydrology of the area. A typical service station may have from two to six monitoring wells and a large industrial tank site may have twenty or more. Further, soil borings are required prior to the installation of monitoring wells to ascertain whether contamination presently exists. Under the Ordinance, if a leak is found to exist, in most circumstances the tank must be removed. If it is replaced, it must be replaced under new construction standards.

With the monitoring devices also comes the requirement of a monitoring program. Included in this program are the specifying of monitoring frequency by the responsible company and recordkeeping of inspections and their results. Further, immediate notification of the enforcing jurisdictions is required for any known or assumed leak. A "whistle-blower" clause was included in the Ordinance to protect employees who bring a potential violation to the attention of the enforcing agency (Sunnyvale Municipal Code, 1983). In several cases the appropriate enforcement agency has been contacted about a potential ordinance violation by a company employee who was not authorized by the company to do so. In no case has the "whistle-blower" provision been used, presumably because of its inherent deterrent value.

All new underground installations are required to be fully double contained for both tanks and piping, and are required to have an intricate set of monitoring devices to detect any leak between the primary and secondary containment devices. Double containment of piping has presented unique challenges. In the case of service stations, it resulted in the use of a flexible material used to line an entire excavation, therefore serving as a double containment device for tanks and piping. The predominant method for industry is the use of concrete vaults. The Ordinance is directed toward performance standards as opposed to specifications, and as a result a variety of techniques and technologies are being utilized in new construction (Sunnyvale Municipal Code, 1983).

Finally, the Ordinance requires public agency inspection. These inspections are intended to assure that hazardous materials management plans are being followed and that frequent monitoring results are being maintained. In all cases in Santa Clara County, trained city fire department personnel conduct the inspection at least once annually. Several cities have specified a twice a year frequency.

It became clear throughout the Ordinance development process that the state of the art of underground containment could not preclude leakage from underground tanks and pipes. As a result, the strategies in the Ordinance of monitoring and double containment were intended to place into effect fail-safe mechanisms that hopefully assure, even if a leak moves beyond secondary containment, that it will be rapidly discovered and abatement processes can move forward more rapidly and with far less expense than had been the case before these strategies were enacted.

IV. IMPLEMENTATION OF THE MODEL STORAGE ORDINANCE

The numerous agency requirements called for under the new Ordinance were found to be extraordinary. Although the requirements were enacted in 1983, by 1985 only a small portion of those required to obtain permits had completed the full permit process. Provisional permits were issued for those that were working their way through the Ordinance requirements, but final permits could not be issued until all aspects of the permit requirements were met. It is anticipated by most cities in the valley that all permits will be issued by late in 1986. Entirely new staffs of people have been built in each city to deal with the implementation issue. Highly trained specialists are required to knowledgeably enforce the Ordinance requirements. For example, the City of Sunnyvale, with a population of 112,000 and the largest number of high technology companies in the county, commissioned a six-member hazardous material response team, with team members also used for compliance inspections. In addition, three highly trained hazardous material specialists work out of that city's Public Safety Department's Fire Prevention Bureau to permit the some 700 businesses and industries subject to the ordinance requirements. Approximately 50% of the funding for the Sunnyvale program is derived from hazardous material permit fees with the remainder coming from general tax sources, although no tax increase was required.

It is too early to determine whether the prevention and detection objectives of the Ordinance will be met. Undoubtedly, changes in the Ordinance will occur over time as a result of the bold actions dictated by the Ordinance. What is known is that the new detection standards are capable of detecting leaks that had previously gone unnoticed. The number of detected leaks has dramatically increased as the Ordinance requirements have begun to be enforced. The number of leaks found at service stations in the City of Sunnyvale in particular has been a substantial portion of the overall leaks found. Over 50% of the service stations who have reached the permit step of placing into service monitoring devices have been found to have leaking tanks. This has come as somewhat of a surprise to the oil industry and has had a substantial effect on the abatement caseload. It is likely that the effectiveness of the prevention and detection aspects of the Ordinance will be known only years after all known locations are permitted. Efforts to implement the Ordinance are ongoing. Public attention, however, has now shifted to the abatement issue.

Abatement

As described earlier, governmental roles and responsibilities shift dramatically from the regulation of storage to the abatement of known leaks. Essentially, the responsibility shifts from cities to the Regional Water Quality Control Board, although the State Department of Health Services and EPA also have significant roles to play. Early in the 1980s the role of the Regional Water Quality Control Board changed dramatically. Prior to this time they were primarily concerned with surface water discharges. As a result, their efforts were directed toward assuring that municipal sewage treatment plants met requisite state and federal discharge requirements. Because they were generally responsible for surface and groundwater, however, the detection of underground contamination thrust them into a new and important role for which they were not thoroughly prepared nor, to this day, are staffed to handle. While the issue of prevention and detection is critical to the protection of groundwater sources, the abatement of known contaminants in groundwater is equally significant. In many regards, the dilemma surrounding new regulatory standards became sublimated to the immense difficulties of abatement.

One major problem has been that of cleanup technologies. At this time, the only technology used is that of groundwater extraction. Procedurally, when a leak is found and the responsible party can be identified, the regional board requires a comprehensive study by the responsible party to determine the nature and extent of groundwater contamination. Monitoring wells are required. Frequently this defining process takes over a year or more. Once the extent of an underground plume is defined, a specific abatement plan must be filed with the regional board. The abatement plan is then reviewed by the board's staff for adequacy and then the board holds public hearings on the abatement plan. In most cases, soil removal coupled with extraction wells is specified. Extraction wells are constructed and groundwater is removed. Generally this groundwater is pumped to surface drainage channels to allow evaporation processes to remove contaminated materials as the water proceeds eventually to the San Francisco Bay. Long-term consequences of this abatement procedure are unknown, but considerable concern is being generated by those in proximity to the open drainage channels. The San Francisco Bay Regional Air Quality Control Board at this time believes that the evaporation processes do not pose serious air pollution health risks. Dramatic quantities of underground water are being withdrawn for extraction purposes with the resulting concern of the drawdown of the underground water supply. According to John O'Halloran, general manager of the Santa Clara Valley Water District, IBM alone is extracting and pumping to waste 18 million gallons of groundwater per day. Early results indicate that this drawdown is beginning to create a problem in maintaining the delicate balance of the underground reservoirs, and therefore may in the future create renewed concerns with land subsidence.

Frequently, the regional board has been presented with more difficult dilemmas. In some cases, a responsible party cannot be found because the initial source of the leak has yet to be discovered or cannot be determined in a legally definite manner. Similarly, circumstances are encountered where there are multiple leaks in the same area with companies arguing over lengthy periods of time whether they are or are not responsible. This further delays the process of abatement and increases the risks that contaminated plumes will contaminate private and public drinking water wells.

Further, even when a responsible party is found, it can be the case that the required cleanup activities are of such a cost that they cannot be absorbed by a company and as a result the company goes into bankruptcy. Although state and federal Superfunds exist to help out in these situations, the process is a slow and difficult one. Superfund resources have been targeted toward major toxic dumps nationally, and in California they are taking higher priority than many of the smaller contamination problems found in the Santa Clara Valley.

Compounding the above dilemmas is the fact that the regional board is woefully understaffed by its own admission and carries an immense caseload far beyond the capabilities of available staff. As a result, attention to new leaks must be placed in priority with some leaks known for many years without abatement activities.

The importance of rapid abatement cannot be underscored. This is the case for several reasons. First, early identification of known leaks provides the opportunity to abate a known leak while materials still remain in the soils prior to migrating to the aquifer. Rapid abatement in these circumstances, regardless of the scope of contamination, permits soils removal, an infinitely less costly and more rapid abatement procedure. Second, contaminants found in the Santa Clara Valley are known to migrate and spread throughout the groundwater table. In some areas this migration is believed to be no more than three feet a year, whereas in other areas it can be several hundred feet per year (California Department of Health Services *et al.*, 1984). With this variance in migration, the potential for contamination of public and private water supplies increases dramatically, and the cost of abatement escalates in geometric proportions. In the case of IBM, according to their representative's comment to this author, they have already spent $17 million in abatement procedures, and that abatement program is ongoing.

The costs of abatement in both health risks and burden on the responsible party, however, are only a part of the problem that demands immediate attention. As described earlier, the Santa Clara Valley has no reasonable option to its present groundwater sources. This groundwater is physically, politically, and economically irreplaceable. As a result, rapid abatement is necessary to assure that toxic pollution does not contaminate the basin to such an extent that a major portion of the available groundwater is not usable. The cost of pumping this water as a waste product or of providing treatment of the contaminated water sufficient to meet drinking water standards may be prohibitive. While the prevention and detection efforts of local governments through the Model Ordinance seem to be well under way and hold great promise for the future, the abatement issue remains unresolved.

V. WHO'S IN CHARGE—UNRESOLVED ISSUES

As can be seen in the preceding description of events and actions, the process of planning for groundwater protection and assurance of drinking water quality has not followed a comprehensive planning approach even though such an approach is obviously demanded by an issue of this magnitude. The reason this has not occurred, however, is quite obvious given the intricate intergovernmental relationships that exist within the Santa Clara Valley, the State of California, and in all probability throughout the entire nation. There is no one agency centrally responsible for the protection of groundwater, nor is that likely to be the case in the future. While that fact may be unappealing to those who are most concerned with the issue and are prepared to allocate the necessary resources, it nonetheless remains true. As a result, the continued planning for groundwater protection in the Santa Clara Valley must of necessity be an iterative one and one of shared responsibility. Further, the issues go well beyond the basics of prevention, inspection, and abatement.

A. The Santa Clara Valley Underground Water Task Force

Recognizing the shared jurisdictional responsibility regarding groundwater protection, the Santa Clara County Groundwater Contamination Task Force was formed in mid-1984. The initiative for formation of this task force came from the Santa Clara County executive officer, the chief administrative officer of the county appointed by the County Board of Supervisors. The task force is composed of representatives from the Environmental Protection Agency, State Department of Health Services, Regional Water Quality Control Board, the county executive, the general manager of the Santa Clara Valley Water District, and a city manager representing the City Manager's Association in Santa Clara County. While this task force has no direct jurisdictional authority or responsibility, it has been crucial in the identification of issues that must be resolved and the development of new legislation required to further the protection of groundwater not only in Santa Clara County but throughout the State of California. Though the task force is not political in nature, issues identified have been provided to elected officials in a variety of ways to enhance the possibility of legislative action. The number of issues identified and raised by this task force is massive, but they can be consolidated into a limited number of categories.

1. **Consolidated responsibility and authority at the state level.** The Santa Clara County case study has provided ample evidence that in California at the present time the capability for action, particularly with abatement strategies, is confounded by a lack of clear strategy and coordination at the state level. For example, while the Regional Water Quality Control Board is directly responsible for abatement, Superfund assistance must be coordinated through the State Department of Health Services and the Federal Environmental Protection Agency. Although the Regional Water Quality Control Board is the abatement agency, it is the State Department of Health Services that is directly responsible for the quality of water that is provided through public water systems. As a result, the local members of the Groundwater Contamination Task Force have called for the consolidation of state planning, standard setting, and implementation. In early 1986, two central measures are before the state legislature that are designed to take these steps. While each differs substantially in who would receive authority, they both have as their objective this consolidation.

2. **Resources.** Abatement of known leaks has been confounded by lack of resources of the Regional Water Quality Control Board to effectuate abatement as rapidly as technically feasible. As a result, the local members of the task force have recommended that the state sharply increase the staffing levels of the Regional Water Quality Control Board. By some accounts, that increase would have to be upward of 400 to 600% to adequately deal with known leak sites as well as those projected to be found over the next several years. Knowing that staffing increases of this magnitude may be politically impossible to obtain, the local government approach has been to request legislation that has the opportunity to fundamentally change the roles and responsibilities of various levels of government. That proposal would allow local government to step into the abatement process if the requisite state agency is incapable of meeting the objective test of moving abatement forward on each known leak site as rapidly as technically feasible.

3. **Standards.** It has become clear through the work of the Groundwater Contamination Task Force as well as other study forums that standards of cleanup for groundwater as well as standards for drinking

water are not adequate. As a part of the abatement process, the regional board must establish "how clean is clean?" as a part of its abatement orders. Few specific quantitative standards exist in this regard, and argument persists whether the level of cleanup of abatement programs presently under way is overly ambitious, adequate, or not sufficient to assure public health protection. Perhaps more difficult is the fact that, for many of the contaminants that leak from tanks, no federal or state action levels for drinking water quality have been established or health risk assessments conducted. As a result, the legislature has been requested to authorize a study of these materials to establish such action levels. (Editor's note: see Chapter 3 for a discussion of the standard setting process.)

4. Overextraction. Extracting more groundwater than is recharged to the aquifer is also known as overdraft or groundwater mining. Taken to its natural extension, continued extraction for abatement purposes of contaminated plumes of groundwater presents a clear and present danger of overextraction of the groundwater basin, with its accompanying potential of water shortages, permanent loss of some of the storage capacity of the groundwater basin, and land subsidence in certain areas. For these reasons, legislation has been sought to call for technological studies of new abatement methodologies. Proposed studies include the possibilities of wellhead treatment of extraction wells utilized for contamination removal coupled with recharge to the aquifer. (Editor's note: see Chapter 12 for an example of contaminant removal followed by recharge.) Studies of underground bacteriological treatment of contaminated areas are now also under way.

5. Cancer registry program. A cancer registry program is not established on a statewide basis. As a result, long-term evidence has not been retained for scientific study to determine the relationship between contamination and the incidence of cancer. This again has been called for under proposed state legislation.

6. Wellhead treatment. Given the possibility of spreading contamination in the groundwater basin, proposals have been made that technological research funded by the state government should be undertaken to determine the cost and treatment methodologies that may provide the opportunity for wellhead treatment of public drinking wells so that even if contamination reaches them, it will be possible to continue to deliver high-quality water to the end user. One of the anticipated difficulties in this study area is the fact that as a result of available water treatment processes to remove toxic contaminants, such as carbon filtration, chlorination is required, which may produce trihalomethanes (THMs), a group of chemicals that include a known carcinogen and that are already present in the treated surface water provided within the county.

7. **Private wells.** As described earlier, there are an estimated 5200 private wells in Santa Clara County. The majority of them are located in the south county area recharge zone with the remainder penetrating into the shallow aquifers of the groundwater basin. These wells are clearly the most susceptible to immediate contamination and yet are the responsibility of no public agency. Not only are they at risk of groundwater contamination due to leaking underground tanks, but they also suffer a high probability of contamination due to past agricultural contaminants, septic systems, etc. Legislation has been proposed that would establish a pilot testing program to improve understanding of the nature of risk and new strategic approaches to the private well dilemma. This proposed program would be paid for by the State of California under the direction of the Department of Health Services.

This dilemma goes well beyond the fact that no agency is responsible for monitoring the quality of the drinking water supply. A frequent testing program of 5200 wells would present an incredible cost. With an estimate of $500 to $600 per test, and a testing frequency of at least once a year, the cost impacts are indeed staggering. Further, these private wells are presently deriving their water at essentially no cost from the groundwater basin and therefore do not provide an ongoing source of revenue as is the case with public and private water systems that recover their costs through user fees. For this reason, some public officials question the use of public funds for essentially a private purpose. Finally, in those areas where private well owners have access to municipal water systems, it is clear that in the short run, as well as the long run, connection to municipal water systems presents a much lower cost than would an ongoing frequent testing program for each private well.

8. **Abandoned wells.** Although the Santa Clara Valley Water District has a sound management scheme and control over newly developed wells, many wells were developed long before this management strategy was in place. As a result, it is believed that there are as many as 1000 abandoned wells throughout the valley floor. In many cases their locations are unknown, and it is believed that many of them may now be located under buildings. These abandoned wells present a clear risk of acting as conduits for contaminants in the surficial aquifer, enabling

them to reach the deeper aquifers. Following the identification of this problem, the Santa Clara Valley Water District moved forward with an ambitious program to locate and seal abandoned wells. The water district has appropriated $1,000,000 in available funds to initiate the effort. Existing district records and maps, aided by cities reviewing their own records and maps, are used to identify abandoned well locations. In late 1985, 88 wells had been sealed through this program over a 9-month period. Presently, the water district does not have the capability of entering private land and demanding the sealing of wells, even though they have accepted the financial responsibility of doing so.

9. Frequency of water testing. Because of the known problem in the Santa Clara Valley, the Department of Health Services has required more frequent testing by purveyors of water supplies in accordance with AB 1803, which amended *Health and Safety Code* Section 4026.2 and added Section 4026.3 and became effective in 1985. Nonetheless, different agencies are taking different strategic approaches to this issue. Some agencies, at considerable cost, have begun ambitious programs of testing that exceed state requirements, regarding both frequency and the chemicals for which tests are performed. Others have chosen not to do so. Additionally, some water purveyors have taken the stance that they will not deliver water with identified contaminants even if they are well below state and federal action levels, whereas others provide water with contaminants as long as those levels meet federal and state standards. These practices are expected to continue, although additional pressure for more rigorous standards is anticipated.

B. Integrated Environmental Management Project

In 1984, the Federal Environmental Protection Agency selected the Santa Clara Valley as one location for its Integrated Environmental Management Project (IEMP). This project has as its intent a cross-media study of risks due to air, water, and other sources (Ott, 1985). This study, which has heavy participation by local governmental and other interests, uses a landmark approach to the health risk issue. Its goal is to examine how the combination of all sources, land, air, and water, forms a health risk profile in the Santa Clara Valley. One of the outcomes of the review will be the assessment of comparable risk. For example, is the risk of drinking water that already contains high levels of THMs greater or less than the risk of drinking groundwater with certain levels of other toxic contaminants? Another groundwater-related outcome will be a projection based on present information of the number of groundwater wells that will likely be contaminated over the next 70-year period.

Although the above description in no way reflects the full scope and magnitude of the IEMP project, it is anticipated that the results will highlight the degree of health risks resulting from the contamination problem and will have broad applicability throughout the nation. (Editor's note: see Chapter 3 for more discussion of the relative contribution of water contaminants to total human exposure.)

VI. CONCLUSION

While officials in Santa Clara County consider themselves to be at the forefront of the groundwater protection issue, the only unequivocal conclusion that can presently be drawn is that the magnitude of the groundwater contamination problem and the issues that follow are far greater than could have ever been contemplated when the first tank leak was found in 1979. This author has drawn the conclusion that, first, there is no reason to believe the Santa Clara Valley is substantially different from other areas of the country that rely heavily on groundwater sources. Second, few seem to understand the immense economic implications of groundwater contamination. Third, it is unreasonable to expect a rapid and uniform national or statewide approach to the protection of groundwater. The planning and implementation process is out of necessity an iterative one.

Fourth, local action is essential and necessary. The experience in the Santa Clara Valley has given ample evidence that the closer the problem is to home, the more likely that creative and rapid steps will be taken to its resolution. Whether local government is purely interested in public health protection or specifically interested in the economic vitality of its area, protection of groundwater is of no less importance than the basics of providing transportation systems, electric and gas supplies, or adequate shelter. Clean groundwater is the lifeblood of urban communities.

REFERENCES

California Department of Health Services, California Regional Water Quality Control Board No. 2, Santa Clara County Public Health Department, Santa Clara Valley Water District, Environmental Protections Agency (1984). "Groundwater and Drinking Water in the Santa Clara Valley: A White Paper." CDHS, Santa Clara, California.

California History Center, De Anza College (1981). "Water for the Santa Clara Valley: A History." De Anza College, Santa Clara, California.

CH_2M Hill (1985). "South Bay Area, Santa Clara County, California: Community Involvement Plan." Santa Clara County, California.

Industry Clean Water Task Force (1984). "Status Report on Cleanup Progress at 82 Sites in Santa Clara County." Santa Clara County, California.

Ott, W. R. (1985). Total human exposure: An emerging science focuses on humans as receptors of environmental pollution. *Environ. Sci. Technol.* **19**(10), 880–886.

Santa Clara Valley Water District (1978). "The Story of the Santa Clara Valley Water District." Santa Clara County, California.

Stoddard, K. (1985). "Key Components of a Comprehensive Toxics Program for California in Fiscal Year 1985–86: Report and Recommendations by the Special Toxics Budget Task Force to the Assembly Ways and Means Committee." California State Legislature.

Sunnyvale Municipal Code (1983). "Storage of Hazardous Materials," Title 20. Sunnyvale, California.

TRS Consultants, Inc. (1985). "South Bay Groundwater Contamination Task Force: Agency Coordination Project." Tiburon, California.

12

South Brunswick, New Jersey, Case Study

G. WILLIAM PAGE
Department of Urban Planning
and Center for Great Lakes Studies
University of Wisconsin–Milwaukee
Milwaukee, Wisconsin 53201

This chapter examines the planning implications of toxics contamination of groundwater by investigating a toxics pollution problem in the municipality of South Brunswick, New Jersey. South Brunswick is primarily a rural township in central New Jersey that has undergone rapid suburban growth in recent years. Planning for an adequate supply of potable water has become a serious present and future constraint on growth in the town. The contamination of the town's water supply source with toxic substances has introduced a whole series of new and serious planning problems.

I. THE MUNICIPALITY

The Township of South Brunswick is located in the southern portion of Middlesex County, New Jersey (see Fig. 1). In recent years the pressure for suburban industrial, commercial, and residential development reached the southern half of Middlesex County and surrounding areas, especially along the Route 1 corridor in the Princeton area in the immediate vicinity of South Brunswick. In addition to Route 1, the New Jersey Turnpike and Route 130 are major roads that pass through the township. Located along the major northeastern United States transportation corridor, South Brunswick is about a one-hour drive by automobile to both Philadelphia and New York City.

South Brunswick is still largely underdeveloped. The municipality contains 26,254 acres. Cropland, pasture, and woodland comprise 41% of the total land area (South Brunswick Township, 1982). From the turn

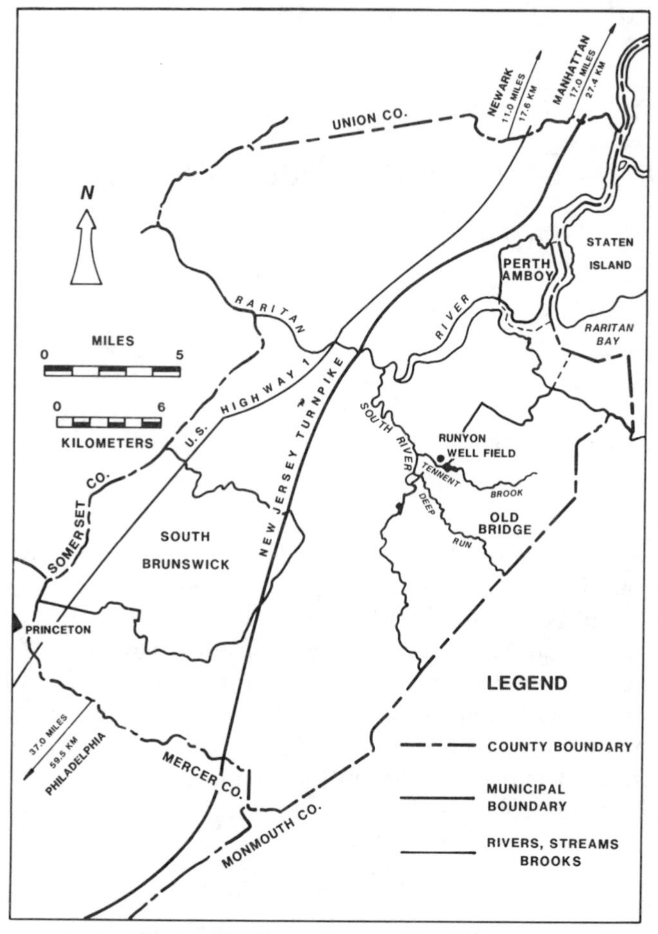

Fig. 1. Middlesex County, New Jersey.

of the century until 1950, the population of South Brunswick grew slowly, but has increased very rapidly since 1950 to a 1980 population of 17,127 (U.S. Bureau of Census, 1981). Its location and population pressures in the area lead to projected populations of 32,566 in 1990 and 48,042 in 2000 (Middlesex County Planning Board, 1984).

Supplying the water requirements of the present population and the projected future population is an important planning consideration in a municipality experiencing rapid population growth. In many respects South Brunswick is in an enviable position in supplying its water needs. The municipality is located in a humid region of the country that averages about 44 inches of precipitation per year (Middlesex County Planning Board, 1979a). It is mostly located within the inner coastal plain physiographic province of New Jersey that has several extensive high-yield and excellent water quality aquifers. The two most important aquifers for South Brunswick, the rest of Middlesex County, and extensive areas to the east and south of Middlesex County are the Old Bridge Sand and the Farrington Sand members of the Potomac–Raritan–Magothy formation. The Old Bridge Sand is the surface formation upon which much of South Brunswick is located. In 1970 the county used a total of 32 million gallons per day (mgd) of water (Geraghty and Miller, Inc., 1976). The Woodbridge Clay formation separates the Farrington Sand from the Old Bridge Sand. In 1974, the county pumped a total of about 18 million gallons per day of water from the Farrington Sand Aquifer (Geraghty and Miller, Inc., 1976). Within South Brunswick there are many private domestic wells drawing from the shallow Old Bridge Sand Aquifer while the large public supply wells and industrial wells draw water from the deeper Farrington Sand.

Despite the relative abundance, the availability of an adequate water supply presents a severe obstacle to future growth. Since 1947, New Jersey has regulated the amount of groundwater that may be withdrawn. Domestic wells are not regulated, but permits are required for pumping more than 100,000 gallons per day of groundwater (Geraghty and Miller, Inc., 1976). The state Water Policy and Supply Council uses the permit power in allocating rights to groundwater. Because municipalities pump more than 18 million gallons per day of water from the Farrington Sand while the estimated optimum yield of the aquifer is only 16.2 million gallons per day, developing municipalities such as South Brunswick are rapidly approaching their maximum diversion right without authorization for additional supplies (Middlesex County Planning Board, 1979a). (Editor's note: see Chapter 1 for a discussion of optimum yield.) South Brunswick has diversion rights of 121 million gallons of water each month, about 4 million gallons per day. In 1975, the municipal water system, entirely dependent on groundwater, used 1.72 million gallons per day while serving 13,372 people (Middlesex County Planning Board, 1979b). In 1983, the actual average daily use of South Brunswick was 2.675 million gallons per day. Projected average daily use in 2000 is 6.866 million gallons per day (Middlesex County Planning Board, 1985).

II. CHRONOLOGY OF TOXICS POLLUTION

South Brunswick operates a municipal water supply system that for a number of years depended on three large-capacity wells drawing water from the Farrington Sand Aquifer. Public supply well number 11 is one of these three wells. It was drilled in 1963 to a depth of 116 feet and has maintained a steady capacity of about 1280 gallons per minute (gpm) over the years (Merk, 1979). Well number 11 is located in the Dayton area within South Brunswick Township near the Jamesburg Road on a small lot purchased by the municipality.

In the fall of 1977, the New Jersey Department of Environmental Protection (DEP) had a sample of water from South Brunswick public supply well number 11 analyzed as part of its innovative monitoring program for toxics in the groundwater of the state (Page *et al.*, 1980; Tucker, 1981). The sample contained high concentrations of several toxic substances, especially 1,1,1-trichloroethane. Several additional samples collaborated the earlier findings. In December 1977, the New Jersey Department of Environmental Protection officially closed South Brunswick public supply well number 11 because of a threat to the public's health.

Over the following months the Department of Environmental Protection and South Brunswick Township sampled and analyzed for toxics many private domestic wells near public supply well number 11. Samples of 27 nearby domestic wells revealed detectable concentrations of 1,1,1-trichloroethane and other toxics. The South Brunswick Health Department at the request of the Department of Environmental Protection closed 11 of the private wells because they had concentrations of 1,1,1-trichloroethane in excess of 500 ppb (Geraghty and Miller, Inc., 1979).

While 1,1,1-trichloroethane was usually the organic microcontaminant found in the highest concentration, samples revealed several other toxic substances in public supply well number 11 and private wells in the immediate vicinity. These toxic substances included 1,1,2,2-tetrachloroethylene, 1,1-dichloroethylene, 1,1,2-trichloroethylene, arsenic, chromium, lead, and zinc. South Brunswick found all these toxic substances in some samples at levels considered a public health threat. Controversy surrounds the public health effects caused by consuming water contaminated with these toxic substances. The ultimate human health effects of consuming these substances at the parts per billion level over a lifetime cannot be confidently predicted. (Editor's note: see Chapter 3 for a discussion of health risks.)

III. RESPONSE TO WELL CLOSINGS

The closing of public supply well number 11 necessitated immediate action by the municipality of South Brunswick. In the short term, the municipality operated the municipal water supply system with only the two remaining wells in operation. In the long term the system needed a third large-capacity well. The municipality decided it would drill a new well because of the size of the area contaminated and the seriousness of the pollution. A new well of 1000 gallons per minute pumping capacity was developed at a cost of about one million dollars and came on line in 1980 (Harris, 1985).

The 11 private domestic wells that had been closed also required immediate action by the municipality. South Brunswick extended water mains into different parts of the contaminated area to supply public water to those residents who had depended on private wells closed because of the toxics pollution. The normal procedure of accessing the cost of extending the water mains to those residents connected to the extended system was waived because of the unique circumstances (Krueger, 1979).

The residents previously dependent on their private domestic wells incurred some expenses because of the contamination. The closing of the 11 private domestic wells because of the toxics pollution in South Brunswick entailed a number of costs. There are difficulties in placing a monetary value on the fear of long-term adverse health effects caused by having consumed a variety of toxic substances for an unknown period of years at dangerously high concentrations. Direct costs include time, effort, and money for using bottled water between the time South Brunswick closed their wells to the time they connected to the municipal water supply system. A substantial cost to the residents affected by the toxics pollution is the cost to the homeowner of connecting their homes with the municipal water mains. Depending on the distance of the home from the street, this cost about $1000 per home (Harris, 1985). Two years after the South Brunswick Health Department ordered their wells closed, two families had still not connected their homes to the public water supply system. Eight years later one family has still not connected to the municipal system and presumably depends on bottled water (Harris, 1985).

IV. CONSULTING STUDIES

South Brunswick paid for several technical consulting studies concerning the toxic substances pollution near public supply well number 11. The initial study, performed by the consulting firm Geraghty and Miller, Inc., for the municipality at a cost of about $5000, gave some indication of the severity of the problem. The second report by this firm cost $52,000 and is the most thorough description of the pollution problem available (Geraghty and Miller, Inc., 1979). An additional contract was signed with Geraghty and Miller, Inc., at a cost of about $120,000 for future study and consulting services connected with this same pollution problem.

The Geraghty and Miller report describes the groundwater characteristics in the vicinity of public supply well number 11. Geraghty and Miller drilled 30 wells for their study. They analyzed the lithologic logs and analytic results of samples from these wells plus 34 industrial, 3 public supply, and 7 private domestic wells. They found that the Woodbridge Clay that separates the Farrington Sand Aquifer from the Old Bridge Sand Aquifer is generally 50 to 90 feet in thickness. The consultant also found a window or discontinuity in the Woodbridge Clay sediments in the vicinity of public supply well number 11. Geraghty and Miller estimate that 1.8 million gallons of water per day flow through this window into the deeper Farrington Sand, the source of water for South Brunswick public supply wells (Geraghty and Miller, Inc., 1979). No organic toxic contaminants have been detected in the Farrington Sand Aquifer in the vicinity except in the area around public supply well number 11 where the Woodbridge Clay is discontinuous.

Having identified the pathway by which toxic contaminants reached well number 11, South Brunswick searched for the source of the pollution. Within the recharge area of well number 11, there are 17 companies and plants that were identified and inspected by representatives of the New Jersey Department of Environmental Protection or by Geraghty and Miller. Among these companies, seven used or stored 1,1,1-trichloroethane (Geraghty and Miller, Inc., 1979). The location of these companies and the results of several hundred chemical analyses from more than 100 wells enabled the identification of three distinct and separate plumes of groundwater contamination by 1,1,1-trichloroethane and other toxic substances.

The largest plume of toxics contamination and the only one of three plumes reaching public supply well 11 originated at the International Business Machines, Inc. (IBM), facility located about 2000 feet west of well 11 (Geraghty and Miller, Inc., 1979). This plume of contamination was 30 feet deep, 1000 feet wide, and 2500 feet long. The consultant estimated that 525 million gallons of water were seriously contaminated with toxic chemicals in the plume. The toxic chemicals included up to 12,000 ppb of 1,1,1-trichloroethane, 1400 ppb of 1,1,2,2-tetrachloroethane, 200 ppb of 1,1-dichloroethylene, 121 ppb of trichloroethylene, and 3150 ppb of zinc (Geraghty and Miller, Inc., 1979). Two of these toxics, 1,1,2,2-tetrachloroethylene and trichloroethylene, are suspected carcinogens (National Research Council, 1977).

The IBM facility has been located in South Brunswick for about 20 years and for many years has been engaged in the printing of holorith computer cards (Krueger, 1979). Employees cleaned the printing equipment with 1,1,1-trichloroethane on a regular basis. Over the years an estimated total of between 500 and 1000 gallons of 1,1,1-trichloroethane have somehow reached the soil on the IBM property (Geraghty and Miller, Inc., 1979). This quantity of chemicals is equivalent to the contents of 10 to 20 55-gallon barrels leaking, being spilled, or being improperly disposed of on the IBM property.

The other plumes of toxic contaminants in the groundwater of South Brunswick did not affect well number 11, but do present serious problems. The plume originating under Mideast Aluminum Industries is also a serious source of contamination. The consultant estimated a plume 10 to 25 feet deep, 1000 feet wide, and 1000 feet long and containing 180 million gallons of water contaminated with dangerous levels of microcontaminants (Geraghty and Miller, Inc., 1979). The plume contained up to 27,000 ppb of 1,1,1-trichloroethane, 500 ppb of 1,1,2,2-tetrachloroethylene, 1200 ppb of 1,1-dichloroethylene, 15,000 ppb of trichloroethylene, 860 ppb of arsenic, and 900 ppb of lead and zinc (Geraghty and Miller, Inc., 1979). At present three domestic wells are affected by the plume of contamination. While public supply well 12 is not affected, the consultant predicted contamination of one of the remaining two public wells in South Brunswick within 2 to 4 years based on the direction and rate of movement of this second plume of toxic contaminants (Geraghty and Miller, Inc., 1979).

V. CORPORATE RESPONSE TO TOXICS CONTAMINATION

The identification of toxic substances contaminating the groundwater of South Brunswick presented a serious threat to two of the corporations with plants in the municipality. South Brunswick quickly identified IBM and Mideast Aluminum as using the toxic chemicals found in groundwater. The IBM corporation was the most threatened by legal action because the Geraghty and Miller report identified the plume of toxic contamination originating at its facility as responsible for the forced closing of public supply well number 11 and 11 private domestic and monitoring wells (Geraghty and Miller, Inc., 1979). IBM hired a consulting firm to study the toxic substances contamination of groundwater under their property as their first action. A Dames and Moore, Inc., consulting report concluded that IBM was not responsible for the toxics contamination of groundwater that cause the closing of wells in the South Brunswick (*Home News*, 1979b). One significant weakness of this report was that its data were limited to those samples collected from wells on the IBM property.

While pleased by the findings of the Dames and Moore report, IBM was not completely satisfied. The denial of responsibility was not convincing when compared to the substantial evidence included in the more thorough Geraghty and Miller report prepared for the municipality. IBM signed a contract for another consulting study for about $200,000 to contest the findings of the Geraghty and Miller report (*Home News*, 1979b).

The second consulting report contracted for by IBM concluded that IBM was indeed responsible for the contamination of groundwater in the vicinity of public supply well number 11 (*Home News*, 1979a). IBM agreed to reimburse many of the costs associated with the closing of the wells, including the township's consultants' fees, the cost of drilling to determine the site of a new well, and the installation of a replacement well. IBM also agreed to reimburse some of the private homeowners for the cost of connecting their homes to the municipal water system (*Home News*, 1979a).

IBM initiated what at the time was an innovative method of treating groundwater contaminated with toxics on its property in South Brunswick. The toxic substances contaminating the groundwater beneath the IBM property are light chlorinated hydrocarbons. These substances are volatile. The treatment approach that IBM used entails (1) pumping the contaminated groundwater from the most down-gradient portion of their property, (2) spraying the water into the air, thereby allowing the organic microcontaminants to volatilize into the atmosphere, and (3) capturing the treated water in ponds located so that they recharge the aquifer on the most up-gradient portion of the property (Krueger, 1979). After several years, the spraying of the contaminated water into the air was replaced by a 40-foot packed column air stripping tower. This facility received an air pollution permit to discharge toxic gases after computer plume dispersion models estimated that the air pollution would not threaten human health. Variations of this system operated from 1978 until September of 1984 when an estimated 90% of the plume of contamination in the Old Bridge Sand Aquifer under the IBM property had been abated (Harris, 1985).

The second major plume of toxic contaminants in the groundwater of South Brunswick Township originating under the Mideast Aluminum property also required action. The municipality closed three private wells because of this contamination (Harris, 1985). Concentrations as high as 2200 ppb of tetrachloroethylene as well as 1,1,1-trichloroethane and trichloroethylene were detected in the plume. Starting in 1980, Mideast Aluminum has been operating a series of air stripping towers and recharging the treated water to the ground. By June 1985 the concentrations of contamination had declined substantially, but were still above levels considered safe (Harris, 1985).

While the sources of the contamination on and under the IBM property could be abated, large portions of the plume of contaminated groundwater had migrated beyond the boundaries of these industrial properties and some of the contamination had passed through the opening in the Woodbridge Clay into the deeper Farrington Sand Aquifer, where it reached well number 11. From 1977 to the summer of 1985 well number 11 was pumped and the contaminated water was discharged directly into the sewer system (pumping to waste) as a means of keeping the plume of contaminated groundwater from migrating and contaminating additional private and public wells. IBM has reimbursed South Brunswick Township for the costs of pumping the well and for the cost of treating these large volumes of water by the Middlesex County Sewage Authority.

In June 1982 the plume of toxic contaminants in the Farrington Sand Aquifer had come dangerously close to one of the two uncontaminated municipal wells in South Brunswick Township. IBM agreed to build a sophisticated water treatment system for the township that could treat the contaminated water from well number 11 and produce water that meets or exceeds all federal and state standards for drinking water. This treatment system will use a 40-foot air stripping tower in conjunction with a 20,000-pound granular activated carbon filtration system designed to produce 11,000 gallons of drinking water per minute with 5 ppb or less of each of the organic contaminants found in the plumes of contamination (Harris, 1985). The municipality constructed the facility on a parcel of land located to minimize connection costs for the other municipal wells if future contamination requires that they also receive this form of sophisticated treatment.

VI. GOVERNMENT ROLES IN DEVELOPING GROUNDWATER PROTECTION POLICIES

Various levels of government have been involved in protecting the groundwater of South Brunswick Township from future contamination with toxic substances. This section will briefly describe the contributions of these government bodies.

The federal government has played a minor role in responding to the groundwater contamination problems in South Brunswick. This role has been largely one of providing technical assistance upon request. This assistance was most useful in evaluating the proposals for and the performance of the treatment technologies for removing toxic chemicals from the contaminated groundwater.

The state government has been involved in the groundwater problems in South Brunswick in a variety of ways. The New Jersey Department of Environmental Protection (DEP) first identified the problem (Tucker, 1981). The Department of Environmental Protection and the N.J. Department of Health have provided technical assistance when requested. U.S. EPA also provided technical assistance in response to the immediate contamination problem. The N.J. Department of Environmental Protection also has contributed to plans for a future supply of water in South Brunswick Township. This has been accomplished primarily through the N.J. Water Supply Management Act of 1981. While these efforts have concentrated on water quantity issues, they have also identified the critical aquifer outcrop areas that should be protected from development or other activities that could introduce contaminants to the soil (Gaston, 1985).

In the years since the discovery of toxic contaminants in the groundwater of South Brunswick Township, the state government of New Jersey has become more involved in groundwater protection issues. Some of these efforts are the N.J. Superfund Program, the N. J. Hazardous Waste Mitigation Program, the A280 groundwater testing requirements, the Toxics Strike Force of the N.J. attorney general's office, and the Toxic Waste Siting Commission.

Middlesex County, primarily through their planning board, has acted to protect groundwater supplies from toxic contaminants. The County Planning Board has taken a regional approach to protecting groundwater supplies that are shared by many municipalities because of the geology of the N.J. coastal plain. The County Planning Board helped South Brunswick Township plan for an adequate future groundwater supply. Making projections of population and water supply requirements and regional solutions to provision of an adequate supply of potable water were especially helpful. The Middlesex County Planning Board also has actively helped South Brunswick protect the quality of its groundwater, although this assistance was greatest in the areas of nonpoint pollution under the 208 planning program of PL 92-500 in the late 1970s.

VII. PROTECTION PLAN ELEMENTS

South Brunswick Township was faced with serious contamination of its groundwater with toxic chemicals in 1977 before this problem was widely known in New Jersey or in other areas of the United States. Few municipalities or other levels of government had established programs for protecting groundwater. South Brunswick used its own resources for analyzing the problem and developing specific programs for ensuring a drinking water supply free of health-threatening levels of toxic chemicals. The groundwater protection plan that has evolved in South Brunswick has several components, some of which are readily transferable to other locations and some that are the result of unique local circumstances.

Municipalities don't want groundwater contamination problems, but when faced with this problem South Brunswick had the good fortune to have involved a large corporation with responsible management and a commitment to being a good corporate citizen. The IBM Corporation provided the township with a sophisticated, state-of-the-art water treatment plant capable of removing a wide range of toxic contaminants from water. (Editor's note: see Chapter 11, the Santa Clara Valley case study, for additional discussion of IBM's involvement in groundwater problems.) This provided South Brunswick Township with one of the most important components of its program for ensuring that its citizens will have a future supply of safe drinking water. The treatment plant began operation in late 1985 and IBM has guaranteed its effectiveness. A treatment facility capable of removing a wide variety of toxic contaminants may be a major component in a plan for providing safe drinking water, but it should not be the entire plan. A municipality should always prefer a clean source of water to treating contaminated water.

One of the first developed and most important components in the groundwater protection plan of South Brunswick Township is its environmental resource inventory. In addition to its many other uses, this document provides detailed information on the geology, topography, soils, and water resources of South Brunswick (South Brunswick Township, 1982). This information provides essential information in the development of groundwater protection plans and in evaluating responses to contamination events. The environmental resource inventory was prepared by the South Brunswick Environmental Commission with the help of municipal officials.

The second component of the South Brunswick Township groundwater protection plan is its surveillance program. Several elements of this ongoing program have been in operation since 1980. One of these elements is the development and enforcement of detailed standards for the storage and handling of hazardous substances (Board of Health Ordinance 1-80). This ordinance provides considerable detail on drum storage areas, loading and unloading areas, requirements for double containment storage tanks, impermeable pans under aboveground tanks, leak detection systems, high-level liquid alarms, overfill and spill detection devices, and other aspects of hazardous materials handling and storage. This ordinance also requires periodic integrity testing of storage tanks and authorizes the Health Department to require industrial or commercial establishments to install observation wells and undertake routine monitoring when they suspect groundwater contamination. Under this program, South Brunswick has identified 63 underground bulk storage tanks (South Brunswick Township, 1982).

A second aspect of South Brunswick Township's surveillance program involves the inspection of hazardous waste-using facilities. The municipality's Planning Department, under its land use regulations, requires a nonresidential occupancy compliance review. Firms must complete a performance standard form (see Fig. 2) that must be approved by the Fire Department, Health Department, and Planning Department prior to the issuance of an occupancy permit. The municipality treats the considerable detail provided on the compliance review of the nonresidential performance standard form as confidential information. South Brunswick retains this information on file where it is available should some form of emergency situation arise at the facility. Existing facilities as well as new facilities are inspected by the Health Department and the Fire Department on a regular basis and by special inspections to ensure that hazardous materials are being stored and handled in accordance with municipal ordinances (Health Department Ordinance 1-78).

The last major component of South Brunswick Township's surveillance program for protecting its groundwater is the water testing program. South Brunswick tests well water in private, commercial, industrial, public supply, and special monitoring wells. The municipality also tests sewage leaving commercial and industrial facilities. The municipal ordinance requires that all private property owners test any wells before the sale of the property can be completed (Health Department Ordinance 2-79). New Jersey requires such an ordinance for the testing of standard water quality parameters such as fecal coliform, hardness, and pH. South Brunswick requires a scan for volatile organic chemicals (VOCs) in addition to the standard water quality parameters. South Brunswick requires this testing by interpreting the state law phrase "other additional chemical determinations as the municipality requires" to include the volatile organic chemical scan (Harris, 1985). The combined volatile organic chemical scan and the standard water quality parameters analysis costs the property owner about $77. All new wells require this testing. Homeowners concerned about the quality of their private well water may have it tested at any time and at their own expense under this program. The volatile organic chemical scan provides identification of 29 purgeable organics at the method selection limit of the equipment. This powerful and low-cost screening device alerts homeowners and public officials to the presence of volatile organic chemicals. The municipality may then begin more thorough sampling

1. Name of owner, mailing address, physical address, and phone number.
2. Name of occupant, mailing address, physical address, and phone number.
3. Proposed use, operation, or process. (Also related operations and flow diagram.)
4. Products, if manufacturing process.
   a. amounts produced (weekly, monthly, yearly)
   b. amounts on hand at any one time (maximum)
5. Chemicals to be used (including but not limited to pesticides, flammables, combustibles, organics, oxidizers, hazardous, etc.).
   a. amounts on hand at any one time (maximum)
   b. method of storage
   c. method of transfer
   d. method of delivery (including delivery point, trace to storage point, and then to point of use)
   e. built-in spill or leak containment features
   f. built-in fire protection features
   g. routine spill prevention and fire prevention measures
   h. emergency action plan for fire, explosion, spill, or leak
   i. special spill containment or firefighting training to be provided to workers
   j. description of a worst case scenario as applicant sees it
   k. special firefighting or spill containment equipment needed and amounts needed in relation to worst case scenario
6. Other items or materials to be utilized or stored.
   a. amounts
   b. method and location of storage
7. Proposed water usage.
8. Description of air pollution control devices including emissions data.
9. S.I.C. number.
10. Types of solid and liquid wastes.
11. Licensed waste hauler and ultimate disposal site for solid and liquid hazardous wastes.
12. N.J. DEP and/or EPA permits required.
13. Location and size of similar operations (including name, address, and phone number of local inspection agencies).
14. Characteristics of waste water effluent (proposed and/or existing facility).

Fig. 2. Information needed for a compliance review of the nonresidential performance standard.

and analysis of the water to confirm the presence of the volatile organic chemicals and to determine the concentration of the microcontaminants present.

South Brunswick Township can require a water quality analysis including a volatile organic chemical scan of any well or sewage line it suspects may be contaminated (Health Department Ordinance 2-79). This ordinance provides that the cost of these analyses be borne by the property owner, but in some instances these fees are waived. The municipality has used this ordinance with considerable success in determining the dimensions of plumes of contaminated groundwater and in identifying sources of contamination. For instance, the municipality used this sampling requirement to identify industries using sewer lines that discharged effluent contaminated with toxic chemicals directly into the ground (Harris, 1985).

South Brunswick also samples and analyzes the water from special monitoring wells. Over 150 monitoring wells are located in South Brunswick as a result of the groundwater contamination problems that have already occurred in the municipality. These wells are sampled on schedules that depend on their location and the concentrations of contaminants present.

Another program used by South Brunswick for protecting its groundwater from toxic contaminants is its local emergency response plan. The municipality developed this plan with the goal of keeping local contamination events from becoming acute. The plan uses information from the surveillance program on file in the municipality to assist in making prompt and informed decisions in the event of an accident, fire, spill, or other emergency involving hazardous substances. The emergency response plan spells out the duties of the Fire Department and other municipal officials and provides some clearly specified procedures to be followed.

The last program involves protecting groundwater from contamination using the land use planning process. Most of South Brunswick Township has highly permeable sandy soil. Large areas in South Brunswick are aquifer recharge zones for the Old Bridge Sand and the Farrington Sand Aquifers that are extensively used for water supply in Middlesex and Monmouth Counties. These aquifer recharge areas comprise a substantial portion of the 10 square miles in the southeastern corner of South Brunswick that are now zoned for industrial development (South Brunswick Township, 1982; Hall, 1985). This large industrially zoned area has experienced some recent development, but remains primarily open space. This area surrounds exit 8A on the New Jersey Turnpike and has locational advantages for the many high technology and other industries that are moving into this area of central New Jersey.

Responsible municipal officials are aware that sound environmental planning principles dictate that aquifer recharge areas and other environmentally sensitive lands be protected from industrial development and other intensive land uses (Hall, 1985; Bittner, 1985). The municipality retains the option of changing the zoning of these aquifer recharge areas to protect them from some forms of development. At present the municipality thinks that it can avoid potential problems by means of careful screening of development proposals at the preapplication stage (Hall, 1985).

South Brunswick Township successfully implemented its groundwater protection program by increasing its budget and staff to accomplish the considerable municipal duties required by this program. Despite the small size of the municipality, South Brunswick added several staff positions and modified or expanded already existing positions to carry out the groundwater protection program. South Brunswick used municipal funds to purchase some sophisticated testing equipment and to pay for an ongoing contractual arrangement for laboratory and consulting services (Harris, 1985). This large increase and reallocation of the municipal budget for environmental programs was achieved through the growth in municipal revenues in a developing municipality and through a small tax increase (Bittner, 1985). The massive publicity about toxic chemical contamination of groundwater in South Brunswick helped create a substantial political consensus on the need for and a willingness to pay for an effective groundwater protection program.

VIII. EFFECTIVENESS OF PLAN COMPONENTS

Measuring the effectiveness of the different components of the groundwater protection program developed by South Brunswick Township is difficult. The following discussion is the opinion of the author based on investigation of the program and discussion of the program with public officials and interested individuals in South Brunswick and neighboring areas of New Jersey. Several components of the groundwater protection program are well conceived and effective. The hazardous material storage ordinance is detailed, thorough, and enforced. The inspection program has proven its effectiveness by identifying previously unknown toxic chemical contamination of groundwater and by alerting citizens and Health Department officials of potential threats to the public's health. The compliance review of the nonresidential performance standard has successfully collected a great deal of important information that industry is sometimes reluctant to reveal. This information is essential for ensuring good design of new facilities and in the development of an emergency response plan.

A major question concerns the effectiveness of the land use planning component of South Brunswick's groundwater protection program. The formal zoning of the township's land use plan does not provide protection of critically important aquifer recharge areas in the township. Instead, South Brunswick relies on informal discussions with potential developers to dissuade inappropriate developments or to ensure the modifications needed for a safe environment. While this has been successful for the past few years, the pressure for industrial development in

South Brunswick remains high and is increasing. There are presently development applications pending by an industry making a plastics-related product, and other industries using products with serious potential to contaminate groundwater are likely to follow (Hall, 1985). The land use planning component of South Brunswick Township's groundwater protection program appears to be the least effective element.

REFERENCES

Bittner, J. (1985). Interview with South Brunswick Township Municipal Administrator, June 24.

Gaston, J. W., Jr. (1985). Letter from the Director, Division of Water Resources, New Jersey Department of Environmental Protection to water users and others concerned in Middlesex, Monmouth and Ocean Counties.

Geraghty and Miller, Inc. (1976). "Task 8 Ground Water Analysis, Lower Raritan/Middlesex County 208 Water Quality Management Planning Program" (unpublished).

Geraghty and Miller, Inc. (1979). "Investigation of Ground Water Contamination in South Brunswick Township, New Jersey." Report issued to the Township of South Brunswick.

Hall, R. (1985). Interview with South Brunswick Township Planning Director, June 24.

Harris, R. (1985). Interview with South Brunswick Township Health Officer, June 24.

Home News, (1979a) June 5, New Brunswick, New Jersey.

Home News, (1979b) July 11, New Brunswick, New Jersey.

Krueger, A. (1979). Chairperson, South Brunswick Environmental Commission (personal interview, December 1).

Merk, L. (1979). South Brunswick Water Supply (telephone interview, December 6).

Middlesex County Planning Board (1979a). "Policies and Practices for Managing Middlesex County's Groundwater Resources," p. 58. MCPB, New Brunswick, New Jersey.

Middlesex County Planning Board, (1979b). "Water Supply Planning Alternatives," p. 135. MCPB, New Brunswick, New Jersey.

Middlesex County Planning Board (1984). "Preliminary Population and Housing Projections, Implied Household Sizes, 1990 and 2000." MCPB, New Brunswick, New Jersey.

Middlesex County Planning Board (1985). "South River Basin Water Supply Feasibility Study." MCPB, New Brunswick, New Jersey.

National Research Council (1977). "Drinking Water and Health." National Academy of Sciences, Washington.

Page, G. W., Greenberg, M., and Tucker, R. (1980). Toxic substances in the groundwater of New Jersey. *Northeast Reg. Sci. Rev.* **10,** 92–113.

South Brunswick Township (1982). "Environmental Resource Inventory." Department of Planning and Development, South Brunswick, New Jersey.

Tucker, R. (1981). "Groundwater Quality in New Jersey: An Investigation of Toxic Contaminants." Office of Cancer and Toxic Substances Research, New Jersey Department of Environmental Protection, Trenton, New Jersey.

U.S. Bureau of Census (1981). "1980 Census of Population and Housing." U.S. Govt. Printing Office, Washington, D.C.

13

Bedford, Massachusetts, Case Study

BONNIE J. RAM
Environmental Consultant
Washington, D.C. 20009
and
HARRY E. SCHWARZ
Environment, Technology and Society Program
Clark University
Worcester, Massachusetts 01610

I. INTRODUCTION

The tradition of home rule in Massachusetts and the decentralized nature of its Department of Environmental Quality Engineering have created a prominent local role in groundwater management. State regulations, such as Title V of the State Environmental Code, are executed through local boards. Towns and cities have the option of formulating controls stricter than federal or state minimum requirements. Since there are no definitive statutes relating to groundwater, the local governments have had to take the initiative in protecting their underground drinking water supplies. Monitoring and sampling of well waters and sewage systems are carried out by the local water superintendents or public works departments in cooperation with state personnel. However, any protection beyond state and federal mandates is very town specific. In most cases, protection measures are devised in response to a crisis or perceived crisis that affects public health. The chemical contamination of groundwaters and other hazardous waste problems have prompted local governments to either seek out state assistance or take matters into their own hands. However, localities are severely restricted in their capabilities to protect their water supplies because of limited funding, staffing, and technical expertise. The following case study of the Town of Bedford, Massachusetts, discusses these issues and describes the response of the contamination of their municipal underground drinking water.

II. BACKGROUND

The Town of Bedford is located 15 miles northeast of Boston. It is surrounded by the towns of Burlington, Lexington, Concord, Carlisle, and Billerica and lies mostly between the Concord and Shawsheen Rivers. Bedford, an area of 14 square miles, is a predominantly residential community. The Bureau of the Census in Massachusetts reported that the population had increased since 1965 by 14.2% and by 1975 was 12,314. The population for 1980 was 13,067. Per capita income in 1975 was $5730, well above the Massachusetts average of $4964. Two highly developed industrial zones, adjacent to Routes 3 and 128, are located in the northwest quadrant of the town. The 1980 fiscal year budget totaled about $12 million and was almost evenly divided between education and other municipal functions such as police, fire, rubbish collection, water, and sewers.

III. HYDROGEOLOGY OF THE TRIBUTARY WATERSHED

The hydrogeologic conditions in the area were summarized in the report of the consulting firm Camp, Dresser & McKee, Inc. (1979). Figure 1 presents an aquifer diagram of the tributary watershed. The bedrock formations, primarily metamorphic rock, are overlayed by unconsolidated glacial deposits. Till, mostly layed down during the last major ice advance, contains bedrock fragments up to large boulder size. Stratified drift in the area consists of layers of silty sand and sand, as well as sand and gravel, pebbles, and cobbles. Thin layers of relatively permeable river alluvium and swamp deposits occur in valleys and other depressions.

Bedrock formations in the area are relatively impervious, yielding small quantities of water recharged by percolation through the overlying glacial deposits. Till soils have low permeability and specific yield and are considered to be moderately impervious boundaries to groundwater flows. Stratified drift deposits are relatively permeable and most of the groundwater that exists in the area flows through these deposits.

Groundwater originates from precipitation filtered through the surficial sediments and from transfer of water in either direction between surface sediments and surface streams. The surface streams are a major discharge system for the local groundwater and have a significant influence on its direction of movement.

IV. INSTITUTIONAL STRUCTURE

Bedford, incorporated in 1729, has a traditional New England local government structure: the Annual Town Meeting acts as the legislative

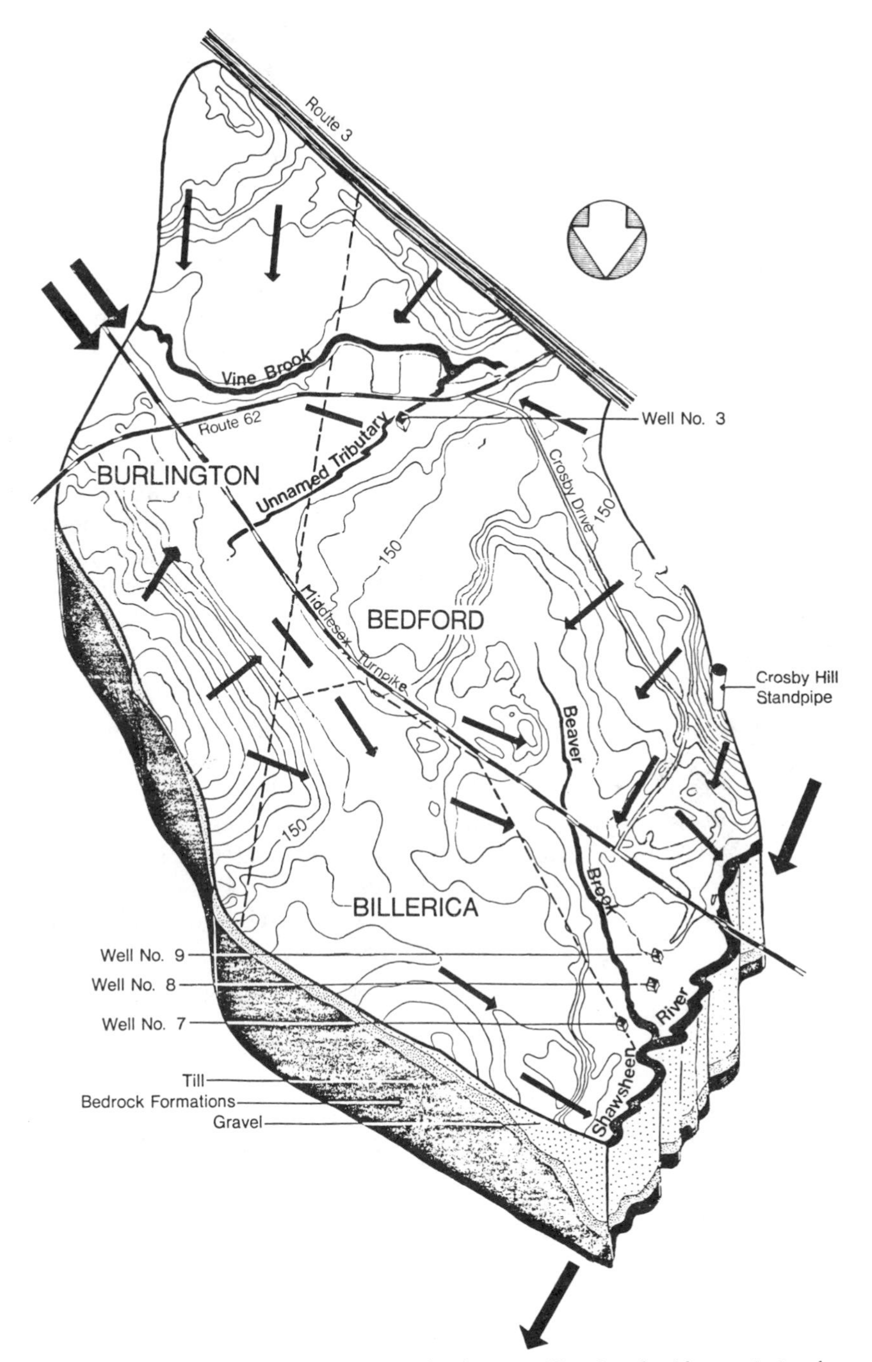

Fig. 1. Beaver Brook–Vine Brook Aquifer diagram. [Reprinted with permission from Camp, Dresser & McKee, Inc. (1979).]

body and the five-member elected Board of Selectmen acts as the executive body. In addition to the Board of Selectmen, the most powerful boards in Bedford with elected positions are the Boards of Health, Planning, School Committee, and Assessors. Appointed citizens serve on the Finance Committee, Conservation Commission, Personnel Board, and Zoning Board of Appeals. Adoption of the town charter in 1974 afforded home rule provisions through the state constitution. The duties of local boards and officials involved with water supply management according to statutory mandate and/or *de facto* policies are briefly described below (see Fig. 2).

Town Administrator

Appointed by the Board of Selectmen, the town administrator is charged with overall management of the town's business. He or she has the authority to set policy and make decisions regarding water supply with the approval of the selectmen. This position was recently created to address the day-to-day demands of town responsibilities.

Board of Selectmen

This elected five-member board is the central policy-making body that manages and plans for the quantity and quality of drinking water supplies among many other duties. It supervises the Public Works Department, the official purveyor of water supply, and enforces regulations ensuring the quality of groundwaters and establishes policies concerning the protection of this resource. The duties of the Public Works Department include: general management, such as metering and billing;

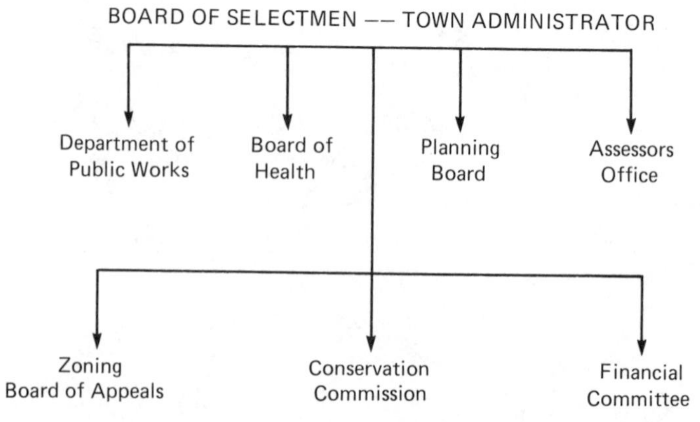

Fig. 2. Local government structure of the Town of Bedford.

maintenance of nine town production wells; monitoring of the municipal sewerage system; and periodic testing for bacterial and mineral constituents in well water in accordance with Massachusetts Drinking Water Regulations (310 CMR 22.00) and the Safe Drinking Water Act (P.L. 93-523). Testing for organics in groundwater was not mandatory before 1981, but the State Department of Environmental Quality Engineering EPA-SPOT program required samples and analysis by the state laboratory in Lawrence or an authorized private lab.

Local Board of Health

The Local Board of Health, pursuant to Chapter 111 of the Massachusetts General Laws (MGL), has broad powers concerning the monitoring of wells. Traditionally, the Bedford Local Board of Health has been primarily concerned with the delivery of health care services. The Local Board of Health has recently assumed broader responsibility regarding the health impacts of environmental pollution including contamination of drinking water.

Planning Board

The local Planning Board is empowered to enact zoning regulations that "conserve health, facilitate the adequate provision of . . . water and encourage the most appropriate use of land" (Chapter 40, MGL). This statutory mandate extends to "eminent domain" powers whereby the board may acquire and purchase land and groundwater sources for municipal supply within town boundaries. The Planning Board has potential authority to protect water and land resources through use of subdivision control and appropriate building and design codes. They can also act as a board of appeals in zoning cases (Chapter 41, MGL).

The Board of Selectmen is the central decision-making body for ensuring an adequate supply of drinking water. In the past decade, the Board of Selectmen has revamped Bedford's future water supply plans. The plans have begun to address the periodic shortages due to increasing demand and declining well yields. Indigenous groundwater supplies have met the average daily demand of approximately one and one-half million gallons per day, yet summer peak demands often exceed safe yields. As a result, the Board of Selectmen received permission in December 1977 to purchase up to 500,000 gallons of water from the Metropolitan District Commission through a Town of Lexington waterline connection. Emergency water provisions were also prearranged with the neighboring towns of Burlington and Concord. The Board of Selectmen believed the town's future groundwater supplies were adequately assessed and demands could be met with these arrangements. However, future plans were made under the assumption that existing groundwaters would continue to yield potable waters. Unforeseen circumstances such as contaminated water supplies were not considered.

V. GROUNDWATER CRISIS

In late fall of 1977, a physicist, residing in Bedford, was testing a gas chromatograph mass spectrometer data system (GC/MS/DS) under development at his company with a water sample from his own household tap. Unexpectedly this experiment identified several trace organics, including dioxane and trichloroethane, in the tap water. Over the next five and a half months, this resident attempted to investigate the public health risks associated with these chemicals:

> He tried first the State Office of Water Pollution Control and could get no more information on dioxane than he already had. So he tried the regional office of the EPA. Then he called the Washington office, then back to the regional EPA laboratory in Lexington, then to the state laboratory at Lawrence, and so on, until finally, at the suggestion of a colleague, he happened to call a toxicologist at the Harvard Medical School and finally got an indication of the nature of the compounds he had discovered. (Reported in Massachusetts Audubon News, 1980)

In May 1978, the resident contacted the chairman of the Board of Selectmen with the conflicting reports about the potentially harmful chemical constituents in Bedford's groundwater. The Board of Selectmen proceeded to verify these reports because of the possible consequences for the town's water supply. Despite their statutory authority to immediately shut down the town wells, the Board of Selectmen had to make a decision between risking the public health with potentially harmful drinking water or leaving the town defenseless if a major fire occurred (Harvard University, 1980). The town officials also did not have enough information to make a final judgment. They contacted the regional Massachusetts Department of Environmental Quality Engineering office requesting assistance in the evaluation of trace organic chemicals possible affecting almost 90% of the town's water supply. As a result of the initial analyses at the state laboratory, two organic compounds, trichloroethylene and 1,4-dioxane, were positively identified in concentrations up to 124 and 400 parts per billion (ppb), respectively (Camp, Dresser & McKee Inc., 1979).

The Board of Selectmen pursued more detailed information concerning the health effects of ingesting these chemicals in drinking water through contacts with the Region I EPA office. The EPA Division of Water Supply did not have established health standards for these chemicals but was able to provide the Suggested No Adverse Response Levels (SNARLs) of 10 ppb for TCE and 20 ppb for 1,4-dioxane. (Editor's note: see Chapter 3 for a discussion of standards.) Realizing that the town's four major production wells contained concentrations in excess of the SNARLs, the Board of Selectmen unilaterally ordered the shutdown of the wells on May 20, 1978. Within a week of receiving new EPA interim standards for the chemical substances, the Department of Environmental Quality Engineering regional office issued an order to discontinue use of these wells for public water supplies. These wells have not been used for public water supply since that time (Camp, Dresser & McKee, Inc., 1979). Fortunately, the town administrator had acted prudently by prearranging emergency water supplies with the Metropolitan District Commission and neighboring towns. These interim agreements provided enough water to meet average daily demands, but the situation remained precarious.

The town administrator and the Board of Selectmen made a concerted effort to secure multiple sources of water supplies by acquiring additional emergency allocations from the Metropolitan District Commission and treated water from the neighboring towns of Billerica and Burlington. BSAF, a local industry suspected to be a source of chemical pollution, also donated $60,000 for water purchases and installed a new pump in Burlington to increase the capacity in Bedford. These supplies have since been temporarily interrupted because of intertown emergency needs, discoloration of the Billerica supply due to manganese content, and discovery of trichloroethylene in Burlington wells. Despite the chain of events, the town of Bedford has had a consistent flow of municipal drinking water because of effective and prudent crisis management by the Board of Selectmen, the town administrator, and the Public Works Department.

Three days following the well closures, the Board of Selectmen instituted a number of measures to curb water usage. These actions included:

1. A 6-month emergency ban on nonessential water usage, e.g., watering lawns and washing cars.
2. A moratorium on new water connections.
3. Authorization through town meetings to revise the flat rate per cubic feet of water to a progressive structure that charged higher prices for large-volume users.
4. Authorization to the town administrator of "eminent domain" powers in order to construct a water connection for the Metropolitan District Commission supplies through Lexington. This action permitted

the town administrator to supersede the traditional bidding proceedings.

These emergency measures effectively reduced the consumption of water in the town, thereby averting any critical water shortage and reducing the volume of water that had to be purchased from outside sources. Bedford acted on its own initiative to manage the crisis.

Community reaction to the groundwater contamination incident and public health threat was moderated by the smooth management and reassurance of the Board of Selectmen and, more importantly, by the consistency of alternate municipal water service. The concern over the health effects of suspected carcinogens and the absence of maximum contaminant levels (MCLs) for the organic chemicals provoked questions rather than panic. The Local Board of Health publicly acknowledged it had neither the resources nor the information to conduct a study of the relationship between the contamination and local health patterns (Bedford Minuteman, 1979a).

At the request of the Bedford Local Board of Health and other Massachusetts towns with similar concerns, the State Department of Public Health compiled statistical data on mortality rates. The Bedford analysis revealed that between 1974 and 1978 ovarian and breast cancers among Bedford women exceeded the statistical expectation by more than 150%. However, the State Department of Public Health asserted that the raw data did not conclude a direct correlation between mortality figures and the town's drinking water contaminants. The assistent commissioner for environmental health of the State Department of Public Health noted that "there is no known case of breast or ovarian cancer associated with the ingestion of an environmental contaminant" (*Boston Globe,* 1980). Bedford residents, in response to this preliminary cancer study, initiated a public education program for breast self-examination that apparently mitigated immediate fears (*Bedford Minuteman,* 1980d). Significantly, community reaction has been mild considering the imminence of public health threats and water shortages. The town had also asked the federal government's Center for Disease Control in Atlanta to study this preliminary data on cancer-caused deaths in Bedford. Center for Disease Control statisticians will begin to investigate whether the State Department of Public Health data indeed point to a health problem by examining death certificates and researching the case histories of every woman who died from cancer (*Bedford Minuteman,* 1980f).

VI. STATE AND LOCAL INVESTIGATIONS

The Massachusetts Department of Environmental Quality Engineering regional and central offices played ancillary roles in the management of the contamination crises by primarily providing help for testing and analysis. The Division of Water Supply and Division of Water Pollution Control were involved in the final authorization of the well shutdown and the tracing of pollution sources, respectively. These state agencies are charged with carrying out the Massachusetts Safe Drinking Water Act (Chapter 111, MGL) and the Clean Waters Act (Chapter 21, MGL) that mandate their responsibilities relating to groundwaters used for drinking water supplies. Essentially, their responsibilities were fulfilled by conducting the initial well tests that corroborated the evidence of the presence of organic compounds in the wells. Because of the insistence of the local Board of Selectmen and the pervasive case of pollution, the state agencies perservered in their investigation (Department of Environmental Quality Engineering, 1980).

The joint efforts of the regional Department of Environmental Quality Engineering and the Town of Bedford included collection of weekly samples from the four contaminated wells to determine the magnitude of the problem. The local Public Works Department was instructed by the Division of Water Supply in sampling techniques for observation wells and the state laboratory in Lawrence conducted the analyses. These tests, conducted through the summer of 1978, revealed concentrations of dioxane up to 1200 ppb in all three wells and up to 500 ppb of trichloroethylene (Camp, Dresser & McKee, Inc., 1979).

The Division of Water Pollution Control regional engineers conducted a separate investigation to determine the sources of contaminants in the groundwaters. The high concentrations in a small surface water body (Beaver Brook) adjacent to the industrial zone led the regional engineer upstream to a local industry (see Fig. 1). He recounted the discovery:

> Suspecting the source of dioxane from a local industry led us to the manhole of their emergency septic tank which was laden with dead grass. The emergency septic tank had apparently overflowed due to a loose valve in the system that was permitting hazardous wastes to discharge into the tank during power failures and overload periods. The excess waste then flowed into the catch basin of this swampy area, drained into the nearby Beaver Brook and eventually leached into the groundwater. Following our tracing activities, the Division made certain that the overflow was prevented by sealing all pumps that transferred these chemicals. (Regional Office, Division of Water Pollution Control, 1980).

This investigation revealed that solvent wastes were routinely being discharged by several local industries into the municipal sewerage system and into on-site emergency septic tanks. Ostensibly, this sewerage ejection system was regulated to ensure compliance with allowable levels of discharge after proper pretreatment to achieve neutralization. These discharges of concentrated chemical cleaning solutions, heavy toxic metals, grease, and oil were therefore in direct violation of state and local regulations for sanitary sewer systems. In addition, several local industries did not fulfill their obligation to have a "licensed scavenger" collect solvents or chemical wastes that were generated, in accordance with "Rules and Regulations Covering Discharge of Sewerage, Drainage, Substances or Wastes to Sewerage Works within the Metropolitan Sewerage District." These regulations were promulgated in accordance with the Federal Water Pollution Control Act of 1972. At the request of Department of Environmental Quality Engineering, the Metropolitan District Commission Sewerage Division reviewed the waste disposal permits of 19 industrial plant sites located within the watershed area of the production wells. The 1978 Metropolitan District Commission report indicated that 13 of these industries use solvents and other hazardous materials that could contaminate the groundwater if improperly discharged. Seven of these 13 firms were cited for discharge violations. Subsequently, additional samples from sewers and surface waters within the well drainage area uncovered high levels of dioxane and trichloroethylene. Independent analysis of sewage effluent by the local Public Works Department traced the source of dioxane to a local industry. However, no definitive source in a legal sense was established for the trichloroethylene.

These events culminated 5 months later (October, 1978) when the Department of Environmental Quality Engineering investigation was presented to the Board of Selectmen. The state's recommendations included:

1. continue monitoring of contaminated wells;
2. establish a permanent water line connection with the Metropolitan District Commission system;
3. prohibit new water connections;
4. negotiate a zero discharge level with the local industry identified as the source of dioxane;
5. consider litigation proceedings with the attorney general's office;
6. hire a consultant to continue the source investigation for trichloroethylene and to conduct other hydrogeological tests; and
7. develop a long-range town water supply assessment.

The Board of Selectmen has since undertaken some of these recommendations while several were previously implemented through local emergency measures. In spite of the fact that Bedford assumed the major responsibility in managing and planning for their drinking water supplies, the state assumed some necessary technical tasks associated with testing and analysis.

The town did have some misgivings concerning the role of the state, but realized the scope of the contamination crisis was beyond their local capabilities. As one town official commented:

> The contamination problem was clearly not an issue that could be completely resolved at the local level. We were at the mercy of the state people. The state people, it turned out, were somewhat at the mercy of the shallow technical resources available to them (Harvard University, 1980).

The reason underlying the communication problems between the state and local administrators can only be hypothesized through the historical events of 1978. The Department of Environmental Quality Engineering did not have access to maximum contaminant levels (MCLs) for the organic microcontaminants in question and were thus forced to rely on EPA's interim standards, i.e., SNARLs. In addition, the Department of Environmental Quality Engineering regional engineers were reluctant to evaluate the health effects of the chemical concentrations without official public health information, including maximum contaminant levels for dioxane and trichloroethylene that was not available at that time from EPA. Moreover, the Department of Environmental Quality Engineering had not yet approved organic chemical testing at any state laboratories because of the absence of mandatory regulations and the lack of sophisticated equipment. During that year, the Lawrence laboratory acquired a GC/MS/DS that was capable of analyzing organic chemical content (Division of Water Supply, 1980). The number of samples and tests, as well as the hours of work required to ensure reliable results, caused further delays in channeling information to the town. Meanwhile, the Town of Bedford was in a quandary concerning the actual public health risks posed by the microcontaminants discovered in their drinking water. The slow reaction of the state agencies prompted the Board of Selectmen to declare a unilateral shutdown of the municipal wells. Thus independent local initiatives, stemming from statutory and *de facto* authority, played a central role in managing the crisis and securing the flow of drinking water.

VII. RESULTS OF BEDFORD'S TECHNICAL INVESTIGATION

The Town of Bedford hired the consulting firm Camp, Dresser & McKee, Inc., to conduct a detailed investigation of the sources of contamination and a hydrogeological analysis of the aquifer system. Camp, Dresser & McKee took over sampling duties from the Department of Environmental Quality Engineering in October 1978 while the state laboratory continued to conduct the analyses. From late April through September 1979 the comprehensive sampling and analyses of municipal wells and sewers were carried out by the consulting firm with the cooperation of the local Department of Public Works. A three-volume report was completed in November 1979, including detailed aquifer mapping and extensive chemical analysis of over 100 observation wells. The consulting firm's report included the following conclusions.

1. Industrial chemicals were discharged to the tributary watersheds of the four municipal wells with major constituents identified as trichloroethylene, 1,1,1,-trichloroethane, tetrachloroethylene, chloroform, dioxane, and tetrahydrofuran. Sixteen compounds, including industrial solvents and trihalomethanes, were identified in the groundwater tributary to the wells and seven of these were listed as EPA "priority pollutants" (see Table I).
2. Water quality sampling of Vine Brook, Shawsheen River, and Beaver Brook indicated periodic contamination with industrial solvents. These streams provide significant recharge to the municipal production wells.
3. Probable sources of contaminant entry into the groundwater system were identified from an examination of available water quality data, groundwater flow patterns, and transmissivity rates. The sites were related to Bedford's municipal sewer system, a sewer lateral in the Town of Burlington, and private industrial drain systems.
4. Current research establishes that minimal, if any, natural attenuation, except for dilution caused by groundwater dispersion, will occur within the aquifer system.
5. Based on natural flow gradients, major concentrations of contaminants could possibly be discharged from groundwaters within 2 to 4 years and lower concentrations could be eliminated in 5 to 12 years. These estimates were based on the assumption that entry sources of contaminants would be eliminated and no remedial action would be taken.
6. The toxic effects of the chemicals found in Bedford wells are uncertain due to conflicting interpretations of available data and their application to public health criteria for drinking water.
7. Alternative water supplies were suggested, including construction of additional wells in Bedford, rehabilitation and cleaning of existing wells, and arrangements with neighboring towns.

VIII. PROFILE OF GROUNDWATER SUPPLIES

The Bedford water supply contamination incident forced the shutdown of the town's four highest production wells, accounting for almost 90% of total supply. In the 3 years following the discovery, the town had

TABLE I

Monitoring Well Sampling Program

Location	Contaminant concentrations (ppb)					
	Trichloro-ethylene	1,1,1-Trichloro-ethane	Tetrachloro-ethylene	Other volatile organic compounds	Dioxane	Tetrahydro-furan
1. Immediate vicinity of municipal wells nos. 7, 8, and 9	424	103	ND[a]	Chloroform, 449 1,1,-Dichloroethane, 35 1,2-Dichloroethane, 20 Methylene chloride, 588	330	13
2. Beaver Brook watershed from the municipal wells to Middlesex Turnpike	1700	ND	ND	Chloroform, 39 *trans*-1,2-Dichloroethylene, 48	ND	59
3. Along Crosby Drive from Middlesex Turnpike to water standpipe	1146	ND	ND	Chloroform, 27 1,2-Dichloroethane, 80 *trans*-1,2-Dichloroethylene, 142 Methylene chloride, 89	ND	ND
4. Beaver Brook watershed from Middlesex Turnpike to its origin	ND	ND	ND	1,2-Dichloroethane, 37 *trans*-1,2-Dichloroethylene, 57 Ethylbenzene, 490	102	ND
5. Along Middlesex Turnpike from Beaver Brook to Mitre	25	ND	ND	*trans*-1,2-Dichlorobethylene, 25 1,2-Dichloroethane, 50	ND	36
6. Mitre Corporation property	<10	ND	ND			
7. Along unnamed brook to Vine Brook	2129	25	50	Chloroform, 65 1,2-Dichloroethane, 35	ND	ND
8. Adjacent to Vine Brook to aquifer boundaries	97	133	7	Chloroform 133 1,2-Dichloroethane, 15 Methylene chloride, 55	ND	ND

Source: Camp, Dresser & McKee, Inc. (1979).

[a] Not detectable.

spent more than one and a half million dollars to purchase water from outside sources. Reconditioning and reconstruction of existing wells costing $215,000 provided 5% of the supply. Bedford must still purchase 85% of its water supplies or 1.81 million gallons of water per day from the neighboring towns of Billerica (60%) and Concord (25%) (*Bedford Minuteman,* 1980e). Water from these sources costs about $600 per million gallons and this cost is expected to rise. The town of Bedford also had to conduct chemical testing of these waters because neighboring towns do not require testing (*Bedford Minuteman,* 1981a). Water supplies from Billerica were temporarily suspended in February 1980 because of discoloration due to high manganese content (*Bedford Minuteman,* 1980a).

The Town of Bedford, having few surface water possibilities, has evaluated the potential for developing additional groundwater drinking water sources. In consultation with their consultants, various sites have been selected as future sources of underground water supplies. These sites are being planned under the assumptions that the currently contaminated wells will not be of potable quality in the near term and aquifer cleanup would be extremely costly. The development of three new well sites, a key step in Bedford's efforts to regain water supply self-sufficiency, required a $1.5 million bond issue for land purchase and facility construction that was approved by a town meeting in the spring of 1980 (*Bedford Minuteman,* 1980c). Treated water from these new wells was expected to cost the Town of Bedford only $160 per million gallons, which is more economical than importing water from nearby towns and the Metropolitan District Commission (*Bedford Minuteman,* 1981a). In the meantime, Bedford must continue to rely on neighboring towns and the Metropolitan District Commission for the bulk of their municipal water supplies. An editorial in the local newspaper sums up the local scenario:

> Bedford has to hope for a regional answer, perhaps an overhaul of the Metropolitan Water System or construction of treatment facilities. The options the town faces now are finite. . . . It's costly to use water in Bedford; the price of our "imported water" can be a significant deterrent to needless use. (*Bedford Minuteman,* 1979b)

IX. SIGNIFICANT LOCAL INITIATIVES

Summarized below are several other important activities that Bedford had undertaken to address its groundwater contamination problem.

1. In the spring of 1980 the Public Works Department initiated a sewer sampling program to identify local industries disposing of chemical wastes improperly. Inspectors were ordered to carry sampling bot-

tles to make unannounced checks for suspected illegal discharges (*Bedford Minuteman*, 1980b).

2. The Local Board of Health broadened their role concerning public health impacts of environmental pollution by creating the Environmental Advisory Board. Composed of resident professionals including a toxicologist, a professor of public health, and a physicist, the Environmental Advisory Board will address the health implications of road salting, hazardous waste disposal, toxic material handling, and air quality (*Bedford Minuteman*, 1981b).

3. The Bedford Planning Board hired the Interdisciplinary Environmental Planning Group, a private consulting firm, to compile existing hydrogeological data on the town's groundwater system. The $30,000 study will include comprehensive mapping to delineate groundwater aquifers and a synopsis of legislation to protect aquifers. The Planning Board is reviewing model zoning bylaws from regional planning agencies with the intention of controlling development in sensitive recharge areas.

4. The Town of Bedford agreed to consider an Aquifer Flushing Project for three of the contaminated wells, a remedial action recommended by Camp, Dresser & McKee. The wells would be pumped to accelerate the natural aquifer flushing action and to determine if the well water can become potable without treatment. A temporary overland pipe, connected to the wells, would discharge to the Shawsheen River downstream from the production wells. The estimated cost of constructing the temporary piping system is $15,000 and $50,000 for power and analytical costs for the municipal wells. The flushing process would accelerate the dilution and ultimate removal of dioxane that was estimated to take 6 months to 1 year. However, flushing of trichloroethylene would be more difficult (Camp, Dresser & McKee, Inc., 1979). This project must be approved by the Massachusetts Department of Environmental Quality Engineering and may require a National Pollution Discharge Elimination System permit because of surface water discharges. The state is still reviewing the proposal and its technical feasibility.

5. Two general bylaws concerning hazardous waste and subsurface storage tanks are under consideration by the Bedford Planning Board. The first recommended bylaw "would prohibit the discharge, deposit, injection, dumping, spilling, leaking, incineration or placing of any waste material that poses a present or potential hazard to human health or safety on any land or water (including groundwaters) within the town boundaries." Certificates of Compliance would be issued by the Local Board of Health. This local regulation would also be in accordance with the state's Hazardous Waste Management Act, Chapter 21C, MGL. The

second bylaw, for regulating underground fuel and chemical storage, is designed to protect ground and surface waters from contamination by liquid fuels and toxic materials from leaking storage tanks (*Bedford Minuteman*, 1980g).

X. LITIGATION PROCEEDINGS

Bedford retained a private law firm, along with the consulting firms, to assess the possible avenues for legal action against the perpetrators of the contamination. Funding for these services was secured through a special town meeting in February 1979, while the hydrogeological study of the extent and source of the pollutants was in progress. The purpose of the litigation would be primarily to prohibit illegal discharges of hazardous substances and to recoup the financial expenditures incurred by the town for water purchases, consultant fees, and other costs related to the contamination of the four production wells since the discovery of the problem in May 1978.

The Town of Bedford had several options to consider for litigation proceedings.

1. Intervention by the state attorney general's office that would relate to the violations of state statutes and regulations, such as the Hazardous Waste Management Act and sewer regulations;
2. Intervention by the regional EPA counsel and U.S. attorney general that would relate to violations of federal pollution control statutes, such as the "imminent hazard" provision of the RCRA (Section 7003); or
3. An independent local suit relying on common law for compensation of property or individual damages. Tort law provides for strict liability without proving negligence or intent as suggested by legal council.

Following extensive closed-door review by the Board of Selectmen, their private lawyer, and the consulting team, the town of Bedford decided to file suit on its own behalf in Middlesex Superior Court in December 1979. The choice to pursue litigation through tort law rather than through state or federal authority was to ensure that the town maintain complete jurisdiction over the case and indisputable rights to any financial entitlement. The local initiative could also be a legal strategy to guarantee further recourse through state or federal suits in the event that the tort litigation was unsuccessful. Moreover, the courts have established a few precedents regarding groundwater contamination through the application of state and/or federal statutes. (Editor's note: see Chapter 2 for a discussion of applicable laws and precedents.) A precedent was claimed in November 1980, when the U.S. attorney general on behalf of EPA settled a lawsuit against W.R. Grace Company in Acton, Massachusetts, pursuant to RCRA Section 7003 ("imminent hazard"). Grace was charged with the contamination of two underground drinking water sources (40% of total supply) because of improper disposal of hazardous waste. The landmark case was settled through consent decree and is one of the first hazardous waste settlements of its kind in the nation (*Environmental News,* 1980).

XI. BEDFORD VERSUS ADVANCED METAL RESOURCE CORPORATION *ET AL.*

The Superior Court case lists four complaints against six local industries in Bedford that allegedly caused the groundwater contamination. The case is particularly interesting because hydrogeological surveys could only establish a direct link to one of the six local industries as a source of dioxane. Most industries located within the tributary watershed of the four contaminated wells use some amount of the solvents detected in the groundwater and therefore, it is assumed, caused some degree of the current as well as past pollution problems. Because of limited historical data concerning local groundwater flows and unknown effects of dilution and dispersion, it is not possible to determine the length of time that industrial contamination has been occurring (Camp, Dresser & McKee, Inc., 1979). Consequently, all six industries within the watershed area have been implicated as suspected sources of the organic constituents found in the aquifer system. Since more than one industry is named as defendant and the physical proof of causation is questionable, this case will undoubtedly set a legal precedent concerning strict liability. The complaint focuses on the fact that the principal sources of potable water for Bedford residents were considered high-quality waters until the discovery of contamination in May 1978. As a result of this contamination, the Town of Bedford incurred substantial public health risks and financial burdens. The lawsuit entails four complaints against six local industries: (1) nuisance; (2) trespass; (3) strict liability; and (4) negligence.

The Town of Bedford, therefore, requests that the courts both enjoin the defendants from continuing these "acts and/or omissions" and determine the amount of damage sustained and to be incurred. The outcome of this suit is described in Section XIII.

XII. SUMMARY AND CONCLUSION

The groundwater contamination crisis in Bedford illuminated a number of interesting issues and problems in relation to the protection of underground drinking water sources by local communities. The Town of Bedford demonstrated a sophisticated approach to resolving its groundwater supply problems by taking initiative with remedial actions and long-term strategies to provide quality and quantity of drinking water (see Table II). The loss of approximately 90% of their total water supply confronted the local officials and residents with the reality of the fragile balance of their land and water resources. Bedford has since made groundwater protection one of its foremost objectives by attempting to control, not prohibit, growth within its boundaries and protect the quality of existing underground drinking water sources.

This case study clearly demonstrates the major role local governments could play to reinforce the existing and future state and federal programs concerning groundwater resources. Local control of land use decisions such as industrial siting and zoning have a direct impact on the viability of aquifer systems and thus affect the implementation of a state or federal groundwater policy. However, crisis experiences, such as occurred in Bedford, reveal the need for greater coordination of legal and institutional mechanisms to achieve policy and/or management objectives.

Crisis management following the contamination discovery in Bedford's four production wells was enhanced by the professional caliber of the local administration. The *de facto* and statutory powers of the Board of Selectmen and their town administrator provided the necessary authority for efficacious decision making. As a result, local water supplies were uninterrupted, water-sharing agreements were formalized with neighboring towns, adequate financial expenditures were approved, and emergency measures to conserve water supplies were enacted. The legal and institutional mechanisms were able to serve the town government's immediate needs and purposes without extensive state or federal involvement. As mentioned previously, this case study should not be generalized for all local experiences. The Town of Bedford is a high-income suburban community with a number of professional and academic representatives serving in local government. As a result, their institutional capabilities in coping with the contamination of the municipal drinking water supplies are above average.

The town of Bedford has begun to develop an impressive range of nonstructural approaches to protect their remaining groundwater resources that can be used as public drinking water supplies. Zoning

TABLE II

Chronology of Bedford Groundwater Contamination Incident and Institutional Involvement

	Local	State	Federal
December 1977			
Planning for future water supplies	x		
May 1978			
Discovery of contamination and unilateral shutdown of 90% of local well water supplies	x	x	x
June 1978			
State discontinuance order and investigation of pollutant sources	x	x	x
Summer 1978			
Local remedial water supply measures instituted, e.g., sewer monitoring	x	x	
October 1978			
State final report of investigation and recommendations	x	x	
February 1979			
Funding provided for private engineering and law firms	x		
November 1979			
Final consulting firm report finds 16 trace organic contaminants in drinking water	x		
December 1979			
Litigation filed against six local industries	x		
January 1980			
Statistical study of local cancer death requested	x	x	
April 1980			
State concludes "statistically significant increase" in cancer deaths among local townswomen	x	x	
December 1980			
Planning Board considering zoning bylaws regulating hazardous waste, underground fuel and storage tanks, and aquifer zoning	x		
December 1980			
Center for Disease Control begins investigation of preliminary cancer and mortality data		x	x

bylaws, subdivision controls, and regulations can serve as effective local preventative measures, e.g., Model Local Zoning Bylaw—Aquifer Protection District (Metropolitan Area Planning Council, Boston, Massachusetts) and General Bylaw to Control Toxic and Hazardous Materials in the Town of Barnstable (Cape Cod Planning and Economic Development Commission). These measures could bolster federal and state legislative mandates such as RCRA and the Massachusetts Hazardous Waste Management Act, respectively. In conjunction with strict enforcement capabilities, these legal provisions could provide the necessary tools to alleviate the vulnerability of small public water supply systems.

Legal mechanisms that monitor the fragile balance of groundwater, surface water, and land resources can preserve pristine aquifers and thereby eliminate risks and/or limit point source pollution. Local and state regulations can specifically address different types of aquifers and development scenarios, given the availability of hydrogeologic data. In Massachusetts the initial siting of industries, under the sole statutory purview of the local government (Chapter 40, MGL), can mitigate potential contamination by at least monitoring the sensitive areas for migration of pollutants and at best prohibiting hazardous activities within the zone of influence and/or recharge area of the well waters. In fact, land use control and review of industrial locations might have lessened the chances of contamination in Bedford since a cluster of industries using hazardous materials are located upstream from the well watershed.

Significantly, the discovery of the contaminated well water was not a result of statutory mandates but rather a fortuitous action by a physicist residing in Bedford. The infiltration of the organic compounds was traced to illegal disposal of concentrated hazardous materials into a malfunctioning septic ejection tank and leaky municipal sewers. Tracking these problems, particularly in cities and towns with aging sewer systems and improperly lined landfills, is difficult. This does not dispel the need for regulation, but points to the limitations of passing legislation without providing adequate monitoring and enforcement support.

A serious weakness of both state and local water resource management is the lack of policy and long-range planning concerning water supply sources. Fortunately, Bedford was able to secure alternative water supplies from the Metropolitan District Commission and nearby towns. The issue of alternative water supply sources, however, would not be as easy to resolve for many other towns in Massachusetts because of isolation from water-rich areas, insufficient groundwater resources, or financial constraints. Moreover, rehabilitation programs to increase the pumpage of existing well sources and to patch up leaky sewer systems with the development of proper disposal facilities would be more cost-effective than the loss of existing or future water supply sources. These policy options must be weighed within the local context so that comprehensive water planning will be institutionally effective and financially feasible.

The Bedford case also introduces the issue of intertown cooperation for water supply agreements and prevention of pollution migration. Groundwater certainly does not flow along political boundaries in Massachusetts, therefore, jurisdictional disputes are inevitable. Planning strategies for adjacent towns sharing groundwater supplies and/or having contiguous industrial zones should include protection of recharge areas, monitoring and maintenance of municipal sewer lines, enforcement of proper disposal of hazardous materials, equitable aquifer withdrawal arrangements, and effective land use controls. Intertown or regional planning also requires coordination to promote consensus rather than competition for resources that could result in self-defeating relationships. Particularly sensitive issues, such as aquifers that cross town boundaries or siting of hazardous waste facilities, might necessitate intervention by the state. Thus far, intertown water supply agreements in Bedford have been facilitated by the Metropolitan District Commission and good relations with neighboring town governments.

The assistance of the Department of Environmental Quality Engineering and related state agencies was partly a statutory responsibility pursuant to the Massachusetts Safe Drinking Water Act (Chapter 111, MGL) and to the Massachusetts Clean Water Act (Chapter 21, MGL) to ensure the quality of public drinking water and to control subsurface pollution sources, respectively. The regional offices of the Department of Environmental Quality Engineering in cooperation with the local Department of Public Works, conducted sampling and testing of groundwaters to determine the extent and possible entry points of the contamination. However, extensive testing procedures, belated acquisition of sophisticated state laboratory equipment, and increasing caseloads led to protracted analysis and delays in the final report to the Board of Selectmen. It is difficult to assess whether the respite in the investigations presented any formidable obstacles for the town, but their frustration was evident. This was demonstrated by the unilateral Board of Selectmen decision to close the wells prior to the Department of Environmental Quality Engineering final order that followed one week later. This experience points to the need to improve communication between state and local institutions, particularly during a crisis period. Steps to improve consistent contacts between the authorities could be accomplished by improving the flow of information. Allowing the regional field offices greater accessibility to staff and funds could enhance crisis management and the degree of response to communities.

Federal, state, and local governments' cooperation was also manifest through transfers of technical information concerning public health risk assessment for the organic constituents found in Bedford's drinking water. Local officials could not ascertain the scope of the problem because of the insufficient data from the state Department of Public Health, Department of Environmental Quality Engineering state laboratories, and EPA. Following a prolonged period of correspondence and telephone contacts, local officials obtained the EPA "suggested guidelines" and interim standards for several of the 16 organic compounds. The reliability of this data, however, was the subject of conflicting opinions of the Department of Environmental Quality Engineering and EPA officials, primarily because of the lack of established numerical standards for organic compounds and the enforcement problems of the Suggested No Adverse Response Level (SNARL) recommendations. Moreover, EPA does not have standards on the risk of combined organic compounds in drinking water as a result of lack of data on their synergistic effects. The fact remains that scientific health risk data on volatile organic compounds in drinking water are quite insufficient to yield clear-cut decisions. The EPA and state agencies must actively pursue research in this area in order to provide data on "acceptable risks" to local governments. In this manner, more effective and accurate decisions could be made concerning the quality of municipal groundwater supplies.

XIII. UPDATE TO 1985

Since the contamination discovery and the shutdown of the four major production wells in Bedford (May, 1978), the town has continued to work toward the improvement of their water supply situation. These steps, carried out in collaboration with a private consulting firm and other government agencies, have demonstrated a concerted effort by the Town of Bedford to develop new groundwater sources, upgrade the quality of their degraded aquifer systems, and implement nonstructural protection mechanisms. As discussed below, these efforts have led to both progressive and frustrating results.

The litigation proceedings filed against six local industries that allegedly caused the pollution of four major production wells was resolved out of court in 1983. In deliberation with the Board of Selectmen, the companies agreed to pay the Town of Bedford $2.5 million compensatory fees over the next 10 years. These fees would cover past and future costs incurred by the town as a result of the water supply lost and the consultant charges. The settlement states that the payments do not constitute an admission of "any fault, liability or wrongdoing by any of the parties implicated" (*Bedford Minuteman,* 1983c). Thus the companies are exonerated from any future liability claims.

The Turnpike wells, numbers 7, 8, and 9, which were originally closed, still show traces of volatile organic chemicals 7 years after they were shut down. The proposed aquifer flushing program was never realized though the National Pollution Discharge Elimination System permit was approved by the State Department of Environmental Quality Engineering. The Town of Bedford was not confident that the technological processes would achieve a zero contaminant level. Similar pollutants were found in the neighboring groundwaters of Burlington. State water quality standards could have been met, however, these standards for organic microcontaminants were in a state of flux. Given the uncertainty of "allowable limits" and the associated health effects, the town opted not to disperse the chemicals by the flushing process.

To date, MCLs have been proposed for dioxane and trichloroethylene. These standards, still under debate, are not considered scientifically reliable until the EPA adopts the limits as an MCL (Environmental Protection Agency, 1985). At this point, available analytical techniques cannot detect dioxane below 10 ppb. The town's policy is to insist on 10 ppb because the "allowable limits" are unknown. Moreover, the recurrence of contamination incidences has made local officials as well as the residents more sensitive and cautious about the quality of their water supply. Bedford's approach to the problem, from a nondetectable point of view, will also affect their plans for the type and intensity of further treatment processes or plants.

In 1983, the town submitted a state application to fund a pilot study to analyze the most cost-effective approach for treating the Turnpike wells. This study, approved in June 1985, will be a joint review of treatment systems with the Town of Burlington. Treating the trichloroethylene in the groundwaters is feasible with available technology, however, the treatment of dioxane is more problematic. The latter process is more sophisticated and costly. A proposed regional treatment system will be evaluated. This state grant stems from Chapter 286 of the Massachusetts General Laws that provides funds to local governments for rehabilitating declining or contaminated water systems. The Town of Bedford, ruling out the aquifer flushing program, has thus chosen to investigate treatment options to restore the quality of their wells.

Well number 3, which is on the property of Mitre Corporation and was shut down along with the Turnpike wells has been restored for municipal use in Bedford. Levels of trichloroethylene have stabilized to 20 micrograms per liter and the presence of other chemicals has declined (*Bedford Minuteman,* 1983d). This well is currently pumping 400,000 gallons per day, and the State Department of Environmental Quality Engineering approved the reopening of the well in 1981 in conjunction with a continuous monitoring program. The other major groundwater resources, the Shawsheen wells numbers 2, 4, and 5, continue to feed 600,000 gallons per day into the public supply system.

Local groundwater supplies now total approximately 1 million gallons per day. However, local demand averages 2.2 million gallons per day and over 3 million gallons per day in peak summer season. These consumption rates are fairly stable as a result of continued water restrictions. Since 1978, the Board of Selectmen has voted each summer to reinstate restriction measures including specified evening hours for watering lawns based on even or odd house numbers and weekdays and prohibiting car washing. If these measures are violated, the party is subject to one warning notice and thereafter their water supply is turned off. This warning has been very effective according to the Public Works Department. Water rates have not changed since they were reassessed in 1978. At that time, the declining rate for water use of 0.25 cents per hundred cubic feet (hcf) was replaced by a universal rate of 1 cent per hcf.

As the town officials ensured the flow of public water supplies, three new wells and a treatment plant have been developed along Hartwell Road (well numbers 10, 11, and 12). Costing $2.3 million, including a state grant of $600,000, the project was completed in March 1983. The wells began producing 860,000 gallons per day with a chemical treatment capacity of 750,000 gallons per day. This well development was a major step in improving the local water supply profile and a significant contribution to their long-term planning efforts.

In the fall of 1983, traces of volatile organic compounds and increasing levels of iron, sulfate, sodium, and calcium were discovered in the new wells numbers 10 and 11. Tuberculation or the buildup of metals, calcium, and other soluble salts on the pipe interior can be broken loose by quick changes in the water flow. Apparently, the increasing sodium levels were linked to the sodium hydroxide solution added to the water supplies to decrease corrosivity. The solution was changed to a potassium base to reverse this trend. The metals continued to show high levels in the groundwater and the sulfates remained at up to 250 micrograms per liter. At present, no treatment process to remove sulfates is available in Massachusetts. The available processes in the country are technologically sophisticated and very expensive.

The central problem remained, however, as organic chemicals were consistently detected in wells numbers 10 and 11, including traces of trichloroethylene between 15 and 20 ppb (*Bedford Minuteman,* 1984a). By December 1983, both wells were closed for public use. Well number 12 also had rising mineral levels, but volatile organic compounds were not present. Several months later the discovery of benzene at 7 ppb, an amount exceeding the EPA standards, shut down the last Hartwell Road well in the spring of 1984 (*Bedford Minuteman,* 1984b). This new well development, in operation for just over one year, led to another dismal scenario for the Town of Bedford.

Given the closure of the Hartwell Road wells and the renewed threat to local water supplies, Bedford requested supply increases from the Metropolitan District Commission from 1 to 2.5 million gallons per day. This temporary agreement, stemming from another water emergency declaration, was approved recently by the Massachusetts Water Resources Authority, the Metropolitan District Commission and neighboring towns. This agreement also ensures that Bedford will be able to meet summer peak demand, which can exceed 3 million gallons of water per day. Meanwhile, the Local Board of Health was charged once again with assessing public health threats from the drinking water.

In the fall of 1982, investigations at Hanscom Air Force Base uncovered traces of hazardous chemicals such as chloroform, dichloroethylene, trichloroethane, and trichloroethylene ((*Bedford Minuteman,* 1983e). High levels of lead and arsenic were also found in the surrounding soils. This discovery was prompted by an Air Force veteran, residing in Bedford, who recalled an incident where hundreds of barrels of hazardous waste were buried 20 years ago. Unfortunately, the exact contents and location of these wastes are unknown. Groundwater flows in the area are restricted by the bedrock formations to three directions, and one of these points toward the Hartwell Road wells ((*Bedford Minuteman,* 1983b). To date, the intrusion of volatile organic chemicals has not been linked conclusively to the Air Force base. Testing and monitoring at 44 sites around the base continue in conjunction with an environmental assessment that the Air Force is conducting.

In their undaunting efforts to redress local water supply problems, the Town of Bedford has proposed three new well sites and a treatment plant along the Concord River. The Putnam Road Aquifer, expected to yield 2.2 million gallons of water per day, is the last potential well field site in Bedford. Metal sources are found naturally in the groundwaters and high levels of carbon dioxide exacerbate water pipe corrosion. Iron levels were detected at up to 18 parts per million (ppm) and manganese levels at 0.3 ppm, while the recommended safe levels are 0.3 ppm and 0.05 ppm, respectively (*Bedford Minuteman,* 1983a). The treatment plant will chemically condition the groundwater to remove carbon dioxide, which causes corrosion. The water will then be oxidized and filtered to remove any metals. Situated on 5 acres of land, the 6000-square-foot plant is projected to cost $2000 per million gallons of treated water or $1330 per milligram with state aid (*Bedford Minuteman*, 1983a).

The proposed well development, estimated to cost $8 million, needs to be approved by the Massachusetts Water Resources Commission. The commission has jurisdiction over this project because it will result in a permanent water diversion from the Concord River. Since the groundwater is produced by induced recharge from the Concord River watershed and the wastewater will be disposed of in Boston Harbor, this project is also subject to the requirements of the Interbasin Transfer Act. Given the legal, technical, and institutional requirements, the new underground water sources are not expected to come on line before 1988.

Beyond the original contamination discovery, the events of the last few years have not fared well for the Town of Bedford. Their attempts to mitigate supply losses and clean up polluted groundwater resources were unsuccessful—as evidenced by the forced closure of the Hartwell Road wells and the continued shutdown of the Turnpike wells. The uncertainty surrounding the Hanscom Air Force Base has emphasized the precarious state of their existing aquifer system and the potential public health threats. Bedford residents, in interviews conducted in 1985, have shown an understanding about these incidents although there is now greater fear and anxiety. These fears are coupled with hesitations on the part of the town government to encourage further developments. A proposed project for a Boston college to move to Bedford will be denied because it would overwhelm their already strained water supply (*Bedford Minuteman*, 1985). Hence, the existing water supply emergency has prompted the town to prohibit further growth in order to conserve their public underground water supplies.

As a result, the Town of Bedford continued to battle with these groundwater supply problems amid their efforts to maintain the quality and quantity of public drinking water. The ghosts of illegal and/or unintentional chemical disposal continue to haunt the town. There is hope that legal and institutional mechanisms have established restrictions to prevent further contamination. However, the problems persist and must continue to be confronted by local governments, the public consumers, and especially those private and public groups that are responsible for proper waste disposal. One cannot afford to wait until groundwater protection becomes more cost-effective. Waiting and hoping can only lead to the loss of precious public underground drinking water sources.

ACKNOWLEDGMENTS

The authors would like to thank local officials in Bedford, Massachusetts, Robert Weimer of Camp, Dresser & McKee, Inc., Boston, Massachusetts, the U.S. Environmental Protection Agency, and the Massachusetts Department of Environmental Quality Engineering for their assistance and cooperation throughout the study. This chapter includes some material developed by the authors for a report entitled "Analysis of Recent Environmental Legislation as It Affects Public Water Supply" prepared for the U.S. Environmental Protection Agency. The statements and conclusions in this case study are solely those of the authors.

REFERENCES

Bedford Minuteman (1979a). December 27, **18**(9), p. 1.
Bedford Minuteman (1979b). December 27, **18**(9), p. 8.
Bedford Minuteman (1980a). March 6, **18**(18).
Bedford Minuteman (1980b). March 6, **18**(18), p. 2.
Bedford Minuteman (1980c). April 10, **18**(23), p. 1.
Bedford Minuteman (1980d). April 17, p. 10.
Bedford Minuteman (1980e). June 26, p. 1.
Bedford Minuteman (1980f). December 24, **19**(8), p. 1.
Bedford Minuteman (1980g). December 18, **19**(7), p. 1.
Bedford Minuteman (1981a). January 22, **19**(12), p. 3.
Bedford Minuteman (1981b). April 10, **18**(23), p. 3.
Bedford Minuteman (1983a). January 20.
Bedford Minuteman (1983b). March 14.
Bedford Minuteman (1983c). April 18.
Bedford Minuteman (1983d). July 14.
Bedford Minuteman (1983e). September.
Bedford Minuteman (1984a). February 16.
Bedford Minuteman (1984b). April 5.
Bedford Minuteman (1985). October 11.
Boston Globe (1980). April 25, p. 19.
Camp, Dresser & McKee, Inc. (1979). "Report on Industrial Chemical Contamination at Municipal Wells Nos. 3, 7, 8 and 9." CDM, Boston, Massachusetts.
Camp, Dresser & McKee, Inc. (1980). Correspondence with the Massachusetts Department of Environmental Quality Engineering, April 8.
Camp, Dresser & McKee, Inc. (1985). Interview, September 13.
Department of Environmental Quality Engineering (1980). Interview, November 11.
Division of Water Supply (1980). Interview, August.
Environmental News (1980). EPA Region I, December, pp. 4–5, Boston, Massachusetts.
Environmental Protection Agency (1985). "Weekly Report," August 9 and 23. EPA, Washington, DC.
Harvard University (1980). "Massachusetts Hazardous Waste Management—Local Role." Department of City and Regional Planning, Cambridge, Massachusetts.
Massachusetts Audubon News (1980). January. Boston, Massachusetts.
Regional Office, Division of Water Pollution Control (1980). Interview, March 10.

14

Summary

G. WILLIAM PAGE
Department of Urban Planning
and Center for Great Lakes Studies
University of Wisconsin–Milwaukee
Milwaukee, Wisconsin 53201

Many different approaches are needed to contend with the problems of microcontaminants in groundwater. Some approaches must be remedial to contend with existing contamination and other approaches are preventive in nature. The actions necessary to contend with actual and potential groundwater contamination span the gamut from urgent to long term. This book attempts to provide the reader with an overview of the topic, some insight into how groundwater protection programs are developed and implemented, and a description of some of the best groundwater protection programs in the United States.

This summary chapter attempts to identify and elucidate certain important themes that are addressed from diverse perspectives in different chapters of the book. In most cases these themes were first addressed in the early chapters that provide a framework for groundwater protection and discuss the options and problems as well as provide guidance on how to go about planning for groundwater protection. These themes are then picked up from the unique perspectives of the case studies.

Each municipality, county, regional agency, or state that is planning for groundwater protection faces a unique set of circumstances that will require a planning process to develop a groundwater protection program that is tailored to fit their unique characteristics. Physical, institutional, legal, and socioeconomic factors all vary in significant ways among political entities that may be interested in planning for groundwater protection. No single set of groundwater protection measures can be ideal for every location. The case study chapters in this book illustrate many of these differences and describe how groundwater protection programs were developed to fit unique needs.

While the process of developing each successful groundwater protection program may be different in many important respects, there are certain important concerns that each must address. These concerns are themes that run through this book and will be the subject of the remainder of this chapter. These themes are the complexity of groundwater systems, fragmentation of laws and institutions, political support, and significant unknowns.

I. COMPLEXITIES OF THE GROUNDWATER SYSTEM

A. Groundwater as Part of the Hydrological Cycle

Groundwater systems are complex portions of the hydrological cycle about which we usually have only a partial understanding. Their location beneath the surface of the earth makes direct investigation of groundwater difficult and expensive. The integral connection of groundwater to the rest of the hydrological cycle causes important complexities that affect groundwater protection programs. Those aspects of groundwater systems that are unknown or only little understood must be carefully considered in planning for groundwater protection.

Chapter 1 warns us that groundwater is an artificial category. Water in the atmosphere, on the land surface, under the land surface, or in the oceans is all part of a continuous process. Failure to understand these connections can lead to many problems. For instance, substantial quantities of microcontaminants reach waterbodies from fallout from the atmosphere (Lantzy and Mackenzie, 1979; Eisenreich *et al.*, 1981; Galloway *et al.*, 1983).

In most cases, groundwater flows into surface waterbodies. Groundwater constitutes the base flow of most streams. The rivers on Long Island, New York, are described as exposed groundwater. Barton Springs Pool, (Fig. 6 in Chapter 9) in downtown Austin, Texas, is a natural outflow of the Edwards Aquifer that is a major source of drinking water and streamflow, as well as a major recreational resource.

In some instances surface water flows into groundwater and recharges the aquifer. In Wausau, Wisconsin, the Wisconsin River recharges the sand and gravel aquifer only when the pumpage rate and the river stage, the volume of water in the river, are high. In other municipalities, surface water is intentionally managed by induced recharge to augment the supply of groundwater. In the Perth Amboy, New Jersey, Runyon Well Field and in the Bedford, Massachusetts, Putnum Road Well Field surface water from rivers is intentionally used to recharge the aquifers and increase the quantity of groundwater that can be withdrawn for use in the municipal drinking water system. Dade County, Florida, is constructing a new canal to bring fresh water from the Everglades into the county where it will be used to recharge the groundwater supply. In so doing, Dade County hopes to change the groundwater flow gradients so that a plume of contamination does not flow toward their well field.

In some municipalities water from precipitation events is captured and used to recharge groundwater supplies. Storm water runoff is channeled into specially constructed impoundments that are designed to allow the water to percolate into the ground and recharge the aquifer (Tourbier and Westmacott, 1974). Long Island, New York, Santa Clara Valley, California, and Austin, Texas, all use induced recharge of captured storm water runoff to replenish the groundwater supply. Induced recharge has the intended effect of increasing the quantity of groundwater available for use as a drinking water supply.

Runoff in many areas, especially in urban areas and agricultural areas, often contains many pollutants. Runoff often contains high concentrations of both organic and inorganic toxic substances. The case studies of Austin, Texas, and Long Island, New York, discuss special precautions taken in their efforts to develop groundwater protection plans that avoid introducing microcontaminants into their groundwater supplies through the use of induced recharge of storm water.

Induced recharge can have a dramatic effect on the quantity of groundwater available for withdrawal and use in a drinking water supply. This is one of the reasons why attempts to estimate the quantity of groundwater that may safely be withdrawn from an aquifer encounter confusion. Chapter 1 strongly suggests that the often-used term "safe yield" be replaced by a more accurate term such as "maximum stable basin yield" or "optimum yield." While these terms may be better, the quantity of water associated with these terms remains difficult to accurately estimate. Good estimates of the optimum yield are important components of a groundwater protection program.

Withdrawal of more than the optimum yield from an aquifer can lead to serious problems. This condition is often referred to as "groundwater mining" or "overdraft conditions." Santa Clara Valley, California, experienced severe land subsidence because of water withdrawals exceeding the optimum yield of the aquifer. The potentiometric surface of groundwater in the Farrington Sand Aquifer in the Runyon Well Field of Perth Amboy, New Jersey, has declined more than 100 feet because of overdraft conditions. In coastal areas such as Perth Amboy, New Jersey, Long Island, New York, and South Brunswick, New Jersey, overdraft conditions over a number of years have caused salt water from the ocean to be drawn into the aquifer and contaminate the groundwater. This problem is known as "salt water intrusion."

Planning for groundwater protection requires special measures when the need for water approaches the optimal yield of the aquifer. Municipalities that do not have surface water supplies suitable for use as a drinking water supply must look for sources of drinking water from outside their municipal boundaries. Perth Amboy, Bedford, the Santa Clara Valley, and Dade County all import water from considerable distances to supply or augment indigenous drinking water sources. Perth Amboy developed their own supply source in a distant municipality and also purchases water from a privately owned water company (purveyor). Bedford purchases water from both neighboring municipalities and a regional water commission. Both the Santa Clara Valley and Dade County purchase exogenous water from the regional water authorities.

The availability of institutional arrangements, physical interconnections of the water distribution systems, and an available surplus of water in the region are important considerations in planning a water supply system. Contamination of groundwater with toxic substances can create an immediate crisis that may not allow time to develop new municipal sources of water such as new wells. In both Perth Amboy and Bedford the municipalities were able to continue to provide a water supply for drinking water and for fire protection despite the closing of all municipal wells caused by the threat to the public's health. Planning for groundwater protection should include making any arrangements possible to be available should a crisis occur. Responding to a crisis that might result from severe groundwater contamination with toxic substances would require that the planning be completed in advance. Portable filtration units using granular activated carbon, described in Chapter 4, are also a possible short-term response to a crisis. Wausau, Wisconsin, used such a unit until a long-term reponse to groundwater contamination in the city could be put in place.

B. Groundwater Uncertainties

Heterogeneities in aquifers produce many uncertainties that affect our ability to understand groundwater systems and our ability to plan for groundwater protection. Our understanding of how water moves through an aquifer that is not a homogeneous material is severely limited, especially because we can never know all of the locations and extents of the nonhomogeneous material. The situation in South Brunswick, New Jersey, presents a good example of one of these problems. The Woodbridge Clay unit acts as an aquitard separating the surface aquifer from the deeper aquifer that was used as a source of municipal drinking water. The municipality learned that the Woodbridge Clay unit was discontinuous. This window in the clay layer allowed toxic contaminants in the surface aquifer to reach the deeper aquifer and contaminate public supply wells. Figure 2 in Chapter 11, the case study of the Santa Clara Valley in California, illustrates a similar groundwater system with discontinuous clay layers separating aquifers.

Modeling the flow of groundwater is exceedingly difficult when heterogeneities are present. Some heterogeneous material is almost always found in aquifers. Chapter 1 showed how knowledge of the heterogeneities present in the subsurface is necessary to predict the movement of contaminants in the saturated zone. Yet, knowledge of the distribution and composition of heterogeneous materials in the subsurface is expensive and time-consuming to obtain. Options for obtaining geohydrological data on this topic are discussed in Chapter 5. In most instances, only the vaguest knowledge of heterogeneities is available.

Despite uncertainties in most of the case studies, modeling of the groundwater flows was undertaken. In some cases, the model predictions were judged to be sufficiently accurate to be useful. In Dade County, many of the groundwater protection measures instituted are based on the groundwater model estimates of how long it will take groundwater to flow from certain points to the nearest public water supply well. The extensive groundwater modeling completed in Wausau, Wisconsin, under the Superfund emergency response action was unable to determine groundwater flows sufficiently to determine the source of contamination for some of the contaminated public supply wells. In the karst terrain around Austin, Texas, groundwater modeling was not attempted because groundwater flow through the Edwards Aquifer is too complicated for modeling efforts. Chapter 1 briefly describes several groundwater models used in making planning decisions and stresses that they be used only by individuals well trained in groundwater hydrology and the use of models, and that the limitations and uncertainties inherent in the model be fully recognized.

The process of groundwater modeling is often more valuable to groundwater protection planning than the results. The monitoring and other data collection activities required to develop mathematical groundwater models provide important information and insights to the site-specific hydrogeological conditions. Chapter 1 tells us that "It is the thought process needed when applying a model that should lead to a decision, not necessarily and certainly not exclusively the answers generated by the model."

Attempts using structural approaches to manage water resources have often caused serious problems with groundwater. Our lack of knowledge concerning conditions in the subsurface has often caused unanticipated problems. Deepening a canal in New Jersey exposed the Farrington Sand Aquifer to brackish water and led to salt water intrusion problems (Chapter 10). The extensive system of canals and levees constructed to control flooding in Dade County, Florida, also contributed to salt water intrusion problems as well as causing substantial and serious disruption to the ecology of the Everglades.

Structural approaches may be important components of groundwater protection plans. Dade County is presently constructing a canal that will supply fresh water that will be used to help keep a plume of contaminated groundwater away from their new well field. Proposed structural components of groundwater protection plans must be carefully evaluated to avoid unforeseen consequences resulting from uncertainties in our knowledge of the subsurface.

II. FRAGMENTATION OF LAWS AND INSTITUTIONS

A. Legislation

In the United States, many levels of government plan for groundwater protection with a multitude of distinct programs authorized by several federal laws, state laws, and municipal ordinances. At all levels of government, fragmentation interferes with effective groundwater protection planning and the implementation of effective groundwater protection programs.

Many federal laws and programs affect efforts to protect groundwater but fail to provide a framework for comprehensive groundwater protection planning. Chapter 2 describes the eight major pieces of federal legislation that affect groundwater. The Office of Technology Assessment (1984) discusses an additional eight federal laws that have a lesser impact on groundwater protection efforts. Despite all this federal legislation, the U.S. Office of Technology Assessment concludes that "not all sources of groundwater contaminants are included, and for the general sources that are included, not all related facilities and/or activities may be covered; not all drinking water supplies are monitored routinely; and standards have not been developed for most contaminants that have already been developed in groundwater."

In May 1986, the House–Senate conference committee voted out a five-year extension of the Safe Drinking Water Act that includes a provision to protect groundwater (Greenhouse, 1986). Such a provision was not present in the Safe Drinking Water Act passed by Congress in 1974. Under this revised legislation, each state is required to develop a program to protect areas around the wells of public water systems. The plans must be submitted to and approved by the U.S. Environmental Protection Agency within three years. The legislation appropriates additional money to the Environmental Protection Agency and authorizes the agency to finance up to 90% of the cost of developing and implementing the state groundwater protection programs. The bill also requires the Environmental Protection Agency to establish Maximum Contaminant Levels (MCLs) for 83 additional contaminants found in drinking water.

Groundwater protection planning efforts at the state level have varied. Some states have been very active in promoting groundwater protection programs (see Chapter 2 for a description of these efforts). Other states have in effect done nothing. In some areas, counties have taken the lead role in groundwater protection, while in other areas municipalities have been most active. The level of activity devoted to planning for groundwater protection in all states is likely to intensify rapidly when the federal government requires and starts financing 90% of the costs of groundwater protection programs under the revisions to the Safe Drinking Water Act.

B. Regional Issues

Planning for groundwater protection often requires a regional approach because plumes of contaminated groundwater don't follow political boundaries. When plumes of contamination extend across political boundaries, groundwater protection programs cannot be effective unless all the political units involved in the contamination event are also involved in the groundwater protection program. Bedford, Massachusetts, had little control with municipal protection measures over the plume of contaminated groundwater that flowed into the municipality from the direction of the Hanscom Air Force Base. South Brunswick, New Jersey, has instituted a very impressive municipal groundwater protection program, but its measures are not effective at protecting South Brunswick from groundwater contamination originating in neighboring municipalities and flowing into South Brunswick's drinking water supply.

Depending on the hydrogeological conditions, aquifers may exist in the subsurface of only small portions of a municipality or of many states. To be effective, groundwater protection programs must be planned for and implemented by the entire region in which the aquifer exists. There are many obstacles to regional approaches in developing a groundwater protection program. One of the most significant obstacles is the fact that states grant to municipalities many of the powers that are most important in groundwater protection plans. In most cases, municipalities zealously guard their power to control local activities. Strong commitments to "home rule" make regional approaches difficult.

The Austin, Texas, case study offers an example of a unique approach used under some circumstances in Texas for regional groundwater protection planning. Texas cities of 100,000 population or more can adopt areawide water pollution abatement programs, including groundwater protection programs, to protect their water supplies from urban runoff pollution in an area of extraterritorial jurisdiction. The area of extraterritorial jurisdiction extends radially from the city for a distance of 5 miles. This is a powerful enhancement of the city's ability to plan for groundwater protection, but it may not be sufficient. The Edwards Aquifer, which supplies Austin's water, has recharge areas beyond the area of extraterritorial jurisdiction to the west of Austin. The city presently has no authority to control growth or environmental degradation in these critical areas.

In many parts of the United States, county governments are taking a leading role in efforts to plan for groundwater protection. County governments have many advantages over municipal governments in planning for groundwater protection because of their size. However, the size of counties is not always optimal for groundwater protection programs. The political boundaries of the county are unlikely to be coterminous with the boundaries of the aquifer. Nassau and Suffolk counties on Long Island, New York, Dade County, Florida, Travis County, Texas, the location of the city of Austin, and Middlesex County, New Jersey, are examples of counties that depend on aquifers that extend beyond county boundaries. In some counties the majority of municipalities may use surface water and be unwilling to implement restrictive groundwater protection programs for which they see no pressing need.

The power of county governments to plan and implement programs varies from state to state. In Marathon County, Wisconsin, as discussed in the Wausau case study, the county government is taking the lead in groundwater protection planning, but it is too early to judge the effectiveness of the implementation. Santa Clara County, California, has succeeded in implementing a groundwater protection program at the county level largely through effective consensus building in the county rather than through the political power of county government. Dade County, Florida, is an example where the county government has functional responsibilities for countywide services such as metropolitan planning, welfare, health, transit, environmental regulation, and the provision of water supply and wastewater services. County governments with primary responsibilities in these important regional services are not common in the United States, but this form of governmental organization has many obvious advantages for resolving regional issues such as groundwater protection.

One of the most notable examples of an attempt to solve a regional environmental problem was Section 208 of the Federal Water Pollution Control Act Amendments of 1972 (P.L. 92-500). Section 208 required a comprehensive areawide planned approach to water pollution problems. The 208 planning programs gathered data at the watershed or aquifer scale. The discussion of data requirements for groundwater protection planning in Chapter 5 identifies the data generated by many 208 planning efforts as the first and most valuable source to investigate when starting work on a groundwater protection program. A committee of the National Research Council has recommended that the federal government fund groundwater protection programs at the basinwide scale (National Research Council, 1986).

Several of the case study locations in this book, for example, Austin, Texas, used the data generated by the 208 planning efforts as a base on which to build groundwater protection programs. On Long Island, New York, the 208 planning efforts was actually used to develop a groundwater protection program. This case study is an interesting critique of the strengths and weaknesses of this significant approach to environmental management. On balance, Section 208 of P.L. 92-500 charted an innovative path for an effective approach to solving regional environmental problems.

In 1986, the amendments to the Safe Drinking Water Act have the potential to make a significant impact on groundwater protection planning. These amendments require the states to develop programs to protect areas around public supply wells and authorize the Environmental Protection Agency to finance part of the cost of developing and implementing these programs (Greenhouse, 1986). These amendments will give the states the opportunity to use some of the lessons learned in 208 planning in developing and implementing their groundwater protection programs.

III. POLITICAL SUPPORT

Building a political constituency to support groundwater protection efforts is probably the most important single step in developing and implementing a successful groundwater protection program. There are two primary reasons why strong political support is essential: (1) developing and implementing groundwater protection programs is expensive and (2) effective groundwater protection programs often require the cooperation of several political entities. Publicizing groundwater problems and efforts to protect groundwater is so important that Chapter 5 suggests that "the early and continuous involvement of a public information specialist may prove as important to the development of a comprehensive local groundwater protection program as the involvement of the community's hydrogeologic consultant." In Wausau, Wisconsin, a public information groundwater specialist was the first additional staff member hired when their groundwater planning effort started.

The desire to control growth in municipalities has been an important factor in generating political support for groundwater management programs. Support for growth management grew in the 1970s around efforts to control suburban sprawl and minimize the negative externalities associated with rapid and unregulated development. The Long Island, New York, case study describes the development of a constituency for groundwater protection that grew out of the public's interest in protecting the natural environment from rapid growth.

Growth management concerns are still a factor in developing groundwater protection programs in some parts of the United States that are experiencing rapid growth. The Austin, Texas, case study is an example of a sunbelt area where the concern to manage growth has been an important political factor in efforts to protect groundwater. Throughout the United States there are some places that are pockets of especially rapid growth. The case study of South Brunswick, New Jersey, describes a municipality in the high technology growth corridor along U.S. Route 1 in the Princeton area. Wherever there are strong pressures for industrial, commercial, or residential development, the political constituency in favor of growth management may be a logical source of support for efforts to develop and implement groundwater protection programs.

In the 1980s, the fear of negative health effects from exposure to microcontaminants in drinking water has become the most important factor in generating political support for groundwater protection programs. Managing toxic and hazardous materials has become the most critical environmental issue in the United States and other economically developed nations. The human health effects of exposure to microcontaminants in drinking water, as described in Chapter 3, are not well understood. Toxic substances management has replaced growth management as a focus for groundwater protection planning. Where groundwater contamination with toxic substances is already a problem, a political constituency for groundwater protection is almost certain to exist. Municipalities that are attempting to plan for the protection of groundwater that is not already contaminated have a greater need of public information programs to help develop a political constituency.

Political support for groundwater protection planning can be developed in different ways. In some situations political support develops from the grass roots. The Austin case study is an example of a city in which private citizen groups provided the impetus to develop groundwater protection programs. The Save Barton Creek Association and the Zilker Park Posse were instrumental in forcing the city council to address the issue. In other situations, the government bureaucracy takes the initiative that helps galvanize public support to the issue of groundwater protection. The Long Island case study is an example of political support being developed from government action. Most places with groundwater protection programs developed political support by some combination of top down and bottom up activity.

IV. SIGNIFICANT UNKNOWNS

Efforts to protect groundwater from contamination by toxic substances are hindered by our lack of knowledge in many areas. There are substantial gaps in our knowledge of which microcontaminants may be present in groundwater, how to predict their movement, how long they will remain a problem, how they will affect human health, the contribution of microcontaminants in drinking water relative to other health risks, the effectiveness of our responses to contamination, and the effectiveness of our efforts to plan for the protection of groundwater.

Our knowledge of which microcontaminants are present in groundwater sources of drinking water is incomplete. On a broad scale, we are unable to accurately predict where groundwater pollution with toxic chemicals will be found (Page, 1986). Toxic materials are used so widely and such small quantities can contaminate groundwater that serious contamination can be found anywhere. Our government has very little knowledge of even the most serious potential sources of contamination. Countless 55-gallon steel drums filled with toxic chemicals are buried in unknown locations throughout the country. These steel drums will inevitably rust and release their contents to the ground where they may reach and contaminate groundwater. Underground storage tanks and their piping are also potential sources of contamination. When the City of Sunnyvale, California, implemented their groundwater protection program, described in the Santa Clara Valley case study, they found that 50% of the gasoline service stations who reached the permit step of monitoring groundwater had leaking tanks. Even when scientists monitor groundwater, they are unable to identify the large majority of the organic compounds present in the water (see the discussion in Chapter 3).

There are significant gaps in our knowledge of groundwater systems that limit our ability to predict the movement of plumes of contaminated groundwater. Chapter 1 describes the importance of heterogeneities in the subsurface and their impact on the movement of water and contaminants. Chapter 1 also discusses the powers and limitations of mathematical models to predict the future and the uncertainties inherent in every modeling exercise. These problems are magnified by the availability of only a small number of trained groundwater hydrologists.

There are significant uncertainties regarding how long contaminated groundwater will remain contaminated. There are very few natural processes that can effectively degrade toxic contaminants once they are in groundwater. Extensive hydrogeological investigations are required to make even an informed estimate of how long the plume of contamination will reamin a threat to human health. Careful study of these problems has not continued long enough to help predict the duration of these problems (Nielsen, 1983). The Bedford, Massachusetts, case study reveals that seven years after the municipality closed their first public supply wells, the concentrations of contaminants in the wells are still too high to use them for a drinking water supply.

The human health effects of exposure to microcontaminants in drinking water, discussed in Chapter 3, is a field with many significant unknowns. Many of the microcontaminants found in groundwater are carcinogens, mutagens, and teratogens, but we do not have a clear understanding of the mechanisms of cancer, mutagenesis, and teratogenesis. We do not know if a threshold exposure to a carcinogen exists or the correct shape of dose–response functions. There are uncertainties about our extrapolations from high dose levels to levels found in drinking water and from animal species to humans. We do not know the health effects of many microcontaminants, such as nonpolar organic compounds, that are often present in water. We know very little about potential synergisms and antagonisms among the complex mixtures of microcontaminants often found in drinking water.

The relative importance of the risk to human health caused by microcontaminants in drinking water is poorly understood and the subject of considerable controversy. Scientists are finding many carcinogens in our environments, and many of them represent a greater threat to human health than the microcontaminants found in drinking water. New research approaches are attempting to assess the total exposure to carcinogens from water, air, skin, and food.

We have learned much from recent experience with remedial efforts to respond to the contamination of groundwater with toxic substances, but significant uncertainties remain. Chapter 4 presents clear estimates of the costs of different technological responses to contamination, but our experience with these technologies extends back only a few years. Countercurrent packed column air stripping towers appear to be the most cost-effective technology to remove volatile organic compounds. The case studies in Wausau and South Brunswick document their use. Air stripping towers do not remove nonvolatile organic compounds nor inorganic toxic substances. In South Brunswick, the municipality built both an air stripping tower and a granular activated carbon filtration facility to reduce to acceptable levels all microcontaminants.

The effectiveness of groundwater protection programs remains uncertain. Planning to protect groundwater has really only begun. The planning process for the first programs to protect groundwater from microcontaminants started in the mid-1970s, and both the development and implementation of plan components are still evolving. It is clear that groundwater protection programs will reduce the quantity of toxic materials reaching our groundwater supplies. The better groundwater protection programs will be more effective than the less thorough and comprehensive programs. The argument to start planning for groundwater protection before there is serious contamination is clear and powerful.

REFERENCES

Eisenreich, S., Looney, B., and Thorton, J. (1981). Airborne organic contaminants in the Great Lakes ecosystem. *Environ. Sci. Technol.* **15**(1), 30–38.

Galloway, J., Eisenreich, S., and Scott, B. (1981). Toxic substances in atmospheric deposition: A review and assessment. *In* "The Potential Atmospheric Impact of Chemicals Released to the Environment" (J. Miller, ed.), EPA 560-80-001. U.S. Environmental Protection Agency, Washington, D.C.

Greenhouse, L. (1986). Bill on drinking water approved by Congress. *New York Times,* May 22, p. 14.

Lantzy, R., and Mackenzie, F. Atmospheric trace metals: global cycles and assessment of man's impact. (1979). *Geochim. Cosmochim. Acta* **43,** 511–525.

National Research Council (1986). "Ground Water Quality Protection, State and Local Strategies." National Academy of Sciences, Washington, D.C.

Nielsen, D. (1983). "Proceedings of the Third National Symposium an Aquifer Restoration and Ground-water Monitoring." National Water Well Association, Worthington, Ohio.

Office of Technology Assessment (1984). "Protecting the Nation's Groundwater from Contamination," OTA-O-233. U.S. Congress, Washington, D.C.

Page, G. W. (1986). Municipal characteristics associated with toxic contaminants in groundwater. *J Plann. Educ. Res.* **5**(2).

Tourbier, J., and Westmacott, R. (1974). "Water Resources Protection Measures in Land Development—A Handbook." University of Delaware, Water Resources Center, Newark.

Index

A

Abatement, 315
  cost of, 317
Absolute ownership, 33, 268
Activated carbon, 83
  granular, 106
  powdered, 107
Adsorption, 11, 105
Advection-dispersion equation, 20
Aeration, 109
  cost, 117
  systems, 114
Airborn pollution, 170
Air stripping, 91
  facilities, 223, 231, 252
American rule, 34
Animal wastes, 174
Aqueduct, 301
Aquifer, 15, 160
  classification, 40
  diagram, 392
  flushing program, 136
  mapping system, 135
  viewpoint, 15
Aquitard, 372
Artesian zone, 264
Atmospheric pressure, 3
Attenuation, 136
Austin Tomorrow Comprehensive Plan, 270

B

Backwashing, 98
Bailer tests, 16
Balcones Fault Zone, 263
Base flow, 3
Bedrock formation, 342
Biodegradation, 13
Biological oxygen demand (BOD), 294
Biotransformation, 13
Biscayne Aquifer, 205, 209
Breakthrough, 106
Buffer zone, *see* Watershed zone

C

Cancer processes, 73
Capillary fringe, 3
Carbon adsorption, 91
Carbonate aquifer, 126
Carcinogenicity, 243
Carcinogens, 73
Cemeteries, contamination from, 174
Cesspool, 142
Chemical oxygen demand (COD), 294
Chlorination, 71, 99
Chlorine dioxide, 100
Citizen participation, 188, 268
Clarification, *see* Sedimentation
Clean Air and Clean Water Acts, 40
Clean-up technique, 316
Clean Water Act, 58
Coagulation, 94
Common law, 33
Compliance review, 336
Comprehensive Environmental Response, Compensation and Liability Act, 28, 57
Condensation, 2
Cone
  depression, 210

Cone (*cont.*)
  influence, 223
  influence thresholds, 224
Configuration, 91
Confined zone, 304
Confining beds, 15, 158
Conservation law, 175
Contaminant and source classification, 40
Contaminant transportation model, 21
Contaminants
  volatic organic (VOCs), 243
Contamination process
  biological, 13
  chemical, 11
  physical, 9
Contaminations
  bacterial, 139
  nonpoint source of, 177
  nontoxics, 291
  storage tanks, 168
  toxics, 294
Convection-dispersion equation, 20
Critical zone, *see* Watershed zone

D

Dalton's Law, 111
Darcy's Law, 4
Deicing salt pollution, 147
Dessicant, 101
Detention time, 97
Differential protection, 39
Diffused air type, air stripping, 115
Diffusion, 10
Diffusion wells, 171
Dilution, 9
Directory of groundwater programs, 43
Discharge
  areas, 5
  industrial waste, 165
Disinfection, 101
Diversion, 10
Diversion right, 327
Dose–response function, 74
Drinking water contamination
  new supply, 88
  technological treatment, 93
Drinking Water Standards, 45

E

Eductor, 116
Edwards Aquifer, 261
Effluents, 261
  sanitary sewers, 165
  sewage treatment plant, 164
Emergency response plan, 338
End-of-pipe techniques, 177
Endogenous water, 288
English Law, 33
English Rule, 34
Environmental Pollution Agency (EPA), 30, 377
Environmental regulations, local, Dade County, FL, 235
Environmental resource inventory, 335
Epidemic research, 77
Evaporation, 2
Exfiltration standards, 233
Exogenous source, 289
Exposure profile, 80

F

Facility storage map, 313
Fall-out, 370
Farrington Sand Aquifer, 327
Fecal streptococci, 266
Federal Insecticide, Fungicide, and Rodenticide Act, 59, 126
Federal Water Pollution Control Act, Amendments of 1972, 377
Fertilizers, contamination of, 167
Field capacity, 8
Filtration, 97
  goal of, 177
  granular activated carbon, 248, 250, 333
  shortcoming of, 178
First flush, 278
First-order decay, *see* Hydrolysis
Flocculation, 95, 253
Flocculation tank, 94
Flood plain, 274
Flow system viewpoint, 15
Fluctuation, 5

G

Gardiners Clay, 161
Gas chromotograph, 70, 183, 346
Geologic well logs, 10
Glacial aquifer, 159
Granular activated carbon, 106
Groundwater
  flow, 2
  flow model, 21, 222
  hydrology, 26
  mining, *see* Overdraft
  monitoring, 13
  planning, 131, 177, 372
  pollution, 17
  potential, *see* Head
  quality protection, 279, 355
  recharge, 7
  travel time, 217
  use, 140
Groundwater protection,
  effectiveness, 339
  land use control, 42
  programs, 38, 376
  user-oriented control, 42
Gt factor, 96

H

Hazardous waste, 53
Head, 3
Henry's Law, 112
Highway deicing, contamination of, 166
Home rule, 376
Hydraulic conductivity, *see* Permeability
Hydraulic gradient, 4
Hydrologic cycle, 1
Hydrolysis, 12

I

Impermeable pans, 335
Impervious cover, 276
  standards of, 277
Incidental discharge, 173
Incinerator quench water, 170
Induced recharge, 370, 371
Infiltration, 7
Ion exchange, 102
Isopleth, 149

K

Karst system, 262, 373

L

Landfills, 166
Landrush, 270
Land subsidence, 371
Land use policy, 234
Lateral study, 139
Leachate
  contamination, 147
  generation of, 166
Leaking underground storage tanks (LUST), 56
Legislation, 374
Lime-soda ash softening, 98
Limited degradation, 39
Lithologic logs, 330
Litigation, 356
Lloyd stratums, 161
Longitudinal study, 139

M

Macrocontaminant, 328
Magothy, 161
Mass flux, 9
Mass spectrometer, 70, 183
Mathematical model, 19
Maximum contaminant level (MCL), 45, 82, 375
Maximum stable basin yield, *see* Safe yield
Metabolic pathways, 76
Microcontaminants, drinking water, 368, 378
Miocene epoch, 263
Moisture content, 6
Monitoring devices, 313
Monitoring program, 254, 313, 328
Morainal setting, 138

Mortality rate, 348
MOUSE program, 21
Multihit model, 74
Multistage model, 74
Municipal wastes, 24
Mutagens, 76, 380
Mutant alleles, 77

N

National Contingency Plan, 58
Natural basin yield, 17
Nested wells, 5
Neutron moisture logger, 6
Nitrate, 81
  contamination of, 150
Nitrite, 81
Nitrosamine, 81
Nondegradation, 89
Nonpoint sources, 18
  urban stormwater, 265

O

One-hit model, 74
Optimal yield, 23, 293, 371
Ordinance, 278, 284
Overdraft, 16, 320, 371
Overdraft conditions, 371
Overland runoff, 2
Ozonation cation exchange reaction, 103
Ozonation system, 101

P

Packed-tower aerator, 115
Parts per billion (ppb), 68
Parts per trillion (ppt), 68
Percolation, 2, 173
Percolation ponds, 302
Perforated casing, 4
Permeability, 4
Pesticide contamination, 167
Pesticide Root Zoon Model (PRZM), 21
Piezometer, 4, 17, 138
Pilot testing, 91
Plume, 18
  toxic contaminant, 330
Plume movement, 90
Point sources pollution, 17
Porosity, 4
Porous material, 3
Potentiometer, *see* Piezometer
Potentiometric surface, 15, 292, 371
Potomac-Raritan-Magothy formation, 291, 327
Powdered activated carbon, 105
Prechlorination, 97
Precipitation, 2, 159
Precipitation exchange, 98
Pressure head, 6
Prior appropriation rule, 34
Pumping tests, 16
Pumping to waste, 333

R

Radon, 81
Rapid mix, 94
Recharge
  areas, 5
  basin, 165
  zone, 304
Redwood slat aerator, 116
Resource Conservation and Recovery Act, 28, 52
Retardation factor, 11
Richard's Equation, 19
River stage, 370
Runoff, 265, 372

S

Safe Drinking Water Act (SDWA), 28, 45, 375
Safe yield, 16
Salinity dams, 228, 371, 374
Salt water intrusion, 292, 372
Sand and gravel mining, 171
Sanitary landfills, 24
Saturated zone, 7
Scavenger waster, 171
Section 208, 377

Sedimentary layer, 262
Sedimentation, 96
Seepage, 2
Sensitive area program, 129
Septic systems, 137
Sludge handling, 97
Softening, 98
Soil moisture characteristic curve, 6
Soil moisture tension, *see* Pressure head
Soil water, 1
Sole source aquifer, 49
Source control program, 129
Spill, 248
  causes, 174
Spray aerator, 116
Standard industrial classification, 42
Storage facilities, 90
Storm water basins, *see* Recharge basin
Storm water runoff control, 278
Stratified drift, 342
Stream runoff, 3
Sub-basin, 303
Subsurface water, 1
Suggested No Adverse Response Levels, 347
Superfund Act, 30, 127, 249
Surface drainage, 159
Surface Mining Control and Reclamation Act, 60
Surface water, 1
Surveillance program, 335
Synergistic effects, 84
Synthetic Organic Chemicals (SOCs), 205

T

Teratogens, 76, 380
Till soil, 342
Total human exposure, 80
Toxic Substances Control Act, 59
Tracer test, 12
Transpiration, 2
Tributary watershed, 342
Trihalomethanes, 71, 352

U

Underflow, 2
Underground injection control program, 48
Uniform protection, 40
United States Coast and Geological Survey, 175
Unsaturated zone, 7
Upland zone, *see* Watershed zone

V

Volatile organic compounds (VOCs), 70, 247

W

Waste
  on-site disposal system, 164
  treatment program, 195
Waste water treatment
  cost, 251
  facilities, 231
Water
  balance equation, 19
  bearing strata, 159
  management, 228
  purveyors, 307
  table, 138
  testing programs, 321, 336
  well, 171
Waterfall type, 115
Watershed, 288
Watershed development standard, 273
Watershed zones, 273
Water treatment
  chemical feed and handling, 99
  cost, 98
Wellhead protection areas, 52

Z

Zero gauge pressure, *see* Atmospheric pressure
Zoning ordinance, 255